A Century of Geneticists

Mutation to Medicine

A Century of Geneticists

Mutation to Medicine

Krishna Dronamraju

CRC Press
Taylor & Francis Group
Boca Raton London New York

CRC Press is an imprint of the
Taylor & Francis Group, an **informa** business

CRC Press
Taylor & Francis Group
6000 Broken Sound Parkway NW, Suite 300
Boca Raton, FL 33487-2742

Printed on acid-free paper

International Standard Book Number-13: 978-1-4987-4866-7 (Paperback)
International Standard Book Number-13: 978-1-138-35313-8 (Hardback)

Visit the Taylor & Francis Web site at
http://www.taylorandfrancis.com

and the CRC Press Web site at
http://www.crcpress.com

I dedicate this book to my loving companion Michele Wambaugh for her unending support, patience, and diligent editorial advice

Contents

SECTION III Biochemical Genetics

SECTION IV Molecular Biology

SECTION V Radiation Genetics

SECTION VI Transposons

SECTION VII Applications of Genetics

Introduction

What can we learn from studying the lives of geneticists? We learn about the successes and failures of those who preceded us. We must avoid the mistakes made in the past. Where progress has been made, we should try to understand the circumstances and how they have contributed to that success. We can evaluate the loss and missed opportunities resulting from the periods of neglect, as in the case of Mendel and Barbara McClintock, which occurred under very different circumstances. We could have made rapid progress had these lapses not occurred at all. Are there similar circumstances today that we are not able to see, but we should make every attempt to uncover?

Are we missing greater opportunities by neglecting or underutilizing the full potential of certain segments of the community, such as women and various minorities and marginal populations? We know from the lives and careers of Mendel and McClintock that their contributions were ignored for prolonged periods because of their circumstances. Mendel's work was ignored mainly because he was not a respected academic, not a faculty member of a university, but a priest who was engaged in totally unexpected experiments in plant breeding.

On the other hand, Barbara McClintock was a respected member of the academic community, possessed excellent credentials, having obtained her doctorate under the direction of Rollins Emerson at Cornell University. But her work on transposons was so far removed from the main stream of genetics that many geneticists found it hard to accept. However, when similar observations were reported by some other investigators (mostly male) in later years, her work was accepted.

Other cases of neglect involved very different circumstances. One such case is the excellent work of Archibald Garrod in human biochemical genetics. His discovery of the inheritance of inborn errors of metabolism appears to have been premature in the sense the scientific community was not sufficiently educated to understand the implications of his work. The science of genetics itself was just rediscovered when Garrod published his paper on alkaptonuria in 1902. They were also not familiar with the metabolic basis of inherited disease that Garrod discovered. His work was ignored until Haldane drew attention to it in 1937.

Another instance of delayed recognition involved Herman J. Muller's discovery of X-ray-induced mutagenesis, which was first reported in 1927. Although it was well recognized within the scientific community, it was not until 1946 that Muller was awarded the Nobel Prize. The reason for the 20-year delay was not due to neglect but lack of public concern regarding the harmful effects of radiation which only became apparent after the atomic bombing of Hiroshima and Nagasaki in 1945.

Beyond genetics, a well-known case of neglect was the lack of interest in the discovery of penicillin by Alexander Fleming in 1929, which was not developed as a therapeutic drug until 1945 by Howard Florey and Ernest Chain. Increasing wartime casualties made it necessary to develop an antibiotic drug to treat the wounded soldiers.

One wonders, how many other opportunities were missed or neglected in the past decades?

In spite of these setbacks, genetics progressed and expanded rapidly by incorporating and accepting concepts and methods from other disciplines such as biochemistry, mathematics, physiology, X-ray crystallography, behavioral sciences, and others.

Multiple disciplines are contributing today to the development of human genome project, complex disease etiology, and human gene therapy. Single-author papers are rare. Most of the publications are authored by multiple authors (some have 50 or more coauthors). Glancing through the contributions of geneticists in this book, one can appreciate the transformation of science. Individuals such as Galton, Bateson, Haldane, Fisher, Muller, Beadle, McClintock, and others, who founded new disciplines while working alone mostly, are now rare. This is generally true of all science. It has become a complex endeavor, requiring large sums of money, multiple technologies, and numerous authors with different technical skills. This makes it even more imperative to pause and appreciate the impact of these giants of the last 100 years.

FURTHER SUGGESTED READING

Fleming, A. (1929) On the antibacterial action of cultures of a penicillium, with special reference to their use in the isolation of B. influenza. *Br. J. Exp. Pathol., 10*: 226–236 (Reprinted in *Rev. Infect. Dis., 2*: 129–139. 1980).

Garrod, A.E. (1902) The incidence of alkaptonuria: A study in chemical individuality. *Lancet, ii*: 1616–1620.

McClintock, B. (1950) The origin and behavior of mutable loci in maize. *Proc. Natl. Acad. Sci. USA, 36*: 344–355.

Muller, H.J. (1927) Artificial transmutation of the gene. *Science, 66*: 84–87.

Acknowledgments

I am pleased to express my gratitude to my partner Michele Wambaugh for much support and valuable editorial assistance throughout the project.

This project has benefitted from my discussions with numerous colleagues, some now deceased, including Ernst Mayr, Joshua Lederberg, James F. Crow, Freeman Dyson, Sir David Weatherall, M.S. Swaminathan, Pushpa Bhargava, Motoo Kimura, Victor McKusick, Elof Carlson, Naomi Mitchison, Cedric A.B. Smith, and Sir Arthur C. Clarke, among others.

Biographical data collection was funded, in part, by the Rockefeller Foundation, The Wellcome Trust, Chemical Heritage Foundation, and the U.S. National Institutes of Health.

I am grateful to Professor M.S. Swaminathan for much valuable information regarding "green revolution" and the photograph with Norman Borlaug.

Professor Ida Stamhuis of the University of Amsterdam has been most helpful for information regarding Hugo de Vries and his photograph.

I thank Loma Kirklins at the Caltech Archives for valuable information and help with the photographs.

Section I

Beginnings

HUGO DE VRIES

Hugo De Vries was a Dutch botanist and one of the rediscoverers of Mendel's work. He is remembered mainly for suggesting the concept of genes, for introducing the term "mutation," and for developing a mutation theory of evolution.

In 1886, he had discovered new forms among a display of the evening primrose (*Oenothera lamarckiana*) growing wild in an abandoned potato field. He found that their seeds produced many new varieties in his experimental garden and introduced the term "mutations" for these suddenly appearing variations. He suggested that evolution might occur more frequently with such large-scale changes. De Vries's theory led T.H. Morgan to study mutations in the fruit fly, until the modern interpretation became the accepted model in later years. However, the primrose variations that de Vries found would not be called mutants today; they turned out to be the result of chromosomal duplications. De Vries was also the first to suggest the occurrence of recombinations between homologous chromosomes, now known as chromosomal crossovers, soon after chromosomes were mentioned in relation to Mendelian inheritance by Walter Sutton.

FRANCIS GALTON

Francis Galton coined the term "eugenics" in 1883 and set down many of his observations and conclusions in a book: *Inquiries into Human Faculty and Its Development*. He pointed out that late marriages of eminent people, and the paucity of their children, were dysgenic. He advocated encouraging eugenic marriages by supplying able couples with incentives to have children. More important from today's point of view of human genetics, Galton's ideas on eugenics have been the driving force that have influenced both the goals and methods in human genetics. His conception of variance, standard deviation, correlation, and regression and others have played a useful role in data analysis. He introduced twin studies for analyzing the relative contributions of genetics and environment in etiology. His introduction of finger print analysis has been most valuable in forensic studies.

WILLIAM BATESON

William Bateson was a British biologist who coined the term "genetics" in 1906 to describe the science of heredity. He also suggested several other terms in genetics, which are used today, such as "zygote," "homozygote," "heterozygote," "allelomorph" ("allele" later), and F1 and F2, respectively, for first and second filial generations.

His 1894 book, *Materials for the Study of Variation*, was one of the earliest formulations of the new approach to genetics. Bateson argued that biological variation exists both continuously, for some characters, and discontinuously for others, and coined the terms "meristic" and "substantive" for the two types. Agreeing with Darwin, he felt that quantitative characters could not easily be "perfected" by the selective force of evolution, because of the perceived problem of the "swamping effect of intercrossing," but proposed that discontinuously varying characters could.

Bateson is well known for his counsel to beginners: "If I may throw out a word of counsel to beginners, it is: Treasure your exceptions! When there are none, the work gets so dull that no one cares to carry it further. Keep them always uncovered and in sight. Exceptions are like the rough brickwork of a growing building which tells that there is more to come and shows where the next construction is to be."

THOMAS HUNT MORGAN

Thomas Hunt Morgan was an evolutionary biologist, geneticist, and embryologist who won the Nobel Prize in 1933 for discoveries elucidating the role that the chromosome plays in heredity. The fruit fly *Drosophila* was Morgan's experimental organism. Morgan and his pupils, A.H. Sturtevant and C.B. Bridges (H.J. Muller joined later), worked in a small crowded room at Columbia University, which later came to be known as the famous "Fly Room." Sturtevant (1965, pp. 46–47) described the room in his book, *A History of Genetics*:

"The possibilities of the genetic study of Drosophila were then just beginning to be apparent; we were at the right place at the right time. The laboratory where we three raised Drosophila for the next seventeen years was familiarly known as 'The Fly Room.' It was a rather small room (16 by 23 feet), with eight desks crowded into it. Besides the three of us, others were always working there—a steady stream of American and foreign students, doctoral and postdoctoral. One of the most important of these was H.J. Muller, who graduated from Columbia in 1910.… There was an atmosphere of excitement in the laboratory, and a great deal of discussion and argument about each new result as the work rapidly developed."

Morgan proposed that the amount of crossing over between linked genes differs and that crossover frequency might indicate the distance separating genes on the chromosome. J.B.S. Haldane (1919) suggested that the unit of measurement for linkage be called the *Morgan*. Alfred Sturtevant developed the first genetic map in 1913. Morgan's fly room at Columbia became world-famous, and he found it easy to attract funding and visiting academics. In 1927, after 25 years at Columbia, and nearing the age of retirement, he moved to establish a school of biology at Caltech in California.

1 Hugo de Vries (1848–1935)

INTRODUCTION

Hugo de Vries (1848–1935), Dutch plant physiologist and geneticist, is the author of the mutation theory of evolution (de Vries 1901–1903).

His interests included the study of heredity and its relation to the origin of species as well as developing a mutation theory. He was especially noted as one of the three rediscoverers of Mendel's laws, the two others being Carl Correns (1864–1933) and Erich von Tschermak (1871–1962). De Vries published three papers on Mendelism in 1900. However, it has also been suggested that he first intended to suppress any reference to Mendel and changed his mind only when he found that Correns (or Tschermak) was going to refer to him (Sturtevant 1965, p. 27).

Hugo de Vries was born in Haarlem, in the Netherlands, on February 16, 1848, the son of Gerrit de Vries and Maria Everardina Reuvens. His father's family had been Baptist ministers and businessmen, and his mother's family, scholars and statesmen. Educated first at a private Baptist school in Haarlem, young de Vries attended gymnasium (equivalent to high school) in The Hague, matriculating in the University of Leiden in 1866. Here, he read two works that greatly stimulated his interest in botany: Darwin's (1859) *Origin of Species* and Julius Sachs' (1868) *Textbook of Botany.* Darwin's book raised de Vries' curiosity about variation and its relationship to the process of evolution, particularly the diversification of species. Sachs' textbook aroused de Vries' enthusiasm of quantitative, experimental work, as opposed to the old-style descriptive taxonomy that made up so much of the field of botany at the time. One of the weakest parts of Darwin's argument for evolution by natural selection had been his lack of coherent understanding of the mechanism of heredity.

EARLY RESEARCH

Pursuing physiological studies at Leiden, de Vries earned his doctorate in plant physiology in 1870 but felt stifled by the university, where conditions for experimental work were crude and where there was open hostility to Darwinism. He therefore decided to continue his education in Germany, first at Heidelberg (1870) and then at Würzburg (1871), with Julius Sachs. Sachs took a great interest in de Vries' career, helping him refine his experimental techniques and nominating him for several important posts over the next few years. Sachs was a strong proponent of experimentation. Under his guidance, de Vries began a series of detailed studies of osmosis, plasmolysis, and the effects of salt solutions on plant cells. He carried out these experiments at Würzburg, then at Amsterdam while teaching in a gymnasium (1871–1877), and finally at the University of Amsterdam, where he was appointed lecturer in plant physiology in 1877 and professor in 1881; he remained in Amsterdam until his retirement in 1918 and later moved to the village of Lunteren.

In the late 1880s, de Vries shifted from experimental work in plant physiology to the study of heredity. His first major publication on this subject was *Intracellular Pangenesis* in 1889 (de Vries 1889), a critical review of the hereditary theories of Darwin, Herbert Spencer, August Weismann, and Carl von Nägeli. All of these writers had proposed some form of particulate theory of inheritance. De Vries added to the list one of his own, the theory of "pangenes" (a term he borrowed from Darwin), unitary particles representing individual traits of an organism and manifesting themselves independently in the adult. De Vries considered the pangene a material unit that could combine and recombine in successive generations, much like atoms in the formation of molecules. Although de Vries' hypothesis cannot be considered a forerunner of the Mendelian-chromosome theory that emerged in the twentieth century, it was an elegant example of the sorts of theories of heredity and evolution that prevailed in later nineteenth-century biological thought.

As a result of his physiological training, de Vries was interested in studying heredity and evolution from a quantitative and experimental, rather than a purely theoretical, point of view. In the early to mid 1890s, he learned of the statistical work on variation being developed by Francis Galton in England. A first cousin of Charles Darwin, Galton measured traits in animal populations and showed that they generally fit into a "normal curve" of distribution. De Vries' studies showed that such curves also existed for many traits in plants. But he also found that many traits showed a bimodal or discontinuous distribution, suggesting that populations are often mixtures of varieties, or races, that can be separated from one another by selection. Crossing several closely related races of poppy, xenia, and other species that differed from one another by only one or a few traits, de Vries arrived independently (by 1896) at what is now known as Mendel's law of segregation. In 1900, he accidently came across and read Mendel's original paper of 1866 and incorporated a discussion of Mendel's results in his own work on the poppy, published in 1900. This publication appears to have triggered both Carl Correns and Erich von Tschermak to read Mendel's work and recognize its importance. The result was to bring to the attention of the scientific world the work of Gregor Mendel, leading to the foundation for modern genetics.

THE MUTATION THEORY

While on a field expedition (in the 1890s) near the town of Hilversum, a suburb of Amsterdam, de Vries observed what he thought were several species of the evening primrose, *Oenothera*, growing together. One seemed to be a parental strain that had given rise to two kinds of offspring, which differed in enough characters to be regarded as distinct species. He transplanted all three types of plants in his experimental garden for an experimental study of evolution. He found that they produced many new varieties and introduced the term "mutations" for these suddenly appearing variations. Although the term "mutation" continued to be used, its meaning changed significantly after T.H. Morgan referred to the eye-color mutants in *Drosophila*. De Vries' theory stimulated T.H. Morgan to launch his studies of mutation in *Drosophila*. Ironically, the large-scale primrose variations turned out to be the result of chromosomal duplications (polyploidy), while the term "mutation" now generally is restricted to *discrete changes in the DNA sequence*. In 1903, de Vries was also the first to suggest the occurrence of recombinations between homologous chromosomes, now known as crossingovers, soon after chromosomes were implicated in Mendelian inheritance by Walter Sutton (Befruchtung und Bastardierung, Veit, Leipzig).

The mutation theory of de Vries was first published in two massive volumes between 1901 and 1903. It was an attempt to explain the origin and inheritance of species differences. De Vries' theory appealed to all biologists from various disciplines, including cytology, embryology, and plant and animal breeding, among others. Although it is called a "mutation theory," it is essentially a concept of heredity and evolution. The timing of de Vries' theory is significant; the Darwinian theory of evolution by natural selection was not yet firmly established. On the other hand, Mendelian genetics was still in the future. De Vries' theory appealed to those who were searching for an alternative to Darwin's theory of natural selection, especially during the period 1903–1913. It was not until five years later that R.A. Fisher (1918) wrote his famous paper, which reconciled the Mendelian and biometric factions in England, and another six years later when Haldane (1924) wrote his first paper on the mathematical theory of natural selection. Soon after, Fisher (1930), Haldane (1932), and Wright (1931) published what became the foundations of population genetics, which drove the last nail in the coffin, burying the theories of de Vries and all others.

Garland Allen (1969) reviewed the reactions to Darwinian theory that were prevailing toward the end of the nineteenth century. In 1885, Darwin's friend and supporter, T.H. Huxley, spoke eloquently of the triumph of Darwinism: "Before twenty years had passed, not only had the importance of Mr. Darwin's work been fully recognized, but the world had discerned the simple, earnest, generous character of the man, that shone through every page of his writings." ("The Darwin Memorial," reprinted in Huxley 1915, pp. 249–250).

However, Darwin's opposition had not been completely eliminated and had, in fact, reappeared at the end of the nineteenth century in various forms. The reception of de Vries' theory must be viewed against this background during the 1903–1913 period.

THEORY OF PANGENESIS

De Vries' theory of pangenesis predisposed him to accept his *Oenothera* findings as examples of the discontinuity that he believed must exist as discrete types in nature.

De Vries believed that new species arose all at once, with no obvious transitional forms between them. De Vries reasoned that such discontinuity in the field implied a discontinuity in the origin of the types themselves.

De Vries' mutation theory was, in fact, a particulate concept of heredity that originated in his own theory of "intra-cellular pangenesis," which was originally published in 1899. Intracellular pangenesis is a discussion arising out of Darwin's hypothesis of pangenesis and Weismann's views on the germplasm.

De Vries referred to pangenes in the sense a chemist refers to his molecules. At the same time, he recognized that hereditary units, unlike the chemist's molecules, could not be isolated and subjected to experimental analysis. He claimed that the characters of an organism are reflected in small hereditary particles, called "pangenes," which are present in all the cells of an organism. They may be active, in the sense they cause the character for which they are responsible to appear, or latent, meaning that they do not produce this effect. They grow and multiply. De Vries claimed that pangenes are usually latent in the germ lines but are quite active in the somatic cells. Organ differentiation occurs in different tissues when particular individual pangenes develop their activity. The development of a character is due to a minimum number of active pangenes of the same kind.

De Vries viewed the *Oenothera* types as discrete, definable types, indicating a discontinuity with no discernible intermediate types. According to de Vries, new species arose all at once, with no obvious transitional types between them. De Vries' theory was in opposition to Darwin's theory of evolution by natural selection, which emphasized that species arose by the action of natural selection on minute, fluctuating variations. In contrast to Darwin, de Vries wrote that species arise by saltation, and the resulting individuals are distinguishable from one another as sharply as most of the so-called small species. De Vries emphasized that his "mutations" are distinguishable from individual, fluctuating variations. He classified the mutations into three types: progressive, retrogressive, and degressive. A progressive mutation need not be visible but could remain latent for a period of time, although it eventually would express itself as a visible mutant character. Retrogressive mutations result in the disappearance of a character that previously existed, such as the loss of variegation in the leaf. Degressive mutations involve the activation of a latent character that previously existed but had been inactive for many generations. De Vries mentioned the occasional appearance of crumpled leaves in *Oenothera laevifolia* as an example of degressive mutations in that species.

Evolutionary Theory in the Late Nineteenth Century

To understand the significance of de Vries's research, it is important to place his investigation in the context of the scientific debates of the period. Charles Darwin's theory of evolution by natural selection was published in 1859. He held that species evolved or changed in form from generation to generation because some members of the species lived for a longer time than others and were able to produce more offspring than their less fit fellows. In the long run, this would result in a species becoming more like the favored variation and less like the unfavored variations. In his *Origin of Species*, Darwin did not establish how variations occurred or how they were inherited. Subsequently, the area of heredity and variation became a recognized field of research for biologists interested in evolutionary theory.

Darwin had put forward the idea that variations between different individuals in a species were usually of a continuous nature. He believed that because of natural selection, certain ranges of this continuous variation would be more favored in the struggle for survival and the species would become changed toward those ranges. However, by the late 1880s and the 1890s, some biologists were becoming convinced that evolution depended on the effect of natural selection on discontinuous variations, not on continuous variations. In the period of de Vries' greatest contributions to science, 1880–1910, he participated vigorously in the debate about the respective roles of continuous and discontinuous variations in the evolutionary process.

Biologists were at the same time involved in much debate and research about the nature of heredity. Darwin realized that one of the gaps in his theory of evolution was an adequate explanation of the mechanism of heredity. To fill this gap, he proposed his theory of pangenesis: Each character in a mature organism was determined by a minute particle, or pangene, passed on from the parental organisms via the sex cells. The pangenes passed from all parts of the parental body through the bloodstream to the sex cells and then determined the character of the appropriate parts of the offspring by similar diffusion as the offspring grew.

One aspect of Darwin's theory of pangenesis caused much debate among biologists. How, they asked, could the pangenes, which were discrete particles, give rise to continuous variations? For this to occur, there would probably have to be some blending of the pangenes from different parents into one pangene. Some biologists preferred to believe that if heredity did depend on the passing of discrete units from parents to offspring, these units would remain discrete in the offspring and give rise to discontinuous variations in the mature offspring. De Vries played an important role in the debate about the process of heredity.

Another area of research was of great importance in the overall picture of evolution. This was the question of the structure of the cell and its nucleus and the analysis of the behavior of cell and nucleus during division. During the last quarter of the nineteenth century, cytologists established a fairly detailed picture of what happened to the nuclear material during cell division. The material was chemically identified, and biologists began to speculate on the connection between the nucleic acids of the chromosomes and the mechanism of inheritance. Again, de Vries played an important role in pointing out the connection between the nuclear material and the particles that controlled the inheritance of characteristics from generation to generation.

TOWARD THE REDISCOVERY

De Vries attempted to support his hereditary theory by studying hybridization as a way to study experimentally. He began conducting hybridization experiments from 1892 onward. New combinations of different characteristics appeared in the offspring, and these new variations bred true. De Vries concluded that characters and their pangenes are independent and can be recombined in various ways, supporting the independent heredity of distinct characters. In 1896, four years before the publication of the rediscovery papers, he hybridized blue and white flowers of *Veronica longifolia* and purple and white flowers of *Aster tripolium*. He tried to interpret these results in a probabilistic framework, approximating a 1.2.1 law. The conclusions are presented in the framework of de Vries' intracellular pangenesis. He used his own terminology of pangenes, central hybrids and old types, corresponding to Mendel's terms—factors, heterozygous, and homozygous organisms. De Vries did not publish the results of his hybridization experiments before his rediscovery papers in 1900. According to the Dutch historian of science, Ida Stamhuis (2015), de Vries's reluctance to publish his results is mostly due to the fact that he was not sure how he could reconcile these data with his theory of *Intracellular Pangenesis*. Quite possibly, he wanted to claim that he had arrived at the same insights as Mendel, independently. In the "rediscovery" papers, he adopted some of the terminology of Mendel. He used Mendel's terms: dominance and recessiveness. But he connected dominance to "the systematically higher character,… or in the case of known descent the older one," which indicated his own approach in which heredity was always connected to evolution. While the rediscovery was in progress, de Vries continued his own work on the *Mutation Theory*. He made it clear that while he recognized Mendel's work, he had discovered the same laws himself and they were in agreement with his own theory (Stamhuis 2015).

De Vries published three papers on Mendelism in 1900, one of which has been overlooked. The first was read by G. Bonnier before the Paris Academy of Sciences on March 26 and was published in the Academy's *Comptes Rendus*. Another paper by de Vries is dated "Amsterdam, March 19, 1900," and was published in the *Revue General de Botanique*, which was edited by Bonnier. It appears likely that these two French manuscripts were sent to Bonnier at the same time. The third paper, in German, was received by the editor of the journal *Berichte der Deutschen Botanischen Gesellschaft* in Berlin on March 14 and was published on April 25. Their order of publication is of some interest. The brief note in the *Comptes Rendus*, the first to be published, does not mention Mendel, although it includes some of Mendel's terminology. The *Revue General* paper is rarely cited. It is longer and mentions Mendel on the last page, with an added footnote referring to the *Berichte* paper and to the papers by Correns and Tschermak. The reference to Mendel is as follows: "This law is not new. It was stated more than thirty years ago, for a particular case (the garden pea). Gregor Mendel formulated it in a memoir entitled 'Versuche uber Pflanzenhybriden' in the *Proceedings of the Brunn Society*, Mendel has there shown the results not only for monohybrids but also for dihybrids…. This memoir, very beautiful for its time, has been known to me and then forgotten."

In the *Berichte* paper, the second to be published but the first to be submitted, the content is same as in the longer French paper, but Mendel is mentioned in several places in the text, concluding with a full recognition of Mendel's contribution.

Even after the rediscovery of Mendel's work, de Vries continued to believe that his own theory of *Intracellular Pangenesis* was basically correct, although he rarely referred to it. Surprisingly, he attached no importance to the Mendelian laws. He waited six months after his rediscovery of Mendel's work before he mentioned it in a letter to his friend in Groningen, Jan Willem Moll (1851–1933). Interestingly, soon after the rediscovery of Mendel's laws, de Vries felt the need to downplay their importance in several different ways. For instance, he always remained vague regarding the exact numbers of pangenes. Second, he claimed that the separation of the hereditary factors in the germ cells does not only occur in the case of Mendelian crossings. He argued that a similar process takes place in other non-Mendelian crosses, and so on.

In summary, he argued that only the *race* characters and not the *species* characters obey Mendel's laws. He appeared to be more interested in the "higher" species than in the "lower" races. He argued further that the occurrence of Mendelian crossings pointed to the loss of the capacity to mutate within a family.

Carl Correns, was a student of Nageli and also of the plant physiologist Pfeffer. It is interesting that both Pfeffer and de Vries were students of Sachs. He studied hybrids of maize and of peas through several generations and arrived at Mendelian results in 1899. In 1900, he published a paper on his work with peas, fully confirming Mendel's results, and repeated the same with a paper in 1901 on his work with maize.

Another rediscoverer of Mendel's work, Erich von Tschermak (1871–1962), was a grandson of Fenzl, under whom Mendel studied systematic botany and microscopy in Vienna. He made crosses of peas and wrote an extensive paper in 1900 after reading the papers of de Vries and Correns.

British biologist William Bateson, who is not usually included among the rediscoverers of Mendel's work, read a paper before the Royal Horticultural Society in London, in May 1900, in which he described the work of Mendel and its confirmation by de Vries. He became an active supporter of Mendel's work and gathered a very active group of workers at Cambridge, which included R.C. Punnett.

BIRTH OF GENETICS

Later, Johanssen deleted "pan" and introduced the word "gene," but it is de Vries who formulated the concept of a definite unit. Darwin's gemmules were extremely minute and distributed in the bloodstream, acting chemically on the reproductive organs. Various papers on specific cases of inheritance followed, and in 1900, de Vries, simultaneously with Correns and Tschermak, called attention to Mendel's paper of 34 years earlier and supported the generalizations from his own work. This was definitely the birth of genetics as a separate science.

In his book on mutation, de Vries recognized the distinction between the minor fluctuations in the expression of a character in the individuals of a species and the larger variations that are heritable. In contrast to Darwin, de Vries and Bateson proposed the role of larger discontinuous steps in evolution. De Vries rejected the Lamarckian idea that environment can determine mutation in a particular direction, however potent it may be in encouraging one or other type of mutation.

During the last several decades, "mutations" have become essential parts of the working theory of genetics and evolution, largely due to the monumental contributions of Nobel laureate H.J. Muller. But these are quite different from the original concept of "mutations" that was introduced by de Vries, who deserves credit for introducing the subject of mutations in our discussions of genetics and evolution from the beginning. Muller (1927) introduced precision in the induction (by radiation) and later measurement of mutation rates, while Haldane pioneered the estimation of human mutation rates and their impact on human populations.

In his preface to *The Mutation Theory*, de Vries wrote: "The origin of species has so far been the object of comparative study only. It is generally believed that this highly important phenomenon does not lend itself to direct observation, and, much less to experimental investigation. This belief has its root in the prevalent form of the conception of species and in the opinion that the species of animals and plants have originated by imperceptible gradations. These changes are indeed believed to be so slow that the life of a man is not long enough to enable him to witness the origin of a new form."

In his book, de Vries intended to show that species arise by saltation and that the individual saltations are highly visible to any one and can be observed like any other physiological process. He wrote further: "In this way we may hope to realize the possibility of elucidating, by experiment, the laws to which the origin of new species conform."

In his book on plant breeding, de Vries wrote: "For the direct observation of the process of mutating, the evening primrose of Lamarck affords, at present, an unequaled opportunity. It produces numerous mutants, and does so in every generation, and almost any sample of pure seed may be used for this study.... All the strains...show the same phenomena of mutability, as far as my experiments go. Where the species is growing in America in the wild condition, is not known, at present, and so it is impossible to decide whether it has acquired the habit of mutating..." He wrote further: "A main point in these observations is that the mutations occur suddenly, without preparation and without intermediates. Nothing indicates on the normal plants what their seeds will produce... The distribution of mutating seeds seems to depend simply upon chance. Nor are there intermediates. Each mutant is as good a representative of its type as its progeny will be."

RECEPTION OF THE MUTATION THEORY

De Vries' period was dominated by the ideas of particulate inheritance and discontinuous variation. Discontinuity in the inheritance of individual characters was for de Vries the logical counterpart of a particulate theory of inheritance, since hereditary particles were themselves discontinuous entities.

Simultaneously, British biologist William Bateson concluded that since species in nature were discontinuous, the variations giving rise to them were also discontinuous. Bateson (1894, 1902) expounded this theme in his survey of natural populations, *Materials for the Study of Variation*. De Vries was inspired by Bateson's work, which influenced his own thinking deeply. Bateson surveyed many natural populations of both animal and plant species, concluding that since species in nature were discontinuous, the variations that produced them were also discontinuous. Furthermore, theories of discontinuous inheritance had been proposed by several others in the preceding years.

The reaction of the biological community to the mutation theory was generally highly favorable. Support came from numerous scientists working in many fields, including several prominent scientists such as C.B. Davenport, T.H. Morgan,

R.R. Gates, Jacques Loeb, Frank Shull, E.G. Conklin, among others. Many were Mendelians. As de Vries himself showed, the Mutation and Mendelian theories were not mutually exclusive. One enthusiastic worker claimed that "no work since the publication of Darwin's *Origin of Species* has produced such a profound sensation in the biological world as...*Die Mutationstheorie* by Hugo de Vries" (Baker 1906). Both R.R. Gates (1911) and C.D. Davenport (1905) paid similar accolades. Morgan's letter to de Vries and his note, which supported de Vries' Mutation theory, are included in the appendix.

De Vries' theory created such an interest that in the early years of the century, several investigators attempted to find mutations in a great variety of organisms besides *Oenothera*. C.B. Davenport, for instance, attempted to find, in 1905–1906, examples of discontinuous variation in animals that would fit under the concept defined by de Vries. Davenport (1906) wrote: "the vast importance of mutations in the origin of species can no longer be questioned. The reviewer is convinced that as good an argument might be made from the zoological side as de Vries has made from the botanical. Undoubtedly, many, if not most of the characteristics of the races of domesticated animals and probably of feral species have arisen by mutation."

Jacques Loeb and T.H. Morgan were interested, independently, in testing de Vriesian mutations in the fruit fly, *Drosophila*.

Thomas L. Casey (1905) explained the sudden appearance of certain genera of mollusks in early Eocene on the basis of the mutation theory, and Schaffner found a variant of *Verbena*, which was growing adjacent to the parent species as a new mutant. Others have tried to repeat de Vries' experiments on other plant species at the New York Botanical Garden.

However, Some other prominent scientists, both in America and Europe, opposed the mutation theory from the very beginning. They considered it naive and lacking in evidence. Plant and animal breeders were generally opposed to the mutation theory. Biometricians in England strongly opposed de Vries' theory of evolution by mutation because, as followers of Darwin, they accepted evolution by continuous variation. Among those opposing were Ludwig Plate and August Weismann in Germany, William Bateson in England, and E. Hart Merriam, C.O. Whitman, O.F. Cook, and others in America. E. Hart Merriam in America, for instance, made a survey of over 1000 species and subspecies of North American mammals and birds to look for differences that might have originated by de Vriesian mutation. He concluded quite firmly: "My own conviction is that the origin of species by mutation among both animals and plants is so uncommon that as a factor in evolution it may be regarded as trivial."

Bradly Davis (1912), in a series of papers, showed that the so-called mutations of *Oenothera* in fact resulted from an unusual hereditary pattern which followed Mendelian laws. Renner (1914) showed that *Oenothera* was a permanent heterozygote between two complexes. In a series of cytological investigations, Cleland showed that the complex of variant forms were not new species at all, but rather complex recombinations. Although de Vries continued to defend his theory until his death in 1935, the mutation concept, as he proposed it, was abandoned by 1915, by the biological community. When Morgan referred to the white-eyed mutant of *Drosophila* in 1910, he was using it in the modern sense which is used today, not the large jumps that de Vries originally referred to in formulating his concept.

DE VRIES VS. DARWIN

The relative merits of the theories proposed by de Vries and Darwin were the focus of much discussion that took place during the early years of twentieth century. Those who were drawn to the mutation theory of de Vries were attracted by several features that seemed to be lacking in the Darwinian theory. These questions are related to the nature of variation, the utility of new characters, the role of selection, the role of the evolutionary time scale, the role of isolation, and the rigor of selection, among others. Major focus was on the nature and source of variations. Questions of interest included the following: What kind of variations were heritable? What factors caused variations? Does selection have a positive and creative role? On what kind of variations does selection act, continuous or discontinuous? De Vries assumed that small, individual variations were probably environmentally controlled and thus not heritable. Drawing attention to degressive and retrogressive mutations to the Mendelian genetics, de Vries showed that these discrete variations were subject to the laws of hybridization.

A major objection against Darwin arose from the origin of variations and their inheritance. The prevailing view in his time was that any new variation would be "swamped" or blended in with the normal variation by cross-breeding. To meet this criticism, Darwin and his supporters assumed that for any variation to persist, it must occur in two or more individuals at a time. However, such a possibility was considered negligible. Until Mendel's laws were established, origin and maintenance of variation continued to be a major problem in understanding evolution.

The mutation theory had one advantage over the Darwinian theory with respect to the "swamping" problem. A mutant is likely to occur several times and is unlikely to be lost in the next generation. Furthermore, a mutant form is unable to produce offspring with its parents and is thus unlikely to be swamped through back-crossing. It served as an important alternative to the Darwinian theory of natural selection, which failed to provide a satisfactory solution to the swamping problem.

The mutation theory raised the whole question of the maintenance of variation and what role selection itself played in the origin of species. While selection acted as a filter that weeded out the unfit, it could not create the fit. As one writer put it, "Natural selection may explain the survival of the fittest, but it cannot explain the arrival of the fittest" (Kellogg 1907). The idea of negativity appeared to have been due to Carl von Nageli, who compared natural selection to a gardener pruning various bushes and trees, without actually creating a new variety.

De Vries correctly saw that selection by itself could not produce evolution. It required a constant source of new, heritable variations. To de Vries, it was new variations (mutations) that provided the creative aspect of the evolutionary process. In this respect, de Vries' theory received strong support from the pure-line experiments of Wilhelm Johannsen (1857–1927) in 1903. Johannsen showed that selection of fluctuating variations only separated out the pure-lines already existing in a heterogeneous population. He noted that selection seemed to produce changes in a population only so long as the selection process was rigorously maintained. But it could not cause the species to transcend that threshold level of variation between one species and another (Johannsen 1903).

Johannsen's work opposed the Darwinian theory in at least two aspects. First, it showed that the selection of fluctuating variation was not able to produce a new species; second, it showed the effects of selection by itself were strictly negative.

In speciation, de Vries and Darwin differed sharply. According to Darwin, selection was essential in the creation of new species. For de Vries, species arose spontaneously, by mutation.

REDISCOVERY OF MENDEL'S WORK

During the 1890s, de Vries carried out many experiments in breeding plants. He crossed plants with different characteristics (for example, hairy and smooth stems) and counted the numbers of plants in succeeding generations that had the different parental characteristics. By the end of the 1890s, he had gathered much evidence to show that there were definite ratios that kept recurring among the offspring (for instance, hairy and smooth stems would occur in a ratio of 3:1). By late 1899, he had obtained similar results in more than 30 different species and varieties. De Vries reasoned that the obtaining of fixed ratios supported his theories of pangenesis and mutation. The pangenes, which determined the characters of the plants, were seen as units that must separate and recombine according to regular patterns during breeding; these regular patterns would give rise to the fixed ratios he had discovered. Mutations would arise from the loss or great change of some of the pangenes.

Sometime in 1900, before de Vries published his new findings about the fixed ratios of characters among the offspring in cross-breeding experiments, he discovered a paper by Gregor Mendel that included an account of the same laws about the regular patterns of inheritance. Mendel's paper had been published in 1866 and had been ignored by the scientific world.

There has been some controversy about de Vries's role in the rediscovery of Mendel's work, including the suggestion that he did not want to acknowledge Mendel's priority in the discovery of the basic laws of genetics. However, it would seem that de Vries never felt that the Mendelian laws were as significant as his own mutation theory, so that his apparent lack of recognition for Mendel could stem from a feeling that biologists were placing too much emphasis on Mendel's laws and not paying enough attention to de Vries's mutation theory.

From 1900 until he retired in 1918, de Vries spent most of his energy trying to find further evidence for his mutation theory. It drew less support as geneticists found more evidence to support Darwin's original theory that the source of evolutionary change was the normal variations that occurred among all numbers of a species. Eventually, the foundations of population genetics, which were laid by Haldane (1924, 1932), Fisher (1918, 1930), and Wright (1931), strengthened both Darwinian natural selection as well as Mendelian genetics. De Vries was still active during these developments.

By the time of de Vries's death in Amsterdam on May 21, 1935, the action of natural selection on ordinary variations had again become the accepted version of evolutionary theory and the term "mutation" was used to apply to any new character of a plant or animal—not only very large and striking variations.

DE VRIES'S POSITION IN THE GENETICS COMMUNITY

Dutch historian Professor Ida Huis presented a critical discussion of de Vries's position in the genetics community, especially in the Dutch genetics community. Initially, he was at the forefront of the international genetics community. He was recognized and respected as a rediscoverer of the Mendelian laws. He became even more famous when he published his two-volume work *Die Mutationstheorie*, in which he presented his argument that large discontinuous variants, which he called "mutations," played a key role in the evolutionary process and these mutations could be studied experimentally. His "mutation" theory made him instantly famous, especially in the United States, resulting in several lecture tours. However, that "successful" period lasted only a few years. With the rise of Mendelian genetics and the modern version of "mutation," de Vries' mutation theory could not stand up to criticism. His colleagues began to see him as a historic relic, one who rediscovered Mendel's work, but no further. Bateson coined the term "genetics" in 1906, discovered linkage, and established the new terminology of genetics which is still used today. Morgan (1910) and his followers established the Mendelian school at Columbia University, redefining "mutation" in its modern sense which we use today (Bowler 1978).

Perhaps embittered by these developments, de Vries gradually shunned participation in genetics gatherings and kept himself aloof from the genetics community. His marginalization was reinforced when the experimental basis of his mutation theory could not be confirmed. Even though the international genetics community wanted to include him, de Vries preferred to keep his distance. He did not attend the first international conference on genetics in 1911 in Paris, nor the international congress of genetics in Berlin in 1927. At the Berlin Congress, he was awarded an honorary position on the Board, but because of his absence, it was given to another Dutch geneticist Tine Tammes (1871–1947), who was actually de Vries's student. Furthermore, in 1922, he did not accept the official invitation to attend the Mendel commemoration in Brunn, the town where Mendel had conducted his important research. De Vries was not a founder of genetics in his native country. His professorship was in physiology and anatomy of plants. His pupil Tine Tammes was appointed extraordinary professor of heredity and variability in Groningen later on. In 1919, a Dutch journal *Genetica* was founded, but de Vries played no part in founding that journal. In 1922, he was invited to contribute to a special issue that was dedicated to Mendel, but he declined.

According to Professor Ida Stamhuis, several of de Vries's anniversaries were celebrated in the Dutch press. In 1916, readers of a prominent newspaper, *De Nieuwe Amsterdammer*, considered de Vries one of the most famous Dutchman of the past 50 years. Although de Vries was not a member of the Dutch genetics community, he was a famous and distinguished scientist in the Netherlands.

REFERENCES

Allen, G.E. (1969) Hugo de Vries and the reception of the "Mutation Theory". *J. Hist. Biol., 2*: 55–87.

Baker, F.C. (1906) Application of de Vries' Mutation Theory to the Mollusca. *Am. Nat., 40*: 327–334.

Bateson, W. (1894) *Materials for the Study of Variation*. New York: Macmillan Co.

Bateson, W. (1902) *Mendel's Principles of Heredity*. Cambridge: Cambridge University Press.

Bowler, P.J. (1978) Hugo de Vries and Thomas Hunt Morgan: The mutation theory and the spirit of Darwinism. *Ann. Sci., 35*: 55–73.

Casey, T.L. (1905) The mutation theory. *Science, 22*: 307–309.

Davis, B.L. (1912) Was Lamarck's evening primrose (*Oenothera larckiana* Seringe) a form of *Oenothera grandiflora* Solander? *Bull. Torrey Botanical Club, 39*: 519.

Darwin, C. (1859) *On the Origin of Species by Means of Natural Selection, or the Preservation of Favoured Races in the Struggle for Life*. London: John Murray.

Davenport, C.B. (1905) Species and varieties, their origin by mutation, by Hugo de Vries (review), *Science, 22*: 369–372.

Davenport, C.B. (1906) The mutation theory in animal evolution. *Science, 24*: 556–558.

De Vries, H. (1889) *Intracellular Pangenesis*. Jena: Fischer.

De Vries, H. (1901–1903) *Die Mutationstheorie*. Leipzig: Von Veit & Co, 2 vols.

Fisher, R.A. (1918) The correlation between relatives on the supposition of Mendelian inheritance. *Trans. R. Soc. Edinb., 52*: 399.

Fisher, R.A. (1930) *The Genetical Theory of Natural Selection*. Oxford: Clarendon Press.

Gates, R.R. (1911) Review of the English translation of *Die Mutationstheorie. Am. Nat., 45*: 254–256.

Haldane, J.B.S. (1924) A mathematical theory of natural and artificial selection. Pt. I. *Trans. Camb. Phil. Soc., 23*: 19–41.

Haldane, J.B.S. (1932) *The Causes of Evolution*. London: Longmans, Green.

Huxley, T.H. (1915) *Darwiniana: Essays*. New York: Appleton.

Johannsen, W. (1903) *Ueber Erblichkeit in Populationen und in reinen linien*. Jena: Gustav Fischer.

Kellogg, V.L. (1907) *Darwinism Today*. New York: Henry Holt.

Morgan, T.H. (1910) Sex limited inheritance in Drosophila. *Science, 32*: 120–122.

Muller, H.J. (1927) Artificial transmutation of the gene. *Science, 46*: 84–87.

Renner, O. (1914) Befruchtung und Embryobildung bei Oenothera Lamarckiana und einigen verwandten Arten. *Flora oder Allgemeine Botanische Zeitung, 107*: 115–150.

Sachs, J. (1868) *Textbook of Botany. Second Edition 1882*. Oxford University Press.

Stamhuis, I.H. (2015) Why the rediscoverer ended up on the sidelines: Hugo de Vries's theory of inheritance and the Mendelian laws. *Sci. Educ., 24*: 29–49.

Sturtevant, A.H. (1965) *A History of Genetics*. New York: Harper & Row.

Wright, S. (1931) Evolution in Mendelian populations. *Genetics, 16*: 97–159.

2 Eugenics and Francis Galton (1822–1911)

Francis Galton founded "eugenics" in 1883 for improving the "human race." He devoted much of his life to exploring variation in human populations and its implications. The term "eugenics" has been used continuously since his death, although not always implying the intent or meaning that Galton originally conceived. With the development of medical genetics, negative eugenics came to be adapted in many medical schools because of its role in eliminating or reducing the frequencies of certain genetically caused diseases and defects.

It is Galton's founding of eugenics as a distinct discipline that justifies his inclusion in this book. By the time of his middle years, Galton had already acquired fame as an explorer, geographer, meteorologist, and travel writer. His lasting and most important contribution to eugenics occurred in his later years. His initial contribution was a paper he wrote in 1865, "Hereditary Talent and Character," which appeared in *Macmillan's Magazine*. He introduced pedigree analysis to study the inheritance of human characters, a method that still remains the standard tool in human genetics. Galton (1869) followed the paper with his book *Hereditary Genius*, which expanded greatly his original theme, and added an important new dimension, the "normal" distribution. Most of *Hereditary Genius* was devoted to the pedigrees of eminent men (see Gillham 2001).

EARLY LIFE

The Galtons were Quakers and businessmen who prospered over generations. Francis Galton was Charles Darwin's half-cousin, both sharing the common grandparent Erasmus Darwin. Francis was born in 1822 and studied medicine at King's College in London. Galton was a child prodigy; he was reading by the age of two; by the age of five, he knew Greek, Latin, and long division; and by the age of six, he was reading poetry and Shakespeare for pleasure. He was educated at King Edward's School, Birmingham, and studied medicine at Birmingham General Hospital and King's College London Medical School. He pursued mathematical studies at Trinity College, University of Cambridge. Soon, his restless mind moved him to explore Africa in the following years under the auspices of the Royal Geographical Society. This became the first European exploration of the northern part of Namibia. His keen interest in quantification was evident from the very beginning; he maintained careful records of longitude, latitude, temperature, and other meteorological details.

He was awarded a gold medal by the Geographical Society in recognition of his quantitative contributions to geography.

Galton then settled in London and became a travel writer. Based on his explorations, he wrote a remarkable book, *The Art of Travel*, first published in 1855 and reprinted in 2001, which provided detailed quantitative instructions on how to plan adequately for an expedition in the bush. Interestingly, he was involved in an acrimonious dispute with Henry Morton Stanley following Stanley's relief of Livingstone, mainly because Galton regarded Stanley as a mere ambitious reporter, not a serious geographer! Galton's interest in meteorology led him to spend many years in weather-related research, resulting in the discovery of the anticyclone.

Francis Galton was a polymath whose restless intellect touched upon several fields of endeavor, often leaving them much enlightened and clarified. He invented pedigree analysis to measure the heritability of human talent and character and was the first to quantify human heredity. He contributed to anthropology, meteorology, biology, psychology, criminology, statistics, geography, human heredity, and eugenics, a term he coined in 1883. Galton devoted much of his life to exploring variation in human populations and its implications. His investigations led to the founding of a research program that encompassed several aspects of human genetics and anthropology, from mental characteristics to variation in human height, from facial images to fingerprint patterns. His work required the invention of novel statistical methods, large-scale collection of population data, and new methods of analysis.

FINGERPRINTS

Galton's interest in personal identification led him to develop the foundations for the fingerprint systems, which are used today by the law enforcement forces throughout the world. This is an excellent example of how a tool for practical application has emerged from basic research. Galton did not set out to develop a method of personal identification which became an indispensable tool for the police forces. On May 25, 1888, Galton presented his fingerprint studies in an impressive Friday Evening Discourse at the Royal Institution entitled "Personal Identification and Description." Galton opened his lecture as follows: "It is strange that we should not have acquired more power of describing form and personal features than we actually possess." Galton reviewed the history when some others have attempted to use fingerprints for personal identification; one of them was William Herschel, who used them to identify pensioners in Calcutta, India; another was Henry Faulds, a young Scottish doctor, working at the Tsukiji Hospital in Tokyo, who noticed the similarity in fingerprints between monkeys and men. On November 27, 1890, Galton read an important paper on fingerprinting before the Royal Society. He presented an extensive analysis of thumbprints from 2500 individuals, including a classification of the ridges, arches, whorls, and interspaces into patterns. Early in his investigations, Galton noted that fingerprints did not vary over time as shown by the data obtained from the same individuals at different time intervals. Galton further wondered if fingerprints might be heritable but lacked sufficient data to reach a firm conclusion. Galton also argued that fingerprints might be useful in the identification of accident victims. Galton emphasized that he was the first to provide evidence for the lifelong persistence of fingerprints, for classifying the complex patterns, and for establishing a consistent methodology for taking fingerprints. He published three books on fingerprinting, including his 1892 monograph, *Finger Prints* (Galton 1892).

ANTHROPOMETRICS

Following his work on *Hereditary Genius*, Galton's study of the human body emphasized both the accumulation of quantitative data on easily measurable characters such as height as well as such psychological characters as human personality and behavior. Galton's goal was to develop methods for characterizing and quantifying human behavior. The final goal is to develop methods for advancing the quality of the human stock, which he discussed in an article titled "Hereditary Improvement."

Galton suggested that moral and intellectual qualities were closely tied together among the nineteenth century English and that "many of the wild instincts of our savage forefathers" had been bred out. Galton attempted to grade them according to their natural ability. However, he found that the worst of the group were the most numerous while the best were the rarest. His goal was to identify the "gifted" and procure for them a socially favored lifestyle that will "bind them together." These ties will be further strengthened by intermarriage, leading to the establishment of an elite group within the population.

To achieve these goals, Galton urged the implementation of three tasks: (a) make continuous enquiries into the facts of human heredity, including the collection of detailed pedigree data; (b) establish an information center on heredity for animal and plant breeders, and (c) establish a separate unit that would evaluate the information gathered. These data would be further augmented in a few years by photographs and physical measurements compiled for a "thousand or more" individuals in each region sampled. Eventually, all schoolboys would be classified according to their natural gifts, both physical and mental. Further research will be made into the genealogies of those showing "remarkable" hereditary qualities. The most promising would be registered at their local centers and will continue to receive special treatment. Galton reasoned that, after a few generations, the selected individuals will become a new center of power themselves. He believed that the number of exceptional individuals would increase by continued breeding and segregation, while the non-gifted would gradually be eliminated from the community. Galton suggested that the least gifted should be actively discouraged from procreation. However, should they procreate, such persons may be considered as enemies to the State and would forfeit "all claims to kindness" by the State.

Galton's "Hereditary Improvement" laid out his entire social agenda. Yet, Galton was not satisfied. He proposed to collect measurements of heights and weights of students from selected schools. Galton concentrated on the data for the boys' heights, showing that they were normally distributed with a mean that increased with age. Based on various measurements on schoolboys, Galton tested his theory that growing up in a city was detrimental to one's health. He found that country boys were "about 1 1/4 inch taller than those in the city group, and 7 lbs heavier," which Galton attributed to "retardation and to total suppression of growth." In later investigations elsewhere, Galton attempted to measure energy, which he defined as "the length of day during which a person is wont to work at full stretch, day by day."

EMPIRICAL TEST OF PANGENESIS AND LAMARCKISM

Galton conducted wide-ranging inquiries into heredity, which led him to challenge Charles Darwin's hypothetical theory of "pangenesis." Darwin had proposed as part of this hypothesis that certain particles, which he called "gemmules," moved throughout the body and were also responsible for the inheritance of acquired characteristics. Galton, in consultation with Darwin, set out to see if they were transported in the blood. In a long series of experiments in 1869 to 1871, he transfused the blood between dissimilar breeds of rabbits and examined the features of their offspring.

Darwin challenged the validity of Galton's experiment, giving his reasons in an article published in *Nature*, where he wrote:

> Galton explicitly rejected the idea of "the **inheritance of acquired characters**" (Lamarckism), and was an early proponent of heredity through selection alone. He came close to rediscovering Mendel's particulate theory of inheritance, but was prevented from making the final breakthrough in this regard because of his focus on continuous, rather than discrete, traits (now known as polygenic traits). He went on to found the biometric approach to the study of heredity, distinguished by its use of statistical techniques to study continuous traits and population-scale aspects of heredity. This approach was later taken up enthusiastically by Karl Pearson and W.F.R. Weldon; together, they founded the highly influential journal Biometrika in 1901. R.A. Fisher would later show how the biometrical approach could be reconciled with the Mendelian approach. The statistical techniques that Galton invented (correlation, regression) and phenomena he established (regression to the mean) formed the basis of the biometric approach and are now essential tools in all the social and biomedical sciences.

HEREDITARY GENIUS

Galton was farsighted in his choice of topics for research. In 1875, in a paper on twin studies, he anticipated the modern field of "behavior genetics." He investigated if twins who were similar at birth diverged in diverse environments and whether twins dissimilar at birth converged when reared in similar environments. Galton concluded that the evidence indicated nature as the dominant force in determining the outcome (rather than nurture).

EUGENICS

One of the enduring fields investigated by Galton in 1883 is what he called *eugenics*, which he discussed in detail in his book *Inquiries into Human Faculty and Its Development*. He observed certain trends in British society which he thought were *dysgenic*. They include the late marriages of eminent people and the paucity of their children. He considered it desirable to encourage able couples with incentives to have children. In 1901, Galton addressed several eugenic issues in his second Huxley lecture at the Royal Anthropological Institute.

A major opportunity to promote eugenics arose when the newly formed Sociological Society met on April 18, 1904, at the University of London with Galton as a featured speaker. His ambitious plan was to introduce eugenics into the national conscience, like a new religion, assuring that humanity shall be represented by the fittest races. Galton wrote: "What nature does blindly, slowly, and ruthlessly, man may do providently, quickly, and kindly." Galton advocated a five-point plan: (a) the newly formed Sociological Society should disseminate knowledge of the laws of heredity and promote their further study, (b) research was needed to determine the extent to which different social classes had contributed to the population over time, (c) records should be collected to establish the circumstances under which large and thriving families of the intellectually adept have most frequently originated, (d) promote social pressure to ensure few eugenically unsuitable marriages would be made, and finally, (e) persistence would be required to realize the success of a eugenic program.

Other participants were more cautious. One speaker expressed skepticism about the assumption that talent and character were heritable. Others wondered about the implications of the newly rediscovered Mendel's laws of heredity, which were being investigated by William Bateson at Cambridge University.

Galton rejected the idea of the inheritance of acquired characters, embracing totally "hard heredity" through selection. He almost discovered Mendel's particulate theory of inheritance; however, he was distracted by his keen interest in continuous, rather than discrete, traits, later called "polygenic" traits.

Galton proceeded to found the "biometric" approach to the study of inheritance, which was based on an extensive use of statistical techniques to investigate continuously varying traits. This method was later followed enthusiastically by Karl Pearson and W.F.R. Weldon. They founded the journal *Biometrika* in 1901, which advanced their approach. The statistical techniques invented by Galton, including correlation and regression (and regression to the mean), formed the foundation for the biometric approach.

In a new book, *Inquiries into Human Faculty and its Development*, Galton (1883) gathered the results of twin studies, ideas in anthropometrics and statistics, as well as psychometrics, psychology, race, and population. His purpose was to touch on various topics more or less connected with that of the cultivation of race or, as we might call it, with eugenic questions. In a footnote, Galton explained that *Eugenics* deals with "questions bearing on what is termed in Greek, eugenes, i.e. good in stock, hereditarily endowed with noble qualities."

Galton's critics were not so kind. Critics from the *Saturday Review*, *Spectator*, and *The Guardian* questioned his noble aim of influencing the "future of humanity," which has been the "grail" of great men since Socrates' time. The *Guardian* critic asked: "Who is to decide whether a man's issue is not likely to be well fitted to play their part as citizens? Do not weak men have strong children, stupid ones wise, wicked good—while, on the other hand, do we not find the weak emanating from the strong, and bad from Good?" (p. 208).

Galton raised fundamental questions regarding human heredity and genetic counseling, which are occupying our attention today. He identified two ethical dilemmas surrounding genetic disease. "Should you inform yourself about whether you carry a genetic disease? And, if you find you have a genetic disease, should you tell other family members? Parents may refrain from doing so through kind motives; but there is no real kindness in the end."

ACTUARIES

Galton was long interested in the occurrence of a disease among the close relatives of a patient, and he was supported by an unlikely source, a well-known actuary W. Palin Elderton, who suggested that life insurances should be helpful in studying the heritability of a disease since when an insured person died, the death certificate was filed in the office along with the original insurance papers. They contained information regarding the causes of death of parents, brothers, and sisters and their ages at death, or their ages if they were alive when the assurance was taken out. The data would facilitate a study of the relationship between inheritance and disease. Galton was advised by Elderton to contact the Institute of Actuaries requesting that they distribute an appropriate circular to insurance agencies, but his plea was rejected.

NOTEWORTHY FAMILIES

In May 1905, upon returning from a vacation in Bordighera on the Italian Riviera, Galton was invited by W.A. Herdman, General Secretary of the British Association (for Advancement of Science) to accept its Presidency. He was nominated by George Darwin and was unanimously supported by the Council. Darwin informed him that Galton need not preside at Council meetings because of his deafness. However, Galton, while appreciating the honor, declined because of frail health. His mental powers, however, were sharp enough to launch a new book, *Noteworthy Families* (Galton and Schuster 1906). He approached the Fellows of the Royal Society (FRS) (listed in the yearbook for 1904) to collect data on their family members and close relatives with respect to their intellectual achievements, seeking evidence for the heritability of high ability. Galton's main conclusion was noteworthiness diminished rapidly as the distance of kinship to the FRS increases, supporting a major point of eugenics that able fathers produce able children in a much larger proportion than the general population. However, Galton did not take into account the very considerable influence (nurture) which fathers exercise upon their sons' careers (Galton 1884, 1889; Galton and Schuster 1906).

One of his last projects was developing a method for measuring "visual resemblance." As one approaches an individual, *general markings* are gradually replaced by *individual features* clearly. Each grade of resemblance was connected with a critical distance, but Galton considered simple distance to be an inadequate measure since one needed to know both distance and size so "the unit is the angle … and size at a distance is expressed as the *angular size*; the distance and area by the *angular area*." He published a detailed explanation in a letter in *Nature* in 1906.

HEREDITY

Galton further reflected on heredity while referring to the publication in 1859 of the *Origin of Species* by his cousin Charles Darwin. We should, however, bear in mind that what was called "heredity" by Galton is, in light of today's knowledge of human genetics, was only a vague notion that refers to a combination of multiple characters and environmental factors. In his *Memories of My Life*, Galton (1909) wrote: "I had been immensely impressed by many obvious cases of heredity among the Cambridge men who were at the University about my own time.... It will be sufficient as an example to give the names of 7 of these Senior Classics, all of whom had a father, brother, or son whose success was as notable as their own" (pp. 288–289).

Galton quoted several individuals who wrote in support of his ideas, including especially a letter from his cousin Charles Darwin, who wrote: "You have made a convert of an opponent in one sense, for I have always maintained that, excepting fools, men did not differ much in intellect, only in zeal and hard work."

However, later research in the twentieth century in human genetics, which includes the mathematical analyses of Bernstein, Fisher, Haldane, Hogben, and Penrose, has led to the destruction of class-bound eugenics. Galton made several assumptions that were not supported by scientific evidence. For instance, he wrote: "The results of the inquiry (of Fellows of the Royal Society) showed how largely the aptitude for science was an inborn and not an acquired gift, and therefore apt to be hereditary.... The paternal influence generally superseded the maternal in early life.... This seemed to afford evidence that the virile, independent cast of mind is more suitable to scientific research than the feminine, which is apt to be biased by the emotions and to obey authority" (*Memories of My Life*, p. 293). No journal today would accept a paper containing such unsupported assertions!

Although Galton was convinced that "nature" is far more important in determining our characters than "nurture" was, he was at a loss to explain the mechanism involved. In his autobiography, although he referred to the recently rediscovered Mendel's laws of inheritance, he failed to grasp their significance. This is surprising because William Bateson (1906) had already discovered the phenomenon of linkage,[1] and Archibald Garrod (1902, 1909) not only discovered the inheritance of metabolic disorders but also explained their recessive inheritance because of parental consanguinity. These events occurred in Cambridge and London, not far from Galton's laboratory in London! Yet, Galton (1909) appeared to be unaware of these remarkable discoveries and wrote in his autobiography: "How far characters generally may be due to simple, or to molecular characters more or less correlated together, has yet to be discovered." He wanted measurements of arm strength, estimates of eye-muscle coordination, sensory stimulation, and persistence of impressions. He wanted these data accompanied by medical histories and photographs.

RACE IMPROVEMENT OR EUGENICS

The last chapter in Galton's (1909) autobiography (*Memories of My Life*) was titled "Race Improvement." Galton advocated what was later termed "negative eugenics." He wrote: "I think that stern compulsion ought to be exerted to prevent the free propagation of the stock of those who are seriously afflicted by lunacy, feeble-mindedness, habitual criminality, and pauperism.... A democracy cannot endure unless it be composed of able citizens; therefore it must in self-defence withstand the free introduction of degenerate stock." Galton drew attention to another practice of humanity: "Many forms of civilisation have been peculiarly unfavourable to the hereditary transmission of rare talent. None of them were more prejudicial to it than that of the Middle Ages, when almost every youth of genius was attracted into the Church and enrolled in the rank of a celibate clergy."

Galton had recorded that the idea of the improvement of the human race was mooted in 1884, and the term "eugenics" was then first applied to it in his *Human Faculty*.

Galton explained in his autobiography his role in approaching the authorities of the University of London to establish a eugenics laboratory. He provided sufficient funds for a small establishment for the furtherance of eugenics, including the salaries of a Research Fellow and a Research Scholar. The University provided rooms and other facilities. The eugenics laboratory was located at University College London, adjacent to Prof. Karl Pearson's biometric laboratory. Galton wrote: "...the young institution promises to be a permanent success" (*Memories of My Life*, p. 321). Indeed it has, under the name of Galton Laboratory.

Galton's lasting impact, as far as the general public is concerned, is in the area of eugenics. The term "eugenics" itself has been used continuously since his death, although not always implying the intent or meaning that Galton originally conceived. Furthermore, it fell into disrepute during the Second World War because of the excessive crimes committed in its name. For that reason, the Galton Professor of Eugenics at University College London, Lionel Penrose, changed the title to Galton Professor of Human Genetics. Nevertheless, the eugenic ideas espoused by Galton were followed and extended by many others, especially J.B.S. Haldane (1923) and Dronamraju (1995) who popularized them in his influential book, *Daedalus, or Science and the Future*. This theme was further continued by Aldous Huxley (1931) in his *Brave New World*, which led to its popularization among the masses, which would have impressed Galton himself! However, it should be pointed out that both books were published long before the Second World War years, when the eugenic crimes were committed. In fact, Haldane became increasingly skeptical of the efficacy of all eugenic programs in his later years.

Havelock Ellis rejected Galton's analogy between animal breeding and human eugenics. Animals were bred for specific purposes "by a superior race of animals, not by themselves." Others have commented that we cannot apply the principles of artificial breeding to man. There is no recognized standard of physical and intellectual perfection. However, there was general support in favor of Galton's advocacy of eugenics, which was partly due to the publication of Galton's speeches in *The Times*.

In his concluding remarks in *Memories of My Life*, Francis Galton wrote: "I take Eugenics very seriously, feeling that its principles ought to become one of the dominant motives in a civilised nation, much as if they were one of its religious tenets.... Man is gifted with pity and other kindly feelings; he has also the power of preventing many kinds of suffering. I conceive it to fall well within this province to replace Natural Selection by other processes that are more merciful and not less effective. This is precisely the aim of Eugenics" (p. 323).

GALTON: A SUMMARY

Francis Galton's contributions ranged from meteorology and exploration to human heredity and eugenics. Although some of his discoveries, such as the use of fingerprinting for personal identification, were very important, his major contribution was *eugenics*.

ENDNOTE

1. Bateson, W. (1906) Bateson first suggested using the word "genetics" to describe the study of inheritance and the science of variation in a personal letter to Alan Sedgwick, dated April 18, 1905. Bateson first used the term "genetics" publicly at the Third International Conference on Plant Hybridization in London in 1906.

REFERENCES

Dronamraju, K.R. (1995) *Haldane's Daedalus Revisited.* New York: Oxford University Press. Foreword by Nobel laureate Joshua Lederberg.
Galton, F. (1869) *Hereditary Genius.* London: Macmillan. Second Edition 1892.
Galton, F. (1883) *Inquiries into Human Faculty and Its Development.* London: Macmillan.
Galton, F. (1884) *Record of Family Faculties.* London: Macmillan.
Galton, F. (1889) *Natural Inheritance.* London: Macmillan.
Galton, F. (1892) *Finger Prints.* London: Macmillan.
Galton, F. (1909) *Memories of My Life.* New York: E.P. Dutton and Company.
Galton, F., and E. Schuster. (1906) *Noteworthy Families.* London: Murray.
Garrod, A.E. (1902) The incidence of alkaptonuria: a study in chemical individuality. *Lancet ii*: 1616–1620.
Garrod, A.E. (1909) *Inborn Errors of Metabolism.* London: Oxford University Press.
Gillham, N.W. (2001) *A Life of Sir Francis Galton.* New York: Oxford University Press.
Haldane, J.B.S. (1923) *Daedalus, or Science and the Future.* London: Kegan Paul.

3 William Bateson (1861–1926)

William Bateson was a British naturalist. He was one of the early Mendelians, whose work helped to establish Mendelian genetics on a firm footing. Bateson was born in Whitby on the Yorkshire coast, the son of William Henry Bateson, Master of St John's College, Cambridge. He was educated at Rugby School and at St John's College in Cambridge, where he graduated with a BA in 1883 with a first in natural sciences.

In 1896, William Bateson married Caroline Beatrice Durham. She was the daughter of a surgeon and very smart. Beatrice would often help Bateson with his work and even worked for income for a short time when financial burdens became a problem, which was pretty much unheard of for females at the time. Letters written by the family throughout the years often had Beatrice holding the family together in times of tragedy and William's bouts of depression (Bateson 1926).

EARLY WORK

William Bateson was keenly interested in embryology and went to the United States to investigate the development of *Balanoglossus*. This worm-like hemichordate led to his interest in vertebrate origins. In 1883–1884 he worked in the laboratory of W.K. Brooks at the Chesapeake Zoölogical Laboratory in Hampton, Virginia. Turning from morphology to study evolution and its methods, he returned to England and became a Fellow of St John's. Studying variation and heredity, he travelled in western Central Asia.

Bateson's major contribution to scientific method was formulated in the lapidary phrase "Treasure your exceptions." He followed this principle, thereby discovering the most important exception to Mendel's laws—linkage.

Besides treasuring his exceptions, Bateson was very skeptical of explanations of many facts that he accepted without question. In particular, he never accepted the word "gene" with its rather wide connotations. Mendel had used the phrase "differendierendes Merkmal," or differentiating character, for his genetical units. If Mendelism had been discovered and accepted in mediaeval Europe, an atomistic theory of substantial forms might have been developed. Bateson used the neutral word "factor."

He was greatly interested in problems of evolution, especially the importance of variation in the evolutionary process. His work on *Balanoglossus* convinced him that meristic variation, that is, variation in the numbers of similar parts in like organisms, occurs in definite steps, with very rare intermediates.

Soon after the rediscovery of Mendelism in 1900, William Bateson and Lucien Cuenot showed that Mendel's principles are applicable to fowl and mice, respectively, thus extending Mendelism to the animal kingdom. We owe the term "genetics" to Bateson, who coined it in 1906 for the study of heredity and variation. Besides "genetics," Bateson also introduced "zygote" for the individual that develops from the fertilized egg (and the fertilized egg itself), and several other terms such as "homozygote" and "heterozygote," "allelomorph" (later shortened to "allele"), as well as "factor," which was replaced by "gene" by Johanssen in 1908.

BALANOGLOSSUS

When Bateson began his studies in science at Cambridge, naturalists were still searching for various clues in tracing the affinities between different branches of the animal kingdom. The most important question of that period was the origin of the vertebrate group. As Punnett (1952) put it: "From what group of invertebrates could they be supposed to descend? Several claimants for the honor had been put forward and argued for, to none had been accorded general recognition." Francis Balfour, then teaching at Cambridge, emphasized the fact that the embryological stages of an animal offered better evidence concerning its affinities than did its adult structure. At that time, *Balanoglossus* was considered a lowly worm of dubious nature with no established affinities to any particular animal group. Through his embryological studies, Bateson showed that *Balanoglossus* must be regarded as the "humblest" member of the vertebrate group (Punnett 1952). Bateson was still in his early twenties at that time and quickly became one of the leading architects of the embryological approach to the study of evolution.

Bateson's zoological studies at Cambridge were influenced by A. Sedgwick and F. Balfour, as well as his contemporary and founder of biometry, W.F.R. Weldon. It was Weldon who encouraged Bateson to study under W.K. Brooks of Johns Hopkins University. Bateson spent the summers of 1883 and 1884 at Hampton, Virginia, and Beaufort, North Carolina, studying the embryology of *Balanoglossus* under W.K. Brooks. Bateson has recorded that it was Brooks who suggested to him to take up heredity as a subject worthy of further research. Two other students of Brooks, T.H. Morgan and E.B. Wilson, also became founders of genetics.

From his studies of *Balanoglossus*, Bateson concluded that the foundations of the theory of evolution by natural selection were far from satisfactory. He believed that the nature of variation itself was poorly understood. He set out to collect materials for further study and discover new facts about the process of biological variation. Those who were accustomed to follow evolution in terms of morphological methods of inquiry warned Bateson that he was embarking on a futile enterprise. However, Bateson, under the influence of Brooks, had already reached the decision that in order to study heredity, one must first understand the nature and causes of variation. To him, this seemed to be the logical course to be followed. However, as subsequent research showed, the effective method was the reverse of this—the origin of variation was better understood only after the mechanisms of segregation and recombination were studied in various species of animals and plants (Sturtevant 1965, p. 25). Indeed, both Bateson and de Vries approached this problem in the same manner—studied variation first and heredity later.

DISCONTINUOUS VARIATION

Bateson's work published before 1900 systematically studied the structural variation displayed by living organisms and the light this might shed on the mechanism of biological evolution and was strongly influenced by both Charles Darwin's approach to the collection of comprehensive examples and Francis Galton's quantitative ("biometric") methods. In his first significant contribution, he shows that some biological characteristics (such as the length of forceps in earwigs) are not distributed continuously, with a normal distribution, but discontinuously (or "dimorphically"). He saw the persistence of two forms in one population as a challenge to the then current conceptions of the mechanism of heredity and says "The question may be asked, does the dimorphism of which cases have now been given represent the beginning of a division into two species?"

MATERIALS FOR THE STUDY OF VARIATION (1894)

In his 1894 book, *Materials for the Study of Variation*, Bateson took this survey of biological variation significantly further. He showed that biological variation exists both continuously, for some characters, and discontinuously for others and coined the terms "meristic" and "substantive" for the two types, respectively. Bateson defined that the Study of Variation is essentially a study of differences between organisms. For each observation of variation, at least two substantive organisms are required for comparison. Bateson wondered: since species are discontinuous, may not the variation by which species are produced be discontinuous too?

Bateson was interested in relating variation and patterns of symmetry. And his attention was particularly focused on two kinds of variation, meristic and substantive. The differences in number and symmetry form an integral and very definite part of the total differences, with special attention being paid to numerical and geometrical differences, which Bateson called "Meristic" changes. These are distinct from variations in the actual constitution or substance in the parts themselves, which Bateson called "Substantive" variations. As Bateson wrote: "These two classes of Variation, Meristic and Substantive, may be recognized at the outset of the study.… An appreciation of this distinction is a first step towards a comprehension of the processes by which the bodies of organisms are evolved" (Bateson 1894, reprinted 1992, p. 23). Among several examples of Meristic variation, Bateson discussed variation in the teeth of vertebrates. Numerical variation may occur in teeth, sometimes by the division of a single member of the series into two and sometimes by a reconstitution of at least a considerable part of the series. The teeth of most vertebrates are differentiated to form a series of organs of differing forms and functions, and the study of variation in teeth may be complicated further by the occurrence of qualitative changes in addition to simply numerical ones.

Bateson added: "The magnitude and Discontinuity of Variation depends on many elements. So far as Meristic Variation is concerned, this Discontinuity is primarily associated with and results from the fact that the bodies of living things are mostly made up of repeated parts—of organs or groups of organs, that is to say, which exhibit the property of 'unity,' or, as it is generally called, 'individuality'. Upon this phenomenon depends the fact that Meristic Variation in number of parts is often integral, and thus discontinuous" (p. 568). Bateson wrote that the phenomenon of serial resemblance is in fact an expression of the capacity of repeated parts to vary similarly and simultaneously. In his concluding reflections, Bateson wrote that it is by the study of variation alone that the problem of evolution can be attacked and that both kinds of variation must equally contribute to its understanding.

One of the questions which Bateson put in the introduction to *Materials for the Study of Variation* has been answered: "The question which the study of variation may be expected to answer, relates to the origin of that Discontinuity of which species is the objective expression. Such Discontinuity is not in the environment; may it not, then, be in the living thing itself." There are at least two reasons for this discontinuity between species and varieties. Living organisms are made up of small and large molecules. Many of the small ones are common to all living things, others to most of them. But the large ones, such as proteins and polysaccharides, are characteristic of species or of genotypes within a species, although some may be found in many different species. Their formation is controlled by large molecules and or sections of large molecules, i.e., genes. These are built up from the ubiquitous types of small molecule and can only vary discontinuously. Many of the discontinuities observed in nature depend on discontinuities in the possible patterns of genes.

SYMMETRY AND MERISTIC REPETITION

Bateson further defined that the study of variation is essentially a study of differences between organisms. For each observation of variation, at least two substantive organisms are required for comparison. Often, it is convenient to compare variation in a parent and offspring. One of the first limitations that Bateson considered is the *magnitude* of variations, which opened the possibility that variation is, to some extent, discontinuous. All organisms are heterogeneous, consisting of organs or parts which in "substance and composition" differ from each other. Bateson further recognized that such structural heterogeneity of material corresponds with a physiological differentiation of function. He emphasized that this differentiation of heterogeneity is found in the bodies of all organisms. It is also found that it is generally orderly and formal, and, according to Bateson, heterogeneity occurs in a particular way and according to geometrical rule. The order of form is generally found in any living body. Heterogeneity is, to some extent, symmetrically distributed around one or more centers. In a majority of cases, these centers of symmetry are themselves distributed about other centers, so that in certain perspectives, the entire body looks symmetrical.

In the simplest possible case of symmetry, there is a series of parts in one direction corresponding to a series of parts in another direction. Symmetry depends on the fact that structures found in one part of an organism are repeated and occur again in another part of the same organism. Symmetrical heterogeneity may be present in a spherical body having a core of different material. This phenomenon of repetition of parts, which generally occurs to form a symmetry or pattern, appears to be a universal character of all living beings. Bateson introduced the term "Merism" to describe this phenomenon.

Bateson (1894) wrote: "Such patterns may exist in single cells or in groups of cells, in separate organs or in groups of organs, in solitary forms or in colonies and groups of forms" (p. 21).

LEAST SIZE OF PARTICULAR TEETH

Continuing his discussion of Meristic variation, Bateson considered the variation in the size of teeth as well as the presence or absence of teeth. He asked: "What is the least size in which a given tooth can be present in a species which sometimes has it and sometime is without it?" In other words, what is the least possible condition, the lower limit of the existence of a given tooth? This question implies an attempt to measure the magnitude or discontinuity of numerical variation in teeth. The least size of a tooth varies for different teeth and different species. It might be expected that any tooth could be reduced to the smallest limits which are histologically conceivable. If we assume that variation is continuous, this would be expected. It is also evident that the size of a tooth in many cases bears a relationship to the size of the adjacent teeth and the general size of the series as a whole. Bateson cited several species where this situation prevails, as well as a few others where exceptions are found. Generally speaking, there is a fairly constant relation between the size of extra teeth and that of the teeth next to which they stand, so that the new teeth are, from the first, of a size and development suitable to their position.

Another phenomenon, the enlargement of the terminal member of a series when it becomes the penultimate, is not only found in teeth but also in ribs, digits, etc., and may be a common feature in a meristic series, so graduated that the terminal member is comparatively small.

PROBLEM OF INTERCROSSING

In common with Darwin, he felt that quantitative characters could not easily be "perfected" by the selective force of evolution, because of the perceived problem of the "swamping effect of intercrossing," but proposed that discontinuously varying characters could. Among other interesting observations he noted were variations in which an expected body part has been replaced by another (which he called "homeotic"). The animal variations he studied included bees with legs instead of antennae; crayfish with extra oviducts; and in humans, polydactyly, extra ribs, and males with extra nipples.

Importantly, Bateson wrote, "The only way in which we may hope to get at the truth [concerning the mechanism of biological Heredity] is by the organization of systematic experiments in breeding, a class of research that calls perhaps for more patience and more resources than any other form of biological enquiry. Sooner or later such an investigation will be undertaken and then we shall begin to know."

In 1897, he reported some significant conceptual and methodological advances in his study of variation. "I have argued that variations of a discontinuous nature may play a prepondering part in the constitution of a new species." He attempts to silence his critics (the "biometricians") who misconstrue his definition of discontinuity of variation by clarification of his terms: "a variation is discontinuous if, when all the individuals of a population are breeding freely together, there is not simple regression to one mean form, but a sensible preponderance of the variety over the intermediates… The essential feature of a discontinuous variation is therefore that, be the cause what it may, there is not complete blending between variety and type." The variety persists and is not "swamped by intercrossing." But critically, he begins to report a series of breeding experiments, conducted by his pupil, Miss E.R.

Saunders, using the alpine brassica *Biscutella laevigata* in the Cambridge botanic gardens. In the wild, hairy and smooth forms of otherwise identical plants are seen together. They intercrossed the forms experimentally, "When therefore the well-grown mongrel plants are examined, they present just the same appearance of discontinuity which the wild plants at the Tosa Falls do. This discontinuity is, therefore, the outward sign of the fact that in heredity the two characters of smoothness and hairiness do not completely blend, and the offspring do not regress to one mean form, but to two distinct forms."

At about this time, Hugo de Vries and Carl Erich Correns began similar plant-breeding experiments. But, unlike Bateson, they were familiar with the extensive plant breeding experiments of Gregor Mendel in the 1860s, and they did not cite Bateson's work. Critically, Bateson gave a lecture to the Royal Horticultural Society in July 1899, which was attended by Hugo de Vries, in which he describes his investigations into discontinuous variation, his experimental crosses, and the significance of such studies for the understanding of heredity. He urges his colleagues to conduct large-scale, well-designed and statistically analyzed experiments of the sort that, although he did not know it, Mendel had already conducted, and which would be "rediscovered" by de Vries and Correns just six months later.

In May 1900, he read a paper before the Royal Horticultural Society in London, in which he described the work of Mendel and its confirmation by de Vries. According to Mrs. Bateson (1928), he first learned of Mendel's work on a train, while going to London from Cambridge to deliver that lecture, and was so impressed by it that he included it in his lecture at once. Bateson (1902) was the leading champion of Mendelism during those early years. In response to an article in *Biometrika* in which Weldon attempted to minimize the importance of Mendel's discovery, Bateson at once prepared a lengthy reply, which was published under the title "Mendel's Principles of Heredity: A Defence" (1902). Punnett (1952) commented: "Weldon's position was remorselessly torn to shreds, and however one may pity the man who stood in his place it cannot be denied that the treatment was well deserved.... Mendel was assured of his hearing, and Bateson began to enrol recruits from among the younger men....at the Cambridge meeting of the British Association in 1904. Bateson became President of the Zoological section... Both Biometricians and Mendelians turned up in full force, and it was before a crowded audience that Bateson delivered his spirited address. The heated debate that succeeded was keenly followed by the packed assembly, and at the end of it there was no mistaking the feeling as to the side with which the victory lay. The Biometrician with his 'Law of ancestral heredity' rapidly faded from the picture, and the was made clear for that remarkable increase in our knowledge of heredity."

LINKAGE

Bateson (1908) reported the first case of incomplete linkage in the sweet pea, involving the gene pairs distinguishing purple flowers from red and long pollen grains from the round. The two dominants, purple and long, were contributed by the same parent, and later in another situation, they noted that one dominant and one recessive of a linked pair came from each parent. These two situations were termed by them as "coupling" and "repulsion[1]," respectively. It was Bateson's passion to study exceptions that led to this discovery.

Among other contributions, mention must be made of Bateson's (1902) suggestion that multiple allelomorphs are involved in determining such phenotypes as stature, and his influence upon Garrod's (1902, 1909) work on the genetic basis of inborn errors of metabolism, in particular interpreting consanguineous parentage of patients with alkaptonuria as evidence of recessive inheritance.

MERISTIC VARIATION

Bateson (1894) laid emphasis on the importance of meristic variation. When the number of like parts such as petals or teeth vary, it is usual to find a whole number of such parts but unusual to find a miniature or incomplete member of the meristic series. Bateson (1894) discussed this problem with reference to the "least size of particular teeth." In Haldane's department at University College London, Gruneberg (1952) studied this phenomenon in the third upper molars of a particular pure line of mice. These teeth are occasionally missing, but when present, they vary in size and can be smaller than the normal, although not rudimentary or incomplete. Gruneberg concluded that the mean size of the tooth rudiment in this line was small and somewhat variable. However, when at a certain critical stage, the rudiment fell below a certain threshold, it regressed or did not develop further. At the other extreme, when a rudiment is too large beyond a certain threshold, it may divide into two or more parts, for instance, an extra limb (Haldane 1958).

Haldane (1958) suggested that the physical principle involved may be similar to surface tension (see Turing 1952). Haldane speculated that the formal physical principles and the mathematical analyses of the formation of vertebrae in a tail may not be unlike that of drops in a liquid filament. Haldane further suggested that other physical causes of discontinuity may lie in the minute particulars of the chemistry and physics of living matter. In his Bateson lecture, Haldane (1958) concluded that this was Bateson's view. Haldane added: "He took the comparison of a zebra's markings with ripple marks quite seriously." Haldane's interpretation of Bateson's views on meristic variation are of great interest. Haldane (1958) wrote: "The forces which hold an atomic nucleus together or disrupt it are like enough to those operative in a drop of water to allow argument from the latter to the former. So, I suggest, are those operative in a tooth or petal rudiment."

Genes that produce meristic variation are not usually constant in penetrance and expression, e.g., polydactyly and ectrodactyly. For instance, Mandeville (1950) reported a gene for absence of upper lateral incisors in man which reduce their size or cause their disappearance. The normal genotype produces a rudiment, which rarely disappears or splits, giving a tooth of fairly uniform size. In the heterozygotes, the rudiment is near the critical size and may either disappear or produce a small abnormal tooth. Meristic variation is seldom strictly Mendelian because of the operation of different causes of discontinuity. Bateson (1894) recognized this heterogeneity and emphasized both heredity and variation. He was particularly interested in emphasizing discontinuous variation at a time when the dominant theme in evolutionary biology was continuous variation, which Darwin had emphasized.

RESEARCH ON INDIAN SPECIES

Haldane's "Bateson lecture" was delivered at the John Innes Horticultural institution in July 1957, shortly before his departure for India. Referring to Bateson's *Materials for the Study of Variation*, Haldane wrote: "when the number of like parts, for example, teeth, vertebrae or petals, can vary, it is usual to find a whole number such parts and unusual to find a miniature or incomplete member of the meristic series… Gruneberg…has studied this phenomenon in the third upper molars of a particular pure line of mice. These teeth are sometimes missing. But when they are present they are variable in size and can be decidedly smaller than the normal, though in no way rudimentary or incomplete…. When, at a certain critical stage, the rudiment fell below a threshold, it regressed or did not develop further. Similarly we may suppose that when a rudiment is too large at a critical stage of development it may divide into two or even more parts, giving an extra limb, for example."

SUBODH KUMAR ROY

Haldane's Indian colleague S.K. Roy pursued several research projects that involved quantitative studies of meristic variation. Roy's research involved the study of variation in like parts in plants, such as petal numbers in flowers in such plants as *Nyctanthes arbor-tristis* and *Jasminum multiflorum* and nipple numbers in goats and cattle. He showed that the petal numbers in *Nyctanthes* vary from four to eight in most flowers. However, when he examined 158,926 flowers, he found that 14 had 9 petals and 1 had 15 petals, with a strong mode at 6 petals. The most surprising finding was that the variance increased towards the end of the flowering season. Haldane commented that if the pots made by a potter became more variable at the end of a day, we should say that he was getting tired, but we do not know what we are to say about a plant. This may well indicate that the developmental canalization (regulation of development) is less rigidly enforced toward the end of the flowering season in plants, but the mechanism that causes it is unknown.

Roy's research was concerned with meristic variation, i.e., variation in like parts, such as teeth and nipples in animals and leaves and petals of plants, which interested Bateson and, later, Haldane. For instance, Roy found the following frequency distribution of petal numbers in three trees of *N. arbor-tristis.*

Number of petals in a flower	4	5	6	7	8	9	15	Total
Number of flowers	451	29,147	109,345	18,956	1,012	14	1	158,926

One of the surprising features Roy found was each tree had its own characteristic mean petal number per flower, and its characteristic variance of petal number, and these statistics generally differed significantly between different trees.

FOUNDING THE DISCIPLINE OF GENETICS

Bateson founded the Genetical Society of Great Britain, coining the term "genetics," which he defined as follows (1906): "the elucidation of the phenomena of heredity and variation: in other words the physiology of Descent, with implied bearing on the theoretical problems of the evolutionist and the systematist, and application to the practical problems of breeders, whether of animals or plants. After more or less undirected wanderings we have thus a definite aim in view." (From Bateson's inaugural address to the third conference on hybridization and plant breeding, 1906).

Bateson became famous as the outspoken Mendelian antagonist of Walter Raphael Weldon, his former teacher, and of Karl Pearson, who led the biometric school of thinking. The debate centered on saltationism versus gradualism (Darwin had represented gradualism, but Bateson was a saltationist). Later, Ronald Fisher and J.B.S. Haldane showed that discrete mutations were compatible with gradual evolution: see the modern evolutionary synthesis.

Between 1900 and 1910 Bateson directed a rather informal "school" of genetics at Cambridge. His group consisted mostly of women associated with Newnham College, Cambridge, and included both his wife Beatrice, and her sister Florence Durham. They provided assistance for his research program at a time when Mendelism was not yet recognized as a legitimate field of study. Since genetics was such an unknown field at the time, Bateson taught very informal classes in Cambridge on the subject. In fact, from 1902 to 1910, he taught his new field at the Newnham College. The women assisted Bateson in his breeding experiments. The results reinforced Mendel's statistics and furthered the field of genetics. The women, such as Muriel Wheldale (later Onslow), carried out a series of breeding experiments in various plant and animal species between 1902 and 1910. The results both supported and extended Mendel's laws of heredity. Hilda Blanche Killby, who had finished her studies with the Newnham College Mendelians in 1901, aided Bateson replicate Mendel's crosses in peas. She conducted independent breeding experiments in rabbits and bantam fowl, as well.

Bateson first suggested using the word "genetics" (from the Greek gennō, γεννώ; "to give birth") to describe the study of inheritance and the science of variation in a personal letter to Adam Sedgwick (1854–1913, zoologist at Cambridge), dated April 18, 1905. Bateson first used the term "genetics" publicly at the Third International Conference on Plant Hybridization in London in 1906. Although this was three years before Wilhelm Johannsen used the word "gene" to describe the units of hereditary information, de Vries had introduced the word "pangene" for the same concept already in 1889, and etymologically, the word "genetics" has parallels with Darwin's concept of pangenesis. Bateson and Edith Saunders also coined the word "allelomorph" ("other form"), which was later shortened to allele.

Bateson codiscovered genetic linkage with Reginald Punnett and Edith Saunders, and he and Punnett founded the *Journal of Genetics* in 1910. Bateson also coined the term "epistasis" to describe the genetic interaction of two independent loci (Bateson and Saunders 1902; Bateson and Punnett 1908).

In his Inaugural lecture for the Professorship of Biology at Cambridge University, Bateson (1908) stated: "...if I may throw out a word of counsel to beginners, it is 'Treasure your exceptions!' When there are none, the work gets so dull that no one cares to carry it further. Keep them always uncovered and in sight. Exceptions are like the rough brick work of a growing building which tells that there is more to come and shows where the next construction is to be." In a wide-ranging lecture, which was primarily addressed to a lay audience, Bateson dealt with segregation and linkage, applications of genetic methods and discoveries, and the physiology of sex determination. Bateson discussed heredity and variation and their bearing on the problem of evolution, stating that the laws of heredity, which were being confirmed and extended in the preceding years, were beginning to put the study of evolution on a firm footing.

HUMAN GENETICS

Bateson's involvement in early human genetics was quite inadvertent. Archibald Garrod, who was a physician at Guy's Hospital in London, observed that there was an increased consanguinity among the parents of patients with certain metabolic disorders, such as cystinuria and alkaptonuria. When Garrod approached Bateson for help in interpreting his results, Bateson deduced that recessive genes were involved because of the obvious connection between parental inbreeding and recessively determined diseases in the offspring. However, Garrod's work (1902, 1909) was ignored for three decades until it was revived, first by Haldane (1930, 1937) and later by Beadle and Tatum (1941). Bateson's participation assured that human genetics had an early start, although it was ignored later.

Bateson (1908) explained explained the meaning of genetic terminology, such as allelomorph, homozygous and heterozygous—terms he invented. He was interested in exploring the connection between consanguinity and recessively inherited diseases. Although clearly outdated by later events, Bateson's lecture was impressive for its historical analysis. He observed that pre-Mendelian methods failed because the methods and data collected were not amenable to analysis. Genetic heterogeneity was emphasized by Bateson with respect to such defects as deaf mutism. He stressed that individual family data must be reported instead of miscellaneous statistics, as was the case with several studies until then.

The most important aspect of Bateson's lecture was the sophistication with which he discussed the subject of genetic etiology, emphasizing not only genetic heterogeneity and inbreeding effects but also sex-linked inheritance as well as complex genetic interaction and linkage.

BATESON'S INTERESTS AND CHARACTER

Bateson was an extremely sensitive boy at school. He described much of his school education as "a time of scarcely relieved weariness, mental retardation, and despair." While discouraged by an outmoded educational system, which emphasized Latin and Greek, Bateson found some comfort in the chemical laboratory. Occasional holidays spent with Linnaeus Cumming (mathematical master and distinguished field naturalist) gave much pleasure. He retained a deep interest in field natural history throughout his life. His early interest in chemistry developed later into full-fledged support for Garrod's inborn errors of metabolism. Indeed, Bateson's genetic interpretation of Garrod's work on biochemical disorders was mainly responsible for the acceptance of that pioneering contribution.

Haldane (1957) knew Bateson well from 1919 until his death. He wrote: "Bateson…could be described as an angry and obstinate old man. But his anger was largely reserved for inaccuracy and loose thinking, and for certain types of injustice. His obstinacy made it difficult to convince him of the truth of theories which had previously been asserted without adequate evidence and were now being substantiated. Correns (1902) in a brilliant guess embodied in a diagram without adequate explanation, had put forward the theory linear arrangement of genes on chromosomes. Bateson, quite rightly, had not accepted the hypothesis. When Bridges and Sturtevant proved it, he was hard to convince, though he was finally convinced of the fact that genes were associated with chromosomes. On the other hand, he instantly accepted new generalisations provided they were statements of fact not involving theoretical superstructures. Thus, he was, I think, the first person to believe my own generalisation about sex-ratio and unisexual sterility in hybrid animals, though not, of course, the rather incoherent explanation of it which I gave. He then displayed a characteristic combination of anger at my ignorance with great generosity in helping me with his immense knowledge of the by-ways of entomological literature. To me, at least, he showed no signs whatever of a senile failure of original thought. On the contrary his last posthumously published paper on the genetics of bolting in root crops initiated a line of research which was later developed by Waddington in his studies on genetic assimilation."

Punnett (1952) had recorded that those who came into contact with Bateson readily felt that they were in the presence of a genius and an intensely virile character, and also a man of intellectual and spiritual power. Bateson was a man of considerable intellectual courage, who did not hesitate to swim against the tide of continuous variation in the post-Darwinian era, studying meristic variation for a period of

10 years. He was well-known as a stubborn man, who was one of the last skeptics to be convinced of the chromosome theory of inheritance.

In summary, Bateson was a man of great vision. As an outstanding biologist, he believed that the study of variation was an essential prerequisite for understanding genetic phenomena. By the time Mendel's work was rediscovered, Bateson reached similar conclusions independently. His study of discontinuous variation ("meristic variation") was undertaken at a time when the predominant theme in biological research was the study of continuous variation as emphasized by Darwin himself. The same stubbornness precluded his acceptance of the chromosomal basis of heredity for several years.

Bateson was a skeptic, who introduced the lapidary phrase—"treasure your exceptions"—into biological research. This attitude led to his discovery of the first exception to Mendelism—linkage. He will be long remembered as a staunch supporter of Mendelian genetics in its early struggling years, especially when under attack by the biometricians in the early decades of the twentieth century.

ENDNOTE

1. With his biochemical background, it was only natural for Haldane (1937, 1954) to introduce "cis" and "trans" into genetic terminology to replace coupling and repulsion, respectively.

REFERENCES

Bateson, B. (1926) *William Bateson: Naturalist.* Cambridge: University Press.

Bateson, W. (1894) *Materials for the Study of Variation.* New York: Macmillan Co.

Bateson, W. (1902) *Mendel's Principles of Heredity: A Defense.* Cambridge: University Press.

Bateson, W. (1908) *Genetics Applied to Medicine.* Huxley's address to the Neurological Society of London. Reproduced in B. Bateson's Memoir (1926).

Bateson, W. (1908) Inaugural Lecture for Chair of Biology, Cambridge University. From B. Bateson's Memoir (1926) and Punnett, R.C. (1952).

Bateson, W., and E.R. Saunders (1902) Experimental studies in the physiology of heredity. *Rep. Evol. Comm. Roy. Soc., 1*: 1–160.

Bateson, W., and R.C. Punnett (1908) Experimental studies in the physiology of heredity. *Poultry Rep. Evol. Comm. Roy. Soc., 4*: 18–35.

Beadle, G.W., and E.L. Tatum (1941) Genetic control of biochemical reactions in neurospora. *Proceedings of the National Academy of Sciences, 27(11)*: 499–506.

Garrod, A.E. (1902) The incidence of alkaptonuria: A study in chemical individuality. *Lancet, ii*: 1616.

Garrod, A.E. (1909) *Inborn Errors of Metabolism.* London: Oxford University Press.

Gruneberg, H. (1952) Genetical studies on the skeleton of the mouse: IV. Quasi-continuous variations. *J. Genet., 51*: 95–114.

Haldane, J.B.S. (1930) *Enzymes.* London: Longmans, Green and Co.

Haldane, J.B.S. (1937) The biochemistry of the individual. In: J. Needham and D.E. Green (eds.), *Perspectives in Biochemistry.* Cambridge: University Press.

Haldane, J.B.S. (1954) *The Biochemistry of Genetics.* London: Allen & Unwin.

Haldane, J.B.S. (1958) The theory of evolution before and after Bateson. *J. Genet., 56*: 11–27.

Johanssen, W. (1908) *Elemente der exakten Erblichkeitslehre.* Jena: Fischer.

Mandeville, L.C. (1950) Congenital absence of permanent maxillary lateral incisor teeth; a preliminary investigation. *Ann. Eugen., 15*: 1–14.

Punnett, R.C. (1952) William Bateson and Mendel's principles of heredity. *Notes Rec. Roy. Soc. Lond., 9*: 336–347. Originally published in *The Edinburgh Review*, 1926, shortly after Bateson's death.

Sturtevant, A.H. (1965) *A History of Genetics.* New York: Harper & Row.

Turing, A.M. (1952) The chemical basis of morphogenesis. *Phil. Trans. Roy. Soc. B, 237*: 37–72.

4 Thomas Hunt Morgan (1866–1945)

Thomas Hunt Morgan was born on September 25, 1866, at Lexington, Kentucky. He was educated at the University of Kentucky, where he took his BS degree in 1886, subsequently doing postgraduate work at Johns Hopkins University, where he studied morphology with W.K. Brooks (McCullough 1969) and physiology with H. Newell Martin.

He showed a keen interest in natural history, and at the age of 10, he started collecting birds, birds' eggs, and fossils during his life in the country. During the summer of 1888, he was engaged in research for the US Fish Commission at Woods Hole. In 1890, he spent the summer at the Marine Biological Laboratory (MBL) in Woods Hole, thus beginning a long-term association with the MBL as a summer investigator and trustee. In 1890, he obtained his PhD degree at Johns Hopkins University. In that same year, he was awarded the Adam Bruce Fellowship and visited Europe, working especially at the Marine Zoological Laboratory at Naples, which he visited again in 1895 and 1900 (Morgan 1896). At Naples, he met Hans Driesch. The influence of Driesch, with whom he later collaborated, created a permanent interest in experimental embryology (Morgan and Driesch 1895).

Morgan was appointed as Associate Professor of Biology at Bryn Mawr College for Women, where he stayed until 1904, when he became Professor of Experimental Zoology at Columbia University, New York. He remained there until 1928, when he was appointed Professor of Biology and Director of the G. Kerckhoff Laboratories at the California Institute of Technology. He remained there until 1945. During his later years, he had his private laboratory at Corona del Mar, California.

During Morgan's 24-year period at Columbia University, his attention was drawn toward the bearing of cytology on the broader aspects of biological interpretation. His close contact with E.B. Wilson offered exceptional opportunities to come into more direct contact with the kind of work that was being actively carried out in the zoological department at that time.

Morgan's early work showed him to be critical of Mendelian conceptions of heredity, and in 1905, he challenged the assumption then current that the germ cells are pure and uncrossed and, like Bateson, was skeptical of the view that species arise by natural selection. In 1909, he began his famous research on the fruit fly *Drosophila melanogaster,* with which his name will always be associated.

W.E. Castle and his associates first used the fruit fly (*Drosophila*) for their research on the effects of inbreeding, and through them, F.E. Lutz became interested in it and the latter introduced it to Morgan, who was looking for less expensive material that could be bred in the very limited space at his command. Shortly after Morgan commenced work with this new material in 1909, a number of striking mutants turned up. His subsequent studies on this phenomenon ultimately enabled him to determine the precise behavior and exact localization of genes. His great success in genetic research using the fruit fly also resulted in an unintended consequence—making *Drosophila* the most famous experimental organism in the history of science, which, incidentally, also earned him a Nobel Prize in 1933. However, Morgan's closest pupil and a successful scientist himself, A.H. Sturtevant (1967), wrote that "even if he had never seen a *Drosophila* in his life, his place in the history of biology would be a high one. Like most biologists—zoologists—of the period, he was trained in comparative anatomy and especially in descriptive embryology. His thesis was done on the embryology of the *Pycnogonidae*, the sea spiders, based on material collected…at Woods Hole. This paper was based on topics of descriptive embryology with the emphasis on phylogeny. This was the custom of the time—this what a zoologist did."

The importance of Morgan's earlier work with *Drosophila* was that it demonstrated that the associations known as *coupling* and *repulsion*, discovered by English workers in 1909 and 1910 using the sweet pea, are in reality two aspects of the same phenomenon, which was later called *linkage*. In 1910, Morgan reported the sex-linked inheritance of white eyes in *Drosophila* and discovered other mutant types as well (Morgan 1910a,b, 1911). One of those, which was later called "rudimentary," was also found to be sex linked. Morgan quickly deduced that one could now test the recombination between genes that lie in the same pair of chromosomes. The result of crosses between white and rudimentary showed that recombination did occur. This was a major step in accepting the chromosome interpretation of Mendelian inheritance. However, it turned out that white and rudimentary lie far apart in the X chromosome; hence, no obvious linkage between them could be observed. But soon, other sex-linked genes were discovered, which made it possible to demonstrate linkage between them. The term "crossing over" was introduced and it became clear that closely linked genes lie close to each other, while more loosely linked genes farther apart. In 1912, Morgan reported the first sex-linked lethal in *Drosophila*. This was also the first lethal in which the heterozygote had no detectable phenotypic effect. This gene had no dominant effect, but males carrying it invariably died. The introduction of marker genes made it possible to locate the lethal on the map of the X. It soon became evident that such recessive lethals constitute the largest single class of mutants in *Drosophila*. They have been very useful in mutation studies and in building complex types of stocks.

Morgan's work also showed that very large progenies of *Drosophila* could be bred. The flies were, in fact, bred by the million, and all the material thus obtained was carefully analyzed. His work also demonstrated the important fact that spontaneous mutations frequently appeared in the cultures of the flies. On the basis of the analysis of the large body of facts thus obtained, Morgan put forward a theory of the *linear arrangement* of the genes in the chromosomes, expanding this theory in his book, *Mechanism of Mendelian Heredity* (1915). Morgan's collaborators in this important work were A.H. Sturtevant, C.B. Bridges, and H.J. Muller. Among these, H.J. Muller was awarded the Nobel Prize in 1946 for his production of mutations by means of X-rays (Carlson 1966, 1981).

E.B. WILSON

T.H. Morgan (1866–1945) and E.B. Wilson (1856–1939) were close friends. Their careers were remarkably similar. Both obtained their PhD at Johns Hopkins, Wilson in 1881 and Morgan in 1890. Both went to Europe, especially to the Biological Station at Naples, where they met several European biologists and made lasting friendships. Both taught at Bryn Mawr College and later joined the Zoology faculty at Columbia University, Wilson in 1891 and Morgan in 1904. Both started their early research in embryology. During summers, both worked at the Woods Hole MBL in Massachusetts, where they were neighbors and their families were very close friends. Yet, according to Sturtevant (1965), their personalities were quite different; Morgan was the idea man, the romantic who dreamed of new ideas and experiments in quick succession, while Wilson was the classic type who was more concerned with the perfection of his product, with setting his ideas in the proper relation to each other and to the main body of science.

At Columbia, Wilson and Morgan working, in proximity, often found intellectual stimulation and comfort. H.J. Muller, who was not accepted at first by Morgan to join his *Drosophila* group, found support from Wilson for some time but was later able to join Morgan's group.

In addition to this genetical work, however, Morgan made contributions of great importance to experimental embryology and to regeneration. So far as embryology is concerned, he refuted by a simple experiment the theory of Roux and Weismann that, when the embryo of the frog is in the two-cell stage, the blastomeres receive unequal contributions from the parent blastoderm, so that a "mosaic" results. Among his other embryological discoveries was the demonstration that gravity is not, as Roux's work had suggested, important in the early development of the egg.

Although so much of his time and effort were given to genetical work, Morgan never lost his interest in experimental embryology and he gave it, during his last years, increasing attention.

To the study of regeneration, he made several important contributions, an outstanding one being his demonstration that parts of the organism that are not subject to injury, such as the abdominal appendages of the hermit crab, will nevertheless regenerate, so that regeneration is not an adaptation evolved to meet the risks of loss of parts of the body. This was discussed in his book *Regeneration*.

Morgan's other books include *Heredity and Sex* (1913), *The Physical Basis of Heredity* (1919), *Embryology and Genetics* (1924), *Evolution and Genetics* (1925), *The Theory of the Gene* (1926), *Experimental Embryology* (1927), and *The Scientific Basis of Evolution* (2nd. ed., 1935), all of them classics in the literature of genetics.

Morgan was made a Foreign Member of the Royal Society of London in 1919, where he delivered the Croonian Lecture in 1922. In 1924, he was awarded the Darwin Medal, and in 1939, the Copley Medal of the Society.

For his discoveries concerning the role played by the chromosome in heredity, he was awarded the Nobel Prize in 1933.

Sturtevant, in his book *A History of Genetics*, wrote: "In 1909, the only time during his twenty-four years at Columbia, Morgan gave the opening lectures in the undergraduate course in beginning zoology. It so happened that C.B. Bridges and I were both in the class. While genetics was not mentioned, we were both attracted to Morgan and were fortunate enough, though both still undergraduates, to be given desks in his laboratory the following year (1910–1911). The possibilities of the genetic study of *Drosophila* were then just beginning to be apparent; we were at the right place at the right time. The laboratory where we three raised *Drosophila* for the next seventeen years was familiarly known as "The Fly Room" (Sturtevant 1959).

FLY ROOM

What was later called "The Fly Room" at Columbia University was in fact a small laboratory (16 by 23 feet), with eight desks crowded into it. Besides the three—Morgan, Sturtevant, and Bridges, who worked there for 17 years—there was a steady stream of American and foreign students. Muller graduated from Columbia in 1910 and spent the winter of 1911–1912 as a graduate student of physiology at Cornell Medical College. He returned later to take an active part in the *Drosophila* work.

Morgan, Sturtevant, Bridges, and (later) Muller worked side by side. Morgan worked while standing, while the others sat at their desks and worked continuously while chatting all the time.

In his book *A History of Genetics*, Sturtevant (1965, pp. 49–50), wrote: "There was a give-and-take atmosphere in the fly room. As each new result or new idea came along, it was discussed freely by the group. The published accounts do not always indicate the sources of ideas. It was often not only impossible to say but was felt to be unimportant, who first had an idea. A few examples come to mind. The original chromosome map made use of a value represented by the number of recombinations divided by the number of parental types as a measure of distances; it was Muller who suggested the simpler and more convenient percentage, the recombinants formed of the whole population. The idea that 'crossover reducers' might be due to inversions of sections was first suggested by Morgan, and this does not appear in my published accounts of the hypothesis. I first suggested to Muller that lethals might be used to give an objective measure of the frequency of mutation. These are isolated examples, but they represent what was going on all the time. I think we came out somewhere near even in this give-and-take, and it certainly accelerated the work."

With reference to his own contribution to producing the first chromosome map, Sturtevant wrote: "In the later part of 1911, in conversation with Morgan…I suddenly realized that the variations in strength of linkage, already attributed by Morgan to differences in the spatial separation of the genes, offered the possibility of determining the sequences in the linear dimension of a chromosome. I went home and spent most of the night (to the neglect of my undergraduate homework) in producing the first chromosome map, which included the sex-linked genes y, w, v, m, and r, in the order and approximately the relative spacing that they still appear on the standard maps" (Sturtevant 1965, p. 47).

However, all was not idyllic within the *Drosophila* group. H.J. Muller, perhaps Morgan's most independent and brilliant student, felt that Morgan had a tendency to use his students' ideas without fully acknowledging them. While recognizing Morgan's unsurpassed abilities as a leader, his fiery and quick imagination, and his frequently penetrating insights, Muller claimed that Morgan was frequently confused about rather fundamental issues involved in the work—such as the theory of modifier genes or the supposed swamping effect of dominant genes in a population. According to Muller, Morgan frequently had to be "straightened out" on such issues by hardheaded arguments with his students—mostly Muller and Sturtevant, with occasional help from E.B. Wilson. Sturtevant concurs with this evaluation at least with regard to the idea of natural selection, which he claims Morgan persisted in misunderstanding until as late as 1914 or 1915.

Morgan and Muller had a difficult relationship. Muller appears to have felt that Morgan made it difficult for him to obtain favorable and permanent teaching positions in the United States. When Muller returned to Columbia as an instructor in zoology in 1918, he hoped to remain there permanently. However, his contract was terminated in 1920, and he had to seek another position at the University of Texas in Austin. Muller thought that Morgan influenced this decision. However, Morgan's leading pupil, Alfred Sturtevant, believed that Morgan not only did not discourage Muller's further appointment but also actively encouraged it. Morgan, who was apparently on sabbatical leave in California at that time, urged Wilson to keep Muller on. However, Wilson, as administrative officer of the department, had found Muller so difficult that he was unwilling to support Muller's promotion. There were several important differences between Morgan and Muller. The most obvious is their age difference. Muller represented the younger generation. They came from widely divergent social classes. Morgan had a relaxed, genial personality of a successful middle class professor, whereas Muller represented the working class, immigrant community, struggling to improve its social and economic well-being. Aggressiveness was alien to Morgan's world. He was laidback and casual, enjoyed cracking jokes with his students and colleagues in a relaxed manner. With his genial background, Morgan must have felt that Muller was occasionally "pushy." Morgan was also uncomfortable with Muller's leftist political attitude. Morgan was reported to have warned some of his colleagues who accompanied Muller to a socialist rally in downtown Manhattan. Morgan and Muller saw the world from different perspectives.

What is clear from an analysis of the reports of many people who worked in the "fly room" during the years 1911–1915 was that Morgan's primary role was that of leader and stimulator. He was constantly coming up with ideas—some wrong, others right—and throwing these out to the eager and brilliant group of young people whom he had working with him. That many of the most far-reaching ideas (such as a quantitative method of making chromosome maps, crossover interference, modifier genes) were first proposed by his students, not directly by Morgan, is also clear. His genius in the development of the *Drosophila* work may have rested more in selecting the right organism, bringing together the right group of people, in working together with them in a democratic and informal way, and in letting them alone, than in producing all the major ideas himself. In fact, it is clear from an analysis of Morgan's published work that he frequently proposed ideas "off the top of his head" and was not always careful to work out their details or implications.

In spite of his aptitude for humor and practical jokes, Morgan always took his work and personal relations very seriously. He was never idle. Whenever he was sitting down, he was always reading and writing. In the afternoon, he went home around 6:00 PM but always carried a brief case full of papers and books that he studied after dinner, often working until midnight or later. In Woods Hole during the summers, he took time to join his family in a picnic or out on E.B. Wilson's sailboat. Then he would work the rest of the afternoon.

J.B.S. HALDANE

Morgan took interest in the mathematical analyses of evolution at the population level, which Haldane carried out in a series of papers from 1924 onward, culminating in his classic work, *The Causes of Evolution*, in 1932. According to Morgan's biographer, Garland E. Allen, Morgan saw the importance of this kind of thinking in clarifying many evolutionary problems, "For example, he saw that such an analysis would help to show conclusively that a dominant gene does not necessarily have any better chance of survival in a population than a recessive. He pointed out that questions about the fate of particular genes in a population could only be answered by an analysis of large amounts of numerical data. Morgan had even begun to recognize that what we today would call 'selection value' had to be assigned to different genotypes."

Morgan wrote: "These theoretical considerations do no more than suggest certain possibilities concerning the theory of natural selection. Before we can judge as to its actual efficiency we must able to state how much of a given advantage each change must add to give it a chance to become established in a population of a given number. Since only relatively few of the individuals produced in each generation become the parents of future generations, numbers count heavily against any one individual establishing itself" (Allen 1978, p. 311). Morgan wrote: "This is a most difficult problem for which we have practically no data, and as yet only the beginning of theoretical analysis has been made of this side of the selection problem. Haldane has developed a partial analysis of the problem for a few Mendelian situations. He points out that the problem is extremely complex and that there is at present not much quantitative information to furnish material for such a study of natural selection by means of gene mutations" (Morgan 1932b, p. 142).

BATESON'S VISIT

William Bateson visited Morgan's laboratory and stayed with him in his home in December of 1921. Bateson wrote of his impressions to his wife in England: "Saw something of Sturtevant yesterday, and thoroughly like him. He and Bridges are quite different from the type I expected. Both are quiet, self-respecting young men.

Sturtevant had more width of knowledge—Bridges scarcely leaves his microscope. I wonder whether they are not the real power in the place. Morgan supplies the excitement in the place. He is in a continual whirl—very active and inclined to be noisy." Bateson's letter would have shocked the members of the *Drosophila* group.

Bateson did not know what to make of Morgan. Unlike European professors, Morgan was on familiar terms with his students. Bateson wrote his impressions of Morgan to his wife: "Nothing that he says is really interesting or original. I like him and I don't. I wish to goodness I had not been made to live in his house. I have to act more affection than I always feel. There is no meanness about him, and I recognize his sincerity—but no size above the ordinary."

On a personal level, Bateson found Morgan difficult to deal with. After his first day at the Morgan house, Bateson wrote home to his wife in England: "The evening and the morning have been the first day. We had a long discursive talk last night, avoiding anything like an actual clash. I fancy more or less that is to be the order of our proceedings. Morgan has a rough good nature that attracts, but I have just the same impression that I got 14 years ago, that he is of no considerable account. His range is so dreadfully small. Off the edge of a very narrow track, he is not merely puzzled, but lost utterly.... He is totally free from pretense—he is almost without shame in his ignorance—I mean of things scientific."

A few days later, Bateson wrote another letter: "I wish I liked Morgan better. I think he has made a great discovery, but I can't see in him any quality of greatness." Professionally, however, Bateson found Morgan's work unassailable. The impressions of the *Drosophila* work made him ultimately reconsider his position on chromosomes. He continued his letter: "Yet, there is no denying the fact that by intensive methods they (Morgan and his group) have got a long way.... Part of each morning is devoted to chromosomes. I can see no escape from capitulation on the main point. The chromosomes must in some way be connected with our transferrable characters.... Cytology here is such a common-place that everyone is familiar with it. I wish it were so with us. Bridges inspires me with complete confidence." A few days later, he wrote again: "I am heartily glad I came. I was drifting into an untenable position which would soon have become ridiculous. The details of the linkage theory still strike me as improbable. Cytology, however, is a real thing—far more important and interesting than I had supposed. We must try to get a cytologist."

TRAINING GROUND

Morgan's laboratory became the training ground for a school of Mendelian genetics—one generation of which emphasized particularly the relationship between genes and chromosomes. Besides Bridges, Sturtevant, and Muller, Morgan's students or postdoctoral associates at Columbia included Alexander Weinstein, E.G. Anderson, H.H. Plough, Theodosius Dobzhansky, L.C. Dunn, Donald Lancefield, and Otto Mohr. These workers, and many others, developed what has come to be called "classical genetics"—that is, genetics at the chromosome level.

Morgan's ideas ranged freely over the broad areas of genetics, embryology, cytology, and evolution. Soon after the *Drosophila* work had gotten under way, he saw that the Mendelian concept could throw considerable light on the problem of natural selection. In 1916, Morgan published his second major work on evolution, *A Critique of the Theory of Evolution* (revised in 1925 as *Evolution and Genetics*), showing clearly his altered views about Darwinian selection. Although he had previously regarded de Vries's mutation theory as an alternative to natural selection, Mendelism now provided a mechanism for understanding the Darwinian theory itself. Mendelian variations (called also "mutations" by Morgan) were not as large or as drastic as those postulated by de Vries. Yet they were more distinct and discontinuous than the slight individual variations which Darwin had emphasized. Most important, they could be shown to be inherited in a definite pattern and were therefore subject to the effects of selection. The Mendelian theory filled the gap which Darwin had left open so long before.

Morgan found it more difficult to make explicit the relationships that he instinctively knew existed between the new science of heredity and the old problems of development (such as cell differentiation or regeneration). In 1934, Morgan attempted to make these connections in a book titled *Embryology and Genetics*. The work proved to be less an analysis of interrelated mechanisms and more a summary of efforts in the two separate fields. Morgan knew well that the time was not ripe for understanding such problems as how gene action could be controlled during development. Yet, *Embryology and Genetics* served an important function of keeping before biologists the idea that ultimately any theory of heredity had to account for the problem of embryonic differentiation. Morgan wisely refrained from drawing conclusions or proposing hypotheses that could not be experimentally verified. One of the most important characteristics of his genius was the ability to restrict the number and kinds of questions that he asked at any one time. For example, by focusing primarily upon the relationships between the Mendelian theory and chromosome structure, he was able to work out the chromosome theory of heredity in great detail. In contrast, other workers, such as Richard Goldschmidt, tried to make those relationships more explicit than the evidence at the time would allow. Consequently, they were often drawn into speculative arguments where no concrete advances could be made.

1925–1945 PERIOD

After the mid-1920s, Morgan's interest shifted away from the specific *Drosophila* work. His new concerns took two forms. One was the attempt to summarize the conclusions deriving from his genetic studies. To this category belong those broader works relating heredity to development and evolution. His other interest turned him to some of the original problems of development and regeneration which had launched his career 35 years previously. During the summers at Woods Hole, and especially after his move to California in 1928, Morgan returned to studies of early embryonic development. The cleavage of eggs, the effects of centrifuging eggs before and after fertilization, the behavior of spindles in cell division, preorganization in the egg, self-sterility in ascidians, and the factors affecting normal and abnormal development were some of the problems in experimental embryology. They represented the type of biological work that Morgan was most interested in. Although he approached the *Drosophila* studies enthusiastically, the mathematics of mapping and many other highly technical problems were less interesting to him than working directly with living organisms. Morgan had the naturalist's love of whole organisms and of studying organisms in their natural environment. He was a good naturalist with a knowledge of many species.

MORGAN'S METHODS

Morgan was a thorough experimentalist. He believed that unbounded speculation was detrimental to the development of sound scientific ideas. However, he believed that the formulation of experiments was essential to developing new concepts and experimental ideas. He thought that the only acceptable hypotheses were those that suggested experimental tests.

By 1909, considerable breeding data suggested that Mendel's laws had wide application. Yet Morgan remained skeptical because he saw no evidence that Mendel's "factors" had any reality. What began to change his mind was the fact that he could test the Mendelian theory by breeding experiments with evidence from cytology in the observed behavior of chromosomes during gametogenesis. As soon as he saw that the white-eye mutation acted as if it were part of the X chromosome, he began to view the Mendelian theory in a completely different light.

Even though Morgan saw Mendel's "factors" as having a possible material basis in the chromosome, he did not automatically accept the idea that genes were physical entities. The physical existence of genes was unnecessary for the validity of the original Mendelian theory and for much of the *Drosophila* work. The Mendelian–chromosome theory stood on its own as a consistent scheme without necessarily being tied to observable physical structures. Until cytological technique materials were developed in the late 1920s, it was impossible to determine a point-by-point correspondence between genetic maps (determined by crossover frequencies) and chromosome structure (determined cytologically). In the preface to *The Mechanism of Mendelian Heredity*, he and his coauthors admitted that Mendel's theories could be viewed independently of chromosomes.

It is significant that the term "gene" did not appear in the first edition of *The Mechanism of Mendelian Heredity.* The authors refer to "factors" just as Bateson had earlier referred to them, although they were aware of the term "gene," which was coined by Wilhelm Johanssen in 1909. The reason they chose to use "factor" at that time was to emphasize the material basis for the mechanism of heredity. Johanssen used the term "gene" at first as a purely abstract concept, consciously disassociated from any of the existing theories of hereditary particles at that time. They were attempting to distance themselves from the viewpoint of Johannsen. Morgan's biological perspective greatly influenced the development of the gene concept, viewing it not simply as a structural but also as a functional unit. There had been a long history of describing Mendelian genes in biochemical and physiological terms. Muller and Altenburg, as well as Sturtevant, were strongly committed to the idea that genes interact through biochemical processes to produce adult characters. Morgan viewed the overall biological, and particularly the developmental, phenomena rather than the specifically chemical relationships in considering the gene as a functional unit. In the absence of biochemical and physiological evidence, Morgan et al. considered

Mendel's law as the adequate scientific explanation of heredity because it fulfills all the requirements of causal explanation (Morgan et al. 1915, p. 227).

However, the chromosomes furnish exactly the kind of mechanism that the Mendelian laws call for, and there is an ever-increasing body of information that points clearly to the chromosomes as the bearers of the Mendelian factors. Preliminary evidence suggested that genes were real entities on chromosomes.

Morgan did not believe that biology should be reduced simply to expressions of physical and chemical interactions. However, he urged other biologists to employ the quantitative and rigorous methodology that had been so successful in experimental embryology. But, reductionism was too simplistic for Morgan. He did believe, however, that biology should be placed on the same footing as the physical sciences: that is, that the criteria for evaluating ideas in biology should be the same as those in physics and chemistry.

Occasionally, Morgan was known to use unorthodox methods, to say the least. Sturtevant (1967) narrated one such instance in his "Reminiscences of Morgan," which he delivered at the MBL in Woods Hole, Massachusetts in 1966:

> He was interested in this period—and for the rest of his life, also—in self-sterility in the ascidian, *Ciona*. For example, if you mix eggs and sperm from the same individual, normally nothing happens. But sometimes self-fertilization does occur. And one of the questions was why? What brings this about? How does it happen? And Morgan had a nice hypothesis: may be the acidity of the water is responsible. Let's see what pH changes will do. But being Morgan, he didn't set up measured amounts of or concentrations. What he did was to take a dish in which eggs and sperm were present and squeeze a lemon over it. And it worked. Then he studied it in more detail after that. This was one of the most successful experiments in the field. (Sturtevant 1967)

Morgan delivered the Presidential address of the Sixth International Congress of Genetics, which was held in Ithaca at Cornell University in 1932 (Morgan 1932a). Morgan outlined the relationships between the new genetics and embryology, cytology, physiology, and evolution. He emphasized the important role of the Mendelian–chromosome theory in understanding the evolutionary process. He emphasized in particular that by mutation of discrete Mendelian genes, small individual variations of the type Darwin had emphasized could be achieved. Through his research on such mechanisms as modifier genes and pleiotropy, Morgan realized that the kind of inheritance that Darwin visualized as the raw material of evolution was in fact produced by various mechanisms that are brought on by Mendelian inheritance. Furthermore, Morgan came to realize the complicated nature of the selection process in the course of evolution when the multiple effects of genes are involved.

Morgan listed what he saw as the outstanding problems of genetics for the immediate future: (a) the physiological processes involved in the replication of genes; (b) the cytology and genetics of the physical events during synapsis and crossing-over; (c) the relations of genes to characters, the way the information of genes is translated physiologically into adult characters; (d) the physical and chemical changes involved in the mutation process; and (e) the application of genetics to horticulture and animal breeding. Amazingly, many of his predictions came true in the following decades. Genetic research utilizing the fruit fly as well as several other species produced much information in response to the challenges posed by Morgan in his Presidential Address.

NOBEL PRIZE

On October 20, 1933, when a phone call informed Morgan of his Nobel Prize, he took it all in his stride with a certain bemused calmness.

His wife, Lilian Morgan, described the day in a letter to their children: "It happened this morning at about 11. Father came into my room (at the Kerckhoff laboratories of Caltech) with the news. He had been called to the telephone for long-distance from New York and thought, of course, that it was one of you, but it was a relay of a cable from Stockholm!… Dr. Mohr told me last summer that many people had wanted this, and their message today is one of the nicest…. The Tolmans came over; in the office it has been a hectic day between the reporters and the messages, not to mention the regular business….I forgot to say that we shall have to go to Stockholm…. Father is quietly sitting in his chair reading Anthony Adverse just as though nothing had happened. I am sure if he could pause a moment he would send his love to you all with that from…. Mother."

Morgan appreciated the significance of the award as an indication that the importance of genetics to medicine was at last being recognized. Both Morgan and his wife were pleased that general biology, through one of its disciplines, namely, genetics, was accorded due recognition as a part of physiology and medicine. What was indeed very pleasing is that Morgan agreed to have his photo taken by the press only if it also includes a group of neighborhood children as well.

NOBEL LECTURE

Morgan's Nobel lecture was titled "The Relation of Genetics to Physiology and Medicine." Morgan emphasized that genetics is closely related to physiology and medicine. He gave several examples of how genes could be viewed as influencing an organism's physiology, embryonic differentiation and biochemical processes in development such as the recent studies of Beadle and Ephrussi on the chemistry of eye-pigment development in *Drosophila*. Morgan explained that a scientific understanding of genetics is helpful in establishing human genetics on a firm footing. For instance, the study of genetics with *Drosophila* had taught a great deal about the interactions that occur between heredity and environment. Morgan emphasized that both heredity as well as environmental factors play important roles in expressing a phenotype such as a human disease. He urged that medical science must take the lead in improving human health by eliminating the causes of human disease.

THE CALIFORNIA INSTITUTE OF TECHNOLOGY

In 1927, George Ellery Hale invited Morgan to come to the California Institute of Technology to establish its first division of biology. After weighing the matter for a short time, Morgan accepted with enthusiasm. Although he had doubts about his abilities as an administrator, the opportunity of heading a new department seemed to far outweigh the possible administrative problems. This move offered several advantages to Morgan, who was then 62. Because the Kerckhoff Laboratory had a generous endowment (from the Kerckhoff family) as well as assistance from the Rockefeller Foundation, Morgan was able from the start to attract a first-rate staff. At Caltech, Morgan developed a modern department based on the concept of biology as he thought it should be studied and taught, where the new experimentalism could play a predominant role. Moving to Caltech also provided Morgan with the opportunity of achieving on a permanent basis the kind of scientific interaction and cooperation that he found so productive first at Naples and later during summers at the MBL, Woods Hole. As he wrote to Hale: "The participation of a group of scientific men united in a common venture for the advancement of research fires my imagination to the kindling point."

In the Caltech period, Morgan's influence in genetics extended beyond the *Drosophila* work and the classical chromosome theory. Although he did not pioneer in the newer biochemical and molecular genetics that began to emerge in the 1940s, he nourished that trend. Both George Beadle, as a National Research Council fellow in 1935, and Max Delbrück, as an international research fellow in biology of the Rockefeller Foundation in 1939, worked with Morgan's group at Caltech; both saw that the next logical questions arising out of the *Drosophila* work were those of gene function. It was their work on the relationships between genes and proteins in simple organisms, such as yeasts and bacteriophages, that prepared the way for the revolution in molecular genetics during the 1950s and 1960s.

Morgan's influence was central to the transformation of biology, in general, and heredity and embryology, in particular, from descriptive and highly speculative sciences arising from a morphological tradition, into ones based on quantitative and analytical methods. Beginning with embryology, and later moving into heredity, he brought first the experimental, and then the quantitative and analytical, approach to biological problems. Morgan's work on the chromosome theory of heredity alone would have earned him an important place in the history of modern biology. Yet in combination with his fundamental contributions to embryology, and his enthusiasm for a new methodology, he can be ranked as one of the most important biologists in the twentieth century.

REFERENCES

Allen, G.E. (1978) *Thomas Hunt Morgan: The Man and His Science*. Princeton: Princeton University Press.

Carlson, E.A. (1966) *The Gene, a Critical History*. Philadelphia: Saunders.

Carlson, E.A. (1981) *Genes, Radiation, and Society: The Life and Work of H.J. Muller*. Ithaca: Cornell University Press.

McCullough, D.M. (1969) W. K. Brooks' role in the history of American biology. *J. History Biol., 2*: 411–438.

Morgan, T.H. (1896) Impressions of the Naples Zoological Station. *Science, 3*: 16–18.

Morgan, T.H. (1910a) Chromosomes and heredity. *Am. Nat., 44*: 449–496.

Morgan, T.H. (1910b) Sex-limited inheritance in *Drosophila*. *Science, 32*: 120–122.

Morgan, T.H. (1911) The origin of five mutations in eye color in *Drosophila* and their modes of inheritance. *Science, 33*: 534–537.

Morgan, T.H. (1912) The explanation of a new sex ratio in *Drosophila. Science, 36*: 718–719.

Morgan, T.H. (1913) *Heredity and Sex*. New York: Columbia University Press.

Morgan, T.H. (1914) Two sex-linked lethal factors in *Drosophila* and their influence and their influence on the sex-ratio. *Jour. Exp. Zool., 17*: 81–122.

Morgan, T.H. (1919) *The Physical Basis of Heredity*. New York: Lippincott.

Morgan, T.H. (1926) Genetics and physiology of development. *Am. Nat., 60*: 489–515.

Morgan, T.H. (1932a) The rise of genetics. *Science, 76*: 261–267 (Presidential Address to the Sixth International Congress of Genetics, Ithaca, New York).

Morgan, T.H. (1932b) *Evolution and Genetics and the Scientific Basis of Evolution*, New York: W.W. Norton.

Morgan, T.H. (1934) *Embryology and Genetics*, New York: Columbia University Press.

Morgan, T.H., and Driesch, H. (1895) Zur Analysis der ersten Entwickelungsstadien des Ctenophoreneies. I. Von der Entwickelung einzelner Ctenophorenblastomeren. II. Von der Entwickelung ungefurchter Eier mit Protoplasmadefekten. *Archiv für Entwicklungsmechanik der Organismen, 2:* 204–215, 216–224.

Morgan, T.H., A.H. Sturtevant, H.J. Muller, and C.B. Bridges (1915) *The Mechanism of Mendelian Heredity*. New York: Henry Holt & Co.

Sturtevant, A.H. (1959) Thomas Hunt Morgan, biographical memoirs. *Natl. Acad. Sci., 33*: 283–325.

Sturtevant, A.H. (1965) The fly room. In: *A History of Genetics*. New York: Harper & Row. pp. 45–50.

Sturtevant, A.H. (1967) Reminiscences of T.H. Morgan. Address delivered at the Marine Biological Laboratory, Woods Hole, MA. August 16, 1967.

Section II

Population Genetics

Population genetics began with the generalization derived by mathematician G.H. Hardy and physician W. Weinberg, who showed, in 1908, that allele and genotype frequencies in a population will remain constant from generation to generation in the absence of other evolutionary influences, such as mutation, selection, genetic drift, and gene flow.

However, the large body of mathematical work that established population genetics as a discipline was due to only three individuals: J.B.S. Haldane, R.A. Fisher, and Sewall Wright. All three, working independently, examined the Darwinian theory of natural selection in terms of Mendelian genetics. Haldane (1924), in particular, examined the consequences of selection in large populations under varying genetic conditions such as dominance and recessivity, mutation pressure, inbreeding and random mating, etc. Haldane (1932) summed up his early work in *The Causes of Evolution.* He was the first to estimate a human mutation rate by balancing natural selection against mutation rates.

Sewall Wright (1931) developed the mathematical theory of random genetic drift, cumulative stochastic changes in gene frequencies that arise from random births, deaths, and Mendelian segregations in reproduction, in small isolated populations. Wright had a long-standing and bitter debate with R.A. Fisher, who argued that most populations in nature were too large for these effects of genetic drift to be important. Wright's name is also associated with the inbreeding coefficient which he developed in 1922. It is the probability of inheriting two copies of the same allele from an ancestor that occurs on both sides of the family.

Fisher (1930) proposed the fundamental theorem of natural selection: "The rate of increase in fitness of any organism at any time is equal to its genetic variance in fitness at that time." Earlier, Fisher (1918) showed in his paper, "The Correlation between Relatives on the Supposition of Mendelian Inheritance," that continuous variation amongst phenotypic traits could be the result of Mendelian inheritance.

In 1968, Motoo Kimura, from Japan, proposed the neutral theory of molecular evolution, which holds that at the molecular level, most evolutionary changes and most of the variation within and between species is not caused by natural selection but by genetic drift of mutant alleles that are neutral. A neutral mutation is one that does not affect an organism's ability to survive and reproduce.

5 J.B.S. Haldane (1892–1964)

John Burdon Sanderson Haldane (JBS) was a man of many contradictions—a scholar of Classics who received no formal education in science but made important contributions to several sciences, a bombing officer in World War I who later embraced nonviolence and Hindu philosophy, a founder of the mathematical theory of evolution who was also a highly skilled popular writer of scientific essays, a Marxist and socialist who was educated at the British elitist schools such as Eton and Oxford, one who admired the Indian caste system, and an apologist for vivisection in mammalian physiological experiments in his youth who could discuss pain in the insect world quite seriously in his later years. Despite such contradictions (or because of them), Haldane remained a great scientist throughout his various transformations, and he is remembered today as one of sciences' greatest champions in the public domain (Dronamraju 1968, 1995, 2011, 2017).

J.B.S. Haldane was born in Oxford on November 5, 1892. Haldane's scientific work covered physiology, genetics, biochemistry, statistics, biometry, cosmology, and other fields. He was a classical scholar who enjoyed quoting Dante, Virgil, and Catullus during his scientific lectures. Haldane's knowledge of multiple disciplines enabled him to cross-pollinate, on the intellectual plane, multiple disciplines. His facility for mathematics introduced a quantitative rigor in his approach to biological problems.

JOHN SCOTT HALDANE

JBS owed much to his father's instruction, foresight, and planning. John Scott Haldane (JSH) trained his son with utmost care from a very early age to become a scientist, especially in physiology. JSH himself was an outstanding physiologist at New College, Oxford, who contributed much to the safety of miners and diving personnel. His early research interest was in testing the impact of air quality on human health. He tested air samples from the slums of Dundee and the London underground, finding high levels of carbon dioxide and carbon monoxide, respectively.

FATHER–SON COLLABORATION

John Scott Haldane involved his son from a very early age in legendary physiological experiments. Both father and son acted as their own "guinea pigs" in physiological experiments involving considerable risk and danger. They certainly involved much pain and discomfort. Obviously, they lived up to their family motto "suffer."

There are not many individuals, perhaps none, who can boast with JBS that his laboratory experience started at the age of two when he used to watch his dad breathe into complex contraptions and draw blood from his arm. In his "An Autobiography in Brief", Haldane (1961) wrote: "I suppose my scientific career began at the age of about two, when I used to play on the floor of his laboratory and watch him playing a complicated game called 'experiments'—the rules I did not understand, but he clearly enjoyed it."

Soon, young Jack joined the experiments. When he was three, his dad started drawing his blood for hemoglobin analysis. His father treated him like an adult and Jack was a quick learner. He already knew the terminology. On one occasion, when he was injured, Jack saw the blood from his injured forehead and asked the doctor, is it "oxyhemoglobin or carboxyhemoglobin?" When Jack was only four, John took him to London, where John was asked to test the air in the Metropolitan Railway building, which was going to be converted into a subway station. Jack helped his father collect air samples in glass tubes.

Jack often accompanied his father into mines. Years later, he recalled a pit in North Staffordshire into which he was lowered in a large bucket on a chain. He was 10 years old at that time. They crawled along an abandoned tunnel, avoiding methane, which gathers near the ceiling. When the party reached a chamber where a man could stand up, one miner raised his safety lamp. It filled with blue flame—pop!—and went out due to lack of oxygen. Methane is flammable, but the miners were saved because they were using Davy safety lamps. The flame was trapped inside the lamp because it was kept in by the wire gauze. Now, JSH used Jack as a guinea pig to demonstrate the effects of breathing methane. He asked Jack to stand up and recite Mark Antony's speech from Shakespeare's *Julius Caesar.* Jack began, "Friends, Romans, countrymen." Later, he said, "I soon began to pant, and somewhere about 'the noble Brutus' my legs gave way and I collapsed on to the floor, where, of course, the air was alright. In this way I learnt that firedamp is lighter than air."

In 1908, John Scott Haldane was invited by the Admiralty to participate in the trials of a new type of submarine. Young JBS, still in his teens, eagerly jumped at the chance of making the first trial dives. He first dived, down to 36 feet, in full gear. But, as he was only 13, the suit was far too large and the cuffs too wide. It rapidly filled with water; however, Jack maintained a cool head and played with the control valves calmly to keep the head above the water level. But when he was pulled up, he was freezing. The crew gave him some whiskey to warm him, and bundled him in a bunk for a long rest.

Haldane was born in Oxford to physiologist John Scott Haldane and Louisa Kathleen Haldane (née Trotter) and descended from an aristocratic intellectual Scottish family. His uncle was Richard Haldane, 1st Viscount Haldane, politician and one time Secretary of State for war. His father was a physiologist, a philosopher, and a Liberal. Younger Haldane once wrote: "My practice as a scientist is atheistic. That is to say, when I set up an experiment I assume no god, angel or devil is going to interfere with its course.... I should therefore be intellectually dishonest if I were not also atheistic in the affairs of the world" (Haldane 1934).

JBS grew up in a household of science, politics, and social reform. From his mother, he inherited her determination and commitment to a cause and refusal to change his mind in the face of adversity, and from his father, he inherited an open mind in scientific inquiries and reach decisions based on evidence and reason. Indeed, JBS was so open minded that he was one of the least dogmatic people I have ever met. It is certainly a rare quality among scientists. About his childhood, Haldane wrote: "As a child I was not brought up in tenets of any religion, but in a household where science and philosophy took the place of faith. As a boy I had very free access to contemporary thought, so that I do not today find Einstein unintelligible, or Freud shocking. As a youth I fought through the war and learned to appreciate sides of human character with which the ordinary intellectual is not brought into contact. As a man I am a biologist, and see the world from an angle which gives me an unaccustomed perspective, but not, I think, a wholly misleading one" (Haldane 1961).

Young Haldane's intellectual versatility, mathematical powers, and prodigious memory were legendary. On one occasion, his father, who had forgotten his log tables on an expedition, was said to have commented: "Never mind. Jack will calculate a set for us." Jack then sat down and completed the task! From an early age, JBS was taught both intellectual and physical courage by his father (Clark 1968, Dronamraju 1985).

Haldane attended Eton and Oxford, where he studied classics, graduating with honors in 1914. As early as 1901, when he was only eight years old, his father took him to the Oxford University Junior Scientific Club for a lecture by A.D. Darbishire on the recently rediscovered Mendel's laws of inheritance. It aroused in him a lasting interest to find out more about the nature of Mendel's laws. Some years later, while he was still in school, Haldane (with his sister Naomi) started breeding guinea pigs and mice on a systematic basis. During that work, Haldane discovered the first case of linkage in vertebrates, which Darbishire himself had overlooked. He reported this discovery before a zoology seminar organized by E.S. Goodrich in 1911, when he was 18 years old; however, when he sought Punnett's advice, he was told to obtain data of his own for independent confirmation. Continuing the experiments, JBS and his sister published their results in 1915, which confirmed their earlier observation of linkage (Haldane et al. 1915). Haldane later remarked that had he published his results in 1911, his discovery of linkage in vertebrates would have been simultaneous with the discovery of linkage in fruit flies by the Morgan School in New York (summarized in Morgan et al. 1915).

HALDANE'S RULE

Haldane's first publications in 1912, however, were in physiology, not genetics. He was a polymath whose scientific work covered physiology, genetics, biochemistry, biometry, statistics, cosmology, and philosophy, among other disciplines. After serving in World War I (1915–1919), Haldane resumed his investigations of linkage, devising the first mapping function and suggesting the measure of chromosome map distance, centimorgan or cM (Haldane 1919), which was later adopted by molecular biologists. Shortly afterward, in 1922, Haldane proposed the generalization concerning the offspring of interspecific crosses, which was termed "Haldane's rule": "When in the first generation between hybrids between two species, one sex is absent, rare, or sterile, that sex is always the heterogametic sex" or, as he put it, "that is to say, the sex which produces two sorts of gametes, namely the male in most animal groups, the female in birds and *Lepidoptera*" (Haldane 1922).

This rule was formulated on the basis of 48 agreements and one exception. Since that time, several other instances have been found to agree with Haldane's rule, which has stood the test of time for over 90 years!

Although JBS was a renowned experimental scientist in physiology, his contributions to genetics were almost entirely theoretical. His mathematical theory of natural selection was a major contribution to twentieth century biology. Haldane's invention of numerous mathematical tests and analyses for treating genetic data from numerous species, especially human pedigree data, have contributed to our knowledge of genetics substantially. His theoretical work has stimulated experimental research by others, especially in the biochemical genetics of plant pigments, cytogenetics, radiation genetics, and linkage studies. In population genetics, Haldane's ideas and methods have stimulated experimental work in mutation research, epidemiology, infectious disease, gene expression, transmission of disease, and several other aspects.

ORIGINS OF LIFE[1]

In the 1920s, Haldane in England and A.I. Oparin in the Soviet Union independently proposed a novel hypothesis about the origin of life on earth. Haldane's ideas were published in 1929, before Oparin's book was translated into English. However, the ideas expressed by both these men on the origin of life were remarkably similar. Haldane proposed that the primordial sea served as a vast chemical laboratory that was powered by solar energy. Complex interaction between carbon dioxide, ammonia, and ultraviolet radiation gave rise to a number of organic compounds in an oxygen-free atmosphere. Haldane described that sea as a "hot dilute soup" containing large populations of organic monomers and polymers that acquired lipid membranes in due course. Further developments ultimately led to the formation of the first living cells. Haldane coined the term "prebiotic soup" to describe the primitive ocean (Haldane 1929).

PHYSIOLOGY

Haldane's initial research in respiratory physiology was conducted in collaboration with his father, John Scott Haldane. Later, JBS continued research in underwater physiology as a subject–investigator. His penchant for somewhat heroic experiments was highly rewarding, because Haldane's rigid scientific discipline prepared him to make detailed and accurate observations according to protocol.

In the 1920s, Haldane conducted several physiological experiments, often employing himself as the guinea pig. He tested the effect of breathing atmosphere containing added carbon dioxide and drinking solutions containing the chlorides of calcium, magnesium, or ammonia on his blood pH and measured the changes in sugar, phosphate, and other components of blood and urine.

In 1927, Haldane investigated the role of carbon monoxide as a tissue poison. His advanced and facile use of mathematics enabled him to describe the kinetics of CO poisoning of enzymes. Later, he turned his attention to problems of high-altitude physiology and other matters.

He carried out other experiments in a closed steel chamber, which simulated conditions in a submarine, when investigating a submarine disaster that occurred off the coast of Liverpool with the loss of 39 lives. In a review of the field of underwater physiology, Case and Haldane (1941) listed six principal problem areas: mechanical effects, nitrogen intoxication, oxygen intoxication, aftereffects of carbon dioxide, bubble formation during decompression, and cold temperature. These topics still remain the areas of concern today.

He followed the Golden Rule, which he learned from his father J.S. Haldane: "To test on ourselves first that which we would have others do."

POPULATION GENETICS

Haldane's most important contribution to genetics was the mathematical theory of natural and artificial selection, which was one of the foundations of population genetics. Haldane and two others, R.A. Fisher (1930) and Sewall Wright, share the honor of being the founders of population genetics. Starting in the 1920s, Haldane published a series of mathematical and theoretical papers that have examined the Darwinian theory of natural selection in terms of Mendelian genetics. Initially, he deduced the number of generations required to cause a specific change in the frequency of a gene under various genetic situations, such as autosomal dominance, recessivity, varying degrees of penetrance, inbreeding and outcrossing, slow selection, rapid selection, isolation, mutation pressure, response to the environment, interaction of selection of one organ with that on another, selection of quantitative traits, very rare characters, and socially valuable but individually disadvantageous characters, among others. For instance, in the simplest case of a dominant gene with a selective advantage of 0.001, a total of 6,920 generations are required to change the gene frequency from 0.001% to 1%, a total of 4,819 generations are required to change from 1% to 50%, a total of 11,664 generations are required to change from 50% to 99%, and a total of 309,780 generations are required to change from 99% to 99.999%.

The mathematics of all three pioneers—Haldane, Fisher, and Wright—agreed essentially despite some differences in their approach. Haldane expressed his results in terms of the ratio, u, of the frequency of the mutant gene (A') to that of its type allele (A) (distribution 1A:uA'). In 1922, Fisher used a function of the gene frequency distribution $(1 - p)$ A:pA', viz., $= \cos^{-1}(1 - 2p)$, which has the property that its sampling variance is constant, $1/2N$, but he shifted later to the gene frequency itself, which was used systematically by Wright from the beginning.

However, all three differed greatly in the application to evolution. Haldane assigned usually constant selective values, but occasionally variable, to each gene or, in some cases each genotype involving two or more interacting loci, deducing deterministically the number of generations required to bring about a specific change in the gene frequency ratio under different types of genetic situations. Fisher, on the other hand, proposed general theorems of evolution, which are applicable to situations of varying complexity. The most important of these was his fundamental theorem of natural selection: "The rate of increase in fitness of any organism at any time is equal to its genetic variance in fitness at that time." Genetic variance was defined as the additive component of the total genotypic variance. This theorem assumes that selection always depends on the net effect of each gene and that there can be no selection among interaction systems between various genes. Evolutionary biologist Ernst Mayr (1959) objected to this assumption as being unrealistic. But both Haldane and Fisher introduced such complications in their later papers.

Both Fisher's and Haldane's approaches were similar in their deterministic character when treating large populations. Both authors recognized that the fate of a single mutation is decided by a stochastic process.

Wright's (1968) approach, on the other hand, was directed toward ascertaining whether some way might exist in which selection could take advantage of the enormous number of interaction systems resulting from a limited number of unfixed loci. He proposed that such a system might exist in a large population, which is subdivided into many, small local populations, isolated enough to facilitate considerable random differentiation of gene frequencies. At the same time, their isolation is limited enough to stop any diffusion of the more successful interaction systems.

NATURAL SELECTION

Haldane's interest in evolution began with studies of natural selection, which Charles Darwin postulated as the main agent of evolutionary change. In 1924, in his first paper in population genetics, Haldane wrote: "In order to establish the view that natural selection is capable of accounting for the known facts of evolution, we must show that not only that it can cause a species to change but that it can cause it to change at a rate which will account for present and past transmutations" (Haldane 1924, reprinted in Dronamraju 1990).

Haldane (1964) cited some examples of his work where mathematical methods yielded deeper insights regarding evolutionary problems which would not have been possible otherwise.

In a spirited defense of his approach, in response to a criticism by Ernst Mayr (1959), Haldane (1964) wrote: "Our mathematics may impress zoologists but do not greatly impress mathematicians. Let me give a simple example. We want to know how the frequency of a gene in a population changes under natural selection. I made the following simplifying assumptions:

1. The population is infinite, so the frequency in each generation is exactly that calculated, not just somewhere near it,

2. Generations are separate. This is true for a minority only of animal and plant species. Thus even in so-called annual plants a few seeds can survive for several years,
3. Mating is at random. In fact, it was not hard to allow for inbreeding once Wright had given a quantitative measure of it,
4. The gene is completely recessive as regards fitness. Again it is not hard to allow for incomplete dominance. Only two alleles at one locus are considered.
5. Mendelian segregation is perfect. There is no mutation, non-disjunction, gametic selection, or similar complications.
6. Selection acts so that the fraction of recessives breeding per dominant is constant from one generation to another. This fraction is the same in the two sexes.

where q_n is the frequency of the recessive gene, and a fraction k of recessives is killed when the corresponding dominants survive."

Haldane acknowledged that H.T.J. Norton gave an equivalent equation in 1910, and Haldane (1924) produced a rough solution when selection is slow; i.e., when k is small. However, he pointed out that even such a simple-looking equation would not yield a simple relation between q and n. Many years later, Haldane and Jayakar (1963) solved this equation in terms of automorphic functions. Haldane (1964) noted that the mathematics are not much worse when inbreeding and incomplete dominance are taken into account. But they are much more complicated when selection varies from year to year and from place to place or when its intensity changes gradually with time. When such problems are solved, the mathematics employed would be truly impressive.

Haldane stated at the outset that he would be dealing only with the simplest possible cases, involving a single completely dominant Mendelian factor or its absence.

When external conditions (such as a change in the temperature or a new disease) change drastically, many genes that have been less favorable to an organism in the past may become more favorable in the new environment. The practical cases to which Haldane applied his theory were ones that involved such change. One of these cases was based on field data assembled by the ecologist Charles Elton on the steady decline in the proportion of silver fox pelts among fox pelts marketed in various parts of Canada in the century preceding 1933. Silver is due to a simple recessive gene. Haldane (1942) calculated that this gene must have been at an average selective disadvantage of about 3% per year or 6% per generation if generation length is taken as two years, compared with its wild type in red.

Haldane's paper, published in 1927, on the treatment of selection in overlapping generations is of special importance because he combined a demographic approach to the study of evolution in the context of Mendelian genetics. Considering a population where the intensity of selection is independent of its size, Haldane addressed the problem of measuring population growth. He calculated the birth rate and death rate, which are not functions of its density. He showed that the oscillations of the population about an exponential function of the time "are either damped or at least increase less rapidly than the population itself." If a population is in equilibrium, oscillations are damped and a stable equilibrium occurs. Haldane concluded that his result was a new proof of the great value of Lotka's (1922) theorem on the stability of the normal age distribution.

Haldane then considered the mode of selection of a dominant autosomal factor in the population, only for female zygotes under the following assumptions: the sex ratio at birth is taken as fixed; the number of dominant and recessive female zygotes is fixed; random mating occurs; and selection and population growth are slow. Haldane then calculated the reproduction rates of the three female phenotypes at a time *t*, and he defined selection as the probability for a female zygote to reach age *x*.

Haldane's approach integrated the mode of selection in demographic structures, showing that selection, when generations overlap, produces an effect that is analogous to that obtained when generations are non-overlapping. He also showed that the mode of selection was the same for the sex-linked factor. Furthermore, Haldane integrated his work with the population growth model of Dublin and Lotka (1925) related to the real structure of the U.S. population in the 1920 census. Haldane computed the differential reproduction rate of different phenotypes at a given time, under the assumption when recessives are rare, when dominants are rare, and when the death rate and the reproduction rate are the same in the two sexes.

DEMOGRAPHIC STRUCTURES

Of the three great pioneers of population genetics, Haldane's attempt was the first to find the solution to genetic problems by integrating them into demographic structures. He was the first to show that a synthesis was necessary between demography and genetics in order to establish the real mode of selection and its processes in a human population. However, Haldane's brilliant interpretation met with no following at that time. French demographer Jean Sutter commented that the integration of genetic problems into demographic structures progressed very slowly for several decades, because of the domination of the deterministic models. After Haldane's death, a breakthrough occurred when Malecot (1966) made use of the chain processes of modern probabilities, which enabled him to set up interesting stochastic models. However, the pioneering contribution of Haldane is still remembered.

Interestingly, one of his correspondents after the Second World War was the well known Italian demographer/statistician Corrado Gini (1884–1965), who developed the "Gini coefficient" for measuring the disparity in the national income of a country. He asked Haldane to send any books and papers that he may have published in the war years. Scientific research was disrupted in Italy during the war years. Communication with foreign countries was difficult. In a letter dated April 30, 1946, Gini wrote: "Here the academic and Scientific life goes on very slowly and with great difficulties. All the public institutions are in very bad financial conditions."

ROLES OF SELECTION AND MUTATION IN EVOLUTION

One of the interesting problems considered by Haldane (1927) was the dynamics of mutation–selection balance to which he returned repeatedly in several later papers. Haldane wrote that if selection acts against mutation, it remains ineffective unless the rate of mutation is greater than the coefficient of selection. Furthermore, it is not selection but mutation that is quite effective in causing an increase in recessives where these are rare in the population. Usually, mutation is also more effective than selection in weeding out rare recessives in a population. Haldane concluded: "Mutation therefore determines the course of evolution as regards factors of negligible advantage or disadvantage to the species. It can only lead to results of importance when its frequency becomes large."

GENE FIXATION

Haldane took up the stochastic treatment of investigating the probability of fixation of mutant genes. When a new mutant appears in a finite population, it either gets lost or becomes established (fixed) in the population after a certain number of generations. This is a fundamental aspect of population genetics that Haldane considered early in his series of pioneering papers in population genetics. In 1927, Haldane investigated the probability of fixation of mutant genes, using the method of generating function suggested by Fisher in 1922. Haldane showed for the first time that a dominant mutant gene having a small selective advantage k in a large random-mating population has a probability of about $2k$ of ultimately becoming established in the population. The calculation of the probability of fixation is more complicated if the advantageous mutation is completely recessive, but Haldane ingeniously showed that it is of the order of k/N, where k is the selective advantage of the recessives and N is the population number. Haldane's results were confirmed by later investigations. He also discussed the situation involving an equilibrium between recurrent mutation and selective elimination and the mechanism by which that equilibrium is reached in the case of complete dominance. Thirty years later, in 1957, Motoo Kimura confirmed Haldane's calculations and extended them to situations involving different levels of dominance. Later, in 1962, Kimura developed a general formula for the probability of eventual fixation, $u(p)$, which was expressed in terms of the initial frequency, p, but also including random fluctuations in selection intensity and random genetic drift in small populations.

When the population is entirely self-fertilized or inbred by brother–sister mating, both dominant and recessive factors have about the same chance as a recessive. Haldane further stated that in situations with partial self-fertilization or inbreeding, an advantageous recessive factor has a finite chance of establishment after one appearance, however large be the population.

METASTABLE POPULATIONS

In Part VIII of his series of papers on the mathematical theory of natural selection, Haldane's (1931) concept of "Metastable populations" stands out. He wrote: "Almost every species is, to a first approximation, in genetic equilibrium; that is to say no very drastic changes are occurring rapidly in its composition. It is a necessary condition for equilibrium that all new genes which arise at all frequently by mutation should be disadvantageous, otherwise they will spread through the population. Now each of two or more genes may be disadvantageous, but all together may be advantageous." Haldane cited the observation of Gonsalez, in 1923, who found that, in purple-eyed *Drosophila melanogaster*, arc wing or axillary speck (both due to recessive genes) shortened life, but together, they lengthened it.

For instance, in the case of a system involving two dominant genes *A* and *B*, he expressed the relative fitnesses of the four phenotypes as follows:

$$\text{AB}1, \text{aaB}1-k_1, \text{Abb}1-k_2, \text{aabb}1+K$$

In an elegantly presented model of the two-locus case, assuming no linkage complications, Haldane plotted the trajectories of points representing the genetic composition of a population in a two-dimensional coordinate system. For a system containing *m* genes, he showed that a population can be represented by a point in *m*-dimensional space. He suggested further that related species represent stable types of the kind described in his model and that the process of speciation may be the result of a rupture of the metastable equilibrium. He added that such a rupture will be specially likely where small communities are isolated. Independently, Sewall Wright (1931) described his theory that evolution is a trial-and-error process that occurs in multidimensional adaptive surfaces and that it tends to occur rapidly in small isolated populations. There was, however, an important difference between Haldane and Wright. While Wright spent much of his life discussing and advancing his theory, Haldane, having made the suggestion, moved on to other subjects.

TYPES OF SELECTION

In 1956, Haldane contributed to our understanding of the various types of natural selection that can impact on the process of evolution. Selection that alters the mean of any character may be called *linear.* If it reduces the variance of a character, which involves elimination of extreme phenotypes, it is called *centripetal.* An example is the effect of selection on human birth weight, which eliminates babies with extremely low and high birth weights. If it increases the variance, it is called *centrifugal.* Kettlewell's (1956) observations on selection for melanism are both linear and centripetal.

Haldane distinguished between the two types of selection: *effective* selection, which changes the gene frequencies, and *ineffective* selection, which does not. If the mean of a character changes in a certain expected direction, it is called *directional* evolution. When the variance is reduced, by eliminating extreme genotypes, then it is called *normalizing* evolution. If its reduction involves elimination of genotypes that vary greatly in diverse environments, it is called *stabilizing* selection. But if the variance is increased, it is called *disruptive* evolution. However, Haldane did not consider a process "disruptive" if it leads to the establishment of a stable polymorphism. The term "normalizing evolution" was proposed by Haldane in this discussion to replace "normalizing selection" because the elimination of phenotypically extreme genotypes may be ineffective and does not necessarily lead to normalization.

It is in fact not possible to know if selection will be effective without the knowledge of principles of genetics. It is also known that at equilibrium (when the intensity of selection is balanced by the rate of mutation), evolution is neither stabilizing nor normalizing. Haldane emphasized that a new vocabulary is noted for the different types of *phenotypic* and *genotypic* selection, e.g., selection at the human Rhesus (Rh) locus due to neonatal jaundice caused by the D antigen, which is phenotypically *disruptive* but genotypically *directional.*

MEASUREMENT OF NATURAL SELECTION IN MAN

In 1954, Haldane proposed to measure the intensity of natural selection by $I = \ln s - \ln S$, where s is the fitness of the optimum phenotype and S is that of the whole population. When he applied this to Karn and Penrose's data on human birth weight distribution, Haldane found that while the death rate in all the babies in their sample was about 4½%, that in the group whose weights lay between 7½ to 8½ lb was only 1½%. Two-thirds of these deaths were phenotypically selective, and the total intensity of selection was 3%. But this was phenotypic selection, not genotypic selection.

Haldane wrote: "...a great deal of phenotypic selection occurs before birth and is hard to measure. Genotypic selection can be measured with some degree of accuracy in certain situations as in the case of hemophilia or certain blood group genes. Haldane suggested that we should try to argue from phenotypic to genotypic selection if adequate data are available. If an environmental influence simultaneously alters a metrical character and raises fitness, there is phenotypic selection in the direction in which that character is altered. For example, if an adequate diet promotes growth in length there is phenotypic selection for length. Even when a metrical character is wholly determined genetically, phenotypic and genotypic selections could be in opposite directions. If we consider a pair of alleles segregating in a population under random mating, Haldane showed that in a population at equilibrium phenotypic selection may be in the direction of the more dominant character. However, when the population is not in equilibrium, phenotypic and genetic selections may be in opposite directions. Haldane concluded that phenotypic and genotypic selections are usually in the same direction when measured over long periods, but not necessarily so in such short periods as three human generations."

STRANGE CASE OF PEPPERED MOTH: INDUSTRIAL MELANISM

Haldane enjoyed applying his theoretical and mathematical analysis to practical situations that involved observations and data collected in the field by other scientists. This was in fact a major part of his contribution to science. His analysis and insights often provided valuable information that was not foreseen by the investigators themselves. One such analysis involved industrial melanism in the industrial districts of England.

Early in the nineteenth century, the peppered moth (*Biston betularia*) was known to most naturalists, including Charles Darwin, as a predominantly white-winged moth liberally speckled with black. While the moths were resting on tree trunks, the lighter form was protected from the predators because it blended with the light colored lichens on the trunks. Later, with the increasing industrialization in the nineteenth century, soot, smoke, and other industrial pollutants from factories darkened the landscape, and the tree trunks where the moths rested were no exception. This sudden change in their environment made them highly vulnerable to birds. However, by the turn of the nineteenth century, the black (*carbonaria*) variant of the moth had largely, if not entirely, replaced the light-speckled (typical) form in the most polluted parts of UK. Experiments by Bernard Kettlewell showed conclusively that the darker melanic forms were eaten much more frequently by birds in the rural areas, where there is less pollution, as compared to the melanic forms in the urban areas, where the tree barks are darker because of greater pollution.

Many years later, with the passage of the 1956 Clean Air Act, the black variant of the moth began to decline and the white form, which was better camouflaged on lichen, reemerged on the trees. Peppered moth evolution is commonly used as an example of industrial melanism. Yet, despite its importance within evolutionary biology, the molecular genetic and developmental control of the *carbonaria*-typical polymorphism is unknown. All that is known on this topic is that the trait is controlled by a dominant gene at a single Mendelian locus.

The following conclusions were reached after extensive experiments and field observations (From: *Industrial melanism in Biston betularia*, Kettelwell 1973).

Number of Moths Eaten by Birds	Pepper	Melanics
Urban (more pollution)	15	43
Rural (less pollution)	164	26

1. Industrial melanism is genetically controlled by a single locus in *B. betularia*.
2. Populations have undergone evolutionary change in color pattern.
3. That change is consistent with the interpretation that it was due to natural selection, in that there is differential survival of the genotypes caused by differential predation on a particular background.
4. Results confirm qualitative prediction of equation for gene frequency change

HALDANE ON INDUSTRIAL MELANISM

Long before the studies of Kettlewell, in 1924, Haldane calculated that the gene for melanism conferred a selective advantage of about 50% on its carriers. It is so high that few biologists accepted his findings at that time. Haldane wrote: "Few or no biologists accepted this conclusion. They were accustomed to think, if they thought quantitatively at all, of advantages of the order of 1 per cent or less. Kettlewell…has now made it probable that, in one particular wood, the melanics have at least double the fitness of the original type."

Haldane also considered analysis of the nearly complete replacement of light-colored moths, *B. betularia*, by a semidominant dark mutant form, which in the end became completely dominant, presumably by direct selection of modifiers. In a paper published in the Proceedings of the Royal Society of London in 1956, Haldane wrote: "Dr. Kettlewell's proof that the dark form *carbonaria* of *B. betularia* has replaced the type, at least in part as the result of selection by bird predators, gives me the right to bring my calculation (Haldane 1924) on this matter up to date." Haldane's analysis of Kettlewell's data led to new and unexpected conclusions: selection against the melanic form was generally much less intense than was found by Kettlewell, there may have been immigration of the lighter form from unpolluted areas into polluted areas, and selection has slowed down (selective advantage) due to some special reason. One such possibility is due to what geneticists call "balanced polymorphism," which involves the existence of two kinds of melanic forms. Although they differ in their underlying genetic basis, they are indistinguishable externally.

BEANBAG GENETICS

The early work in population genetics was characterized as "beanbag genetics" by Ernst Mayr (1959) because he likened the treatment of single genes to the work of early Mendelians who used beans of various colors in bags to study Mendel's laws of genetics. Mayr wrote: "The Mendelian was apt to compare the genetic contents of a population to a bag full of colored beans. Mutation was the exchange of one kind of bean for another. This conceptualization has been referred to as beanbag genetics." Mayr (1959) emphasized that considering genes as independent units is meaningless from the physiological as well as the evolutionary viewpoint. Mayr emphasized that interactions between genes play a crucial role in evolution.

In their early work, both Haldane and Fisher treated genes as noninteracting independent units, whereas Wright considered their interaction (epistasis) as an essential part of evolution. In their later work, both Haldane and Fisher also considered more complicated situations including genic interaction. All three pioneers of population genetics took into account the complications arising from linkage, dominance, and epistasis in their calculations. In retrospect, Haldane commented that Mayr overstated his case. Initially, Haldane treated genes as noninteracting independent units for the sake of simplifying mathematical analyses. I have discussed this subject in detail in my book *Haldane, Mayr and Beanbag Genetics* (Dronamraju 2009):

> There is a lot more to "Beanbag Genetics" than was apparent from Mayr's critique. It goes far beyond counting mendelian ratios or a simple phenotype–genotype relationship. Indeed, Beanbag genetics today encompasses molecular clocks, nucleotide diversity, DNA-based phylogenetic trees, as well as the four major forces: mutation, selection, migration and random genetic drift. Rates of evolution at the nucleotide level can be measured and compared among diverse populations and among species. But most important of all, genetic differences between populations could be measured only by hybridization. Today it is commonplace to compare DNA differences and similarities in diverse species, orders and taxa. Finally, the most important evidence for the concept of "out of Africa" in human ancestry came from nucleotide diversity, in other words, beanbag genetics.

HUMAN GENETICS

Haldane was a founder and an early investigator of the genetic analysis of human pedigrees and populations. His interest in human genetics was an extension of his work on the mathematical theory of natural selection. He investigated various mathematical models of equilibria by balancing natural selection against mutation rates. Extending this principle to human genetics, Haldane (1932, 1935) considered the example of hemophilia. He deduced that as one-third of the hemophilics in each generation are being eliminated by natural selection, an equal number of new cases are arising in each generation by mutation. That was the beginning of what has now become the extensive field of human mutation rates.

Haldane's contributions to human genetics are entirely theoretical (Dronamraju 1989). He developed methods of analysis and tests of significance as the need arose in various situations. However, his ideas have stimulated experimental research of many kinds by other investigators. Much of what is now called "transplantation genetics" began with the discovery of the H-2 locus in the mouse by Peter Gorer, who was pursuing research under Haldane's direction at University College London.

In the 1930s, a small group of individuals (including Haldane 1931) started developing statistical methods to understand the transmission of characters in families. That was the formal beginning of the science of human genetics, which later gave rise to medical and clinical genetics. Earlier, Archibald Garrod (1909), physician at Guy's Hospital in London, investigated the recessive inheritance of several human biochemical disorders, such as cystinuria and albinism; however, the significance of Garrod's pioneering work was not fully appreciated for thirty years until Haldane drew attention to it in an interesting paper on chemical individuality in 1937.

From the 1930s onward, Haldane and others have steadily developed the methods and applications of human genetics. At first, the methods were mainly statistical and mathematical, but later, biochemical and molecular methods have contributed to the rapid growth of human genetics and its many extensions, such as immunogenetics and biochemical genetics.

For the next 30 years, Haldane extended human genetics meticulously, developing methods for the estimation of human mutation rates and mutation impact, linkage and human gene mapping, measurement of the effect of natural selection in human populations, interaction between genetic factors and the environment, estimation of genetic loads and radiation damage, as well as the impact of infectious disease, and other aspects. Haldane's influence was far greater than could be judged simply from reading his numerous published papers because of his readiness to communicate his ideas without reservation to his students and colleagues.

Haldane and Bell (1937) pioneered what later came to be known as the "the human genome project." Using statistical methods, they took the first step in 1937 in his estimation of the distance between the loci for the genes for haemophilia and color blindness on the X chromosome map. From these modest beginnings, the human genome project has evolved into today's multibillion enterprise.

Haldane pioneered several other areas of human genetics, such as the estimation of genetic damage resulting from radiation in atomic bomb testing, the genetic effects of human inbreeding, and the genetic analysis of complex disease traits. Furthermore, he wrote a great deal on the applications of human genetics for improving the human species, but he was opposed to formal eugenic programs mainly because he did not believe that we had sufficient knowledge of human genetics to benefit from them. In his plenary address to the 11th International Congress of Genetics, in 1963, Haldane wrote: "The appalling results of false beliefs on human genetics are exemplified in the recent history of Europe. Perhaps the most important thing which human geneticists can do for society at the moment is to emphasize how little they yet know. This is a thankless task...the influence of genetical ideas, such as that of racial purity, has often been disastrous.... It does not much matter if many people know nothing about hybrid maize or progeny tests for milk yield. It matters a great deal if they know nothing about human genetics, because it is a topic of interest to all human beings, and the gaps in their knowledge will be filled by superstition or intellectually dishonest propaganda."

GENETIC LOADS AND THE IMPACT OF MUTATION

In 1937, Haldane wrote a paper with the deceptively simple title "The Effect of Variation on Fitness," enunciating a basic principle of population genetics that formed the foundation for assessing the impact of mutation on the population. He showed that the effect of mutation on the fitness of a population is independent of how deleterious the mutant phenotype is but is instead determined almost entirely by the mutation rate. The principle he outlined in that paper became a useful tool later in assessing radiation damage, which was produced from the open air testing of atomic bombs. It was adopted by the BEAR committee (Committee on Biological Effects of Atomic Radiation) of the U.S. National Academy of Sciences in 1956 when the impact of testing nuclear weapons became a subject of great political and social importance. Haldane showed in his 1937 paper that the effect of mutation on the reproductive fitness of a population is independent of how deleterious the mutant phenotype is but is instead determined almost entirely by the mutation rate (Haldane 1937). His idea, which was independently used by H.J. Muller and others, provided the first basis for various assessments of the genetic effects of radiation. It also showed that any increase in mutation rate would have an effect on fitness ultimately equal to this increase.

DISEASE AND SELECTION

An ingenious idea of Haldane that led to a great deal of epidemiologic research involved the role played by infectious disease in the evolutionary process. The best known paper of Haldane (1949b) on this subject, titled "Disease and Evolution," was his address to an international conference in Italy in 1949, which was simply titled "Disease and Evolution." Haldane wrote: "the struggle against disease, and particularly infectious disease, has been a very important evolutionary agent, and that some of its results have been rather unlike those of the struggle against natural forces, hunger, and predators, or with members of the same species."

Haldane noted the similarity in geographic distribution of both thalassemia and malaria, especially in the Mediterranean region. The high incidence of thalassemia puzzled every one. Quite intuitively, Haldane saw a connection between the high incidence of thalassemia and malaria, which led to their prevalence in Italy, Greece, and surrounding areas.

Neel and Valentine (1947) thought that the higher incidence of thalassemia was due to its higher mutation rate. They suggested a high mutation rate of 1 in 2500 births. But Haldane (1949a, 1949b) offered a different explanation. He wrote: "I believe that the possibility that the heterozygote is fitter than normal must be seriously considered. Such increased fitness is found in the case of several lethal and sub-lethal genes in *Drosophila* and *Zea*." Haldane stated this explanation clearly in his address to the 8th International Congress of Genetics in Stockholm in 1948, which was published in 1949 (Haldane 1949a).

EUGENICS

Haldane's (1923) first book, a popular version of a futuristic speculation of eugenic applications, stands out for its originality, daring, and scientific predictions. Such freshness and bold approach were never to be seen again in his writings. *Daedalus, or Science and the Future* was a slim octavo volume that was an instant success. It was a daring presentation of what might be forthcoming in science and how it might affect the future human society in health, social and cultural traditions such as marriage and reproduction, religion, and political institutions. Haldane wrote speculative essays on science and eugenics repeatedly throughout his life, but the passion and bold imagination which he displayed in *Daedalus* was never to be seen again. Haldane wrote in his introduction to *Daedalus* that "it is the whole business of a university teacher to induce people to think." Haldane achieved that goal admirably through his writings, lectures, personal contacts, and, above all, the example of his own life and work, which many others have emulated.

Of great interest were Haldane's biological predictions, especially revolutionary advances in molecular biology and reproductive biology, which came true in the 1970s and beyond. His prediction of genetic manipulation and *in vitro* fertilization, resulting in the mass production of "test-tube" babies, deeply influenced several scientists and writers. One of them was his friend Aldous Huxley (1932), who copied these ideas in his *Brave New World*.

Haldane's *Daedalus* served several purposes: his predictions of genetic engineering and mass production of "test-tube" babies have had a significant impact on scientists and writers alike. He warned of the need to be prepared for the ethical dilemmas and new challenges that will be confronting humanity. Perhaps, *Daedalus* was also Haldane's attempt to shock the establishment that undoubtedly delighted him. By his predictions, Haldane was suggesting which lines of research would be desirable. Haldane emphasized that a eugenically minded government would consider proper feeding as an important aspect of its administration. No eugenic measure could have much effect if an inadequate diet causes physical or mental disabilities. Another fact of social inequality is an educational system that does not recognize the large innate inequalities of intelligence. An educational system designed by a biologist would recognize that inequality. An ideal society that is based on biological knowledge will recognize that large innate differences exist and that we must take both genetics and environment into consideration in matters of eugenic planning.

STERILIZATION

Haldane had written on eugenics and related issues throughout his life, adding new ideas and ethical consequences in successive papers. However, the passion and imagination that he displayed in *Daedalus* were never to be seen again. Indeed, with the growth of knowledge of human genetics, he became more and more skeptical of any immediate benefits of eugenic applications. In fact, he was averse to using the term "eugenics" because of its connotation with the racist policies of Nazi Germany before and during the Second World War.

As a mathematical biologist, Haldane was well aware of the slow rate of change that would be brought on by a eugenic program. In this respect, Haldane's position was quite different from that of Julian Huxley and Herman J. Muller, who were more enthusiastic about the prospects of eugenics by germinal choice. Furthermore, Haldane (1934, Norman Lockyer Lecture) noted that "eugenic organizations rarely include a demand for peace in their programmes, in spite of the fact that modern war leads to the destruction of the fittest members of both sides engaged in it." Haldane profoundly disagreed with the views of those who were recommending wholesale sterilization of mentally defective individuals, petty criminals, and the chronic unemployed, the so-called "social problem group."

Even when he seemed to have a common interest initially with a correspondent regarding eugenic methods, Haldane was quick to withdraw support when sterilization was mentioned as one of the means of achieving a eugenically improved society. Such was the case with his exchange with a certain Ethna Donnelly in Los Angeles. At first, he was supportive of what he thought was a project on ectogenesis, but when Donnelly mentioned "sterilization of all below a certain stage of mental and physical fitness," he quickly backed off, saying "you can count me out." Haldane wrote: "I am in complete disagreement with your views on sterilization, which, in my opinion, are to quote your own words, 'actuated by cruel impulses or goaded by fear' which are almost certainly based on ignorance of the facts. If, therefore, you are going to hook-up your project with that kind of propaganda, you can count me out" (letter dated August 18, 1947).

During the heyday of his Marxist association, Haldane viewed earlier eugenic programs in terms of a class struggle, as sterilization and other eugenic measures proposed would impact disproportionately on the economically disadvantaged. An unexpected consequence of the rigor of methodology of human genetics, which was established by the mathematical analyses of Haldane, Hogben, and others, was that it led to the destruction of class-bound eugenics.

ADDRESS TO THE XI INTERNATIONAL CONGRESS OF HUMAN GENETICS, 1963

In 1963, Haldane was invited to address the XI International Congress of Human Genetics, which was held in The Hague in the Netherlands. Haldane pointed out that, for a variety of reasons, men and women find it harder to accept the implications of genetics than those of other scientific disciplines. First, genetic findings may conflict with or reinforce two powerful emotions, the sexual and parental. Second, they may conflict with or reinforce theories that support the class structure of human societies and the divisions between societies or, alternatively, with theories that regard class and national divisions as evil. The findings of genetics may conflict with religious doctrines as well, but it is not as troublesome as the conflicts brought on by the studies of astronomy and evolution. Haldane emphasized that it is impossible to speak on the applications of human genetics without offending someone.

The important point that Haldane emphasized is that generalizations based on race are usually misleading and untrue. The genetic basis of racial comparisons is not meaningful because each may be superior to others in its own environment with respect to a certain trait. Socially important innate characters can be reversed by changed circumstances. Haldane cited the example of Norwegians and Danes, whose ancestors (Vikings) thought it disgraceful to die otherwise than in battle, but today, they are most peaceful people and have, in fact, low murder rates.

PROSPECTS FOR EUGENICS

Haldane had little patience with many eugenic programs. One of his objections is that they do not take into account the extreme slowness of the evolutionary process. He wrote: "…the development of population genetics has begun to show us why evolution is so vastly slower a process than any of Darwin's contemporaries believed. So far from thinking that human or other animal populations are evolving at an observable speed, it is often justifiable to think of them as being in equilibrium. Natural selection, based on genetical differences, is certainly occurring in human populations, but it is very nearly balanced by mutation, segregation, and migration." He pointed out further that no eugenic program (short of the prevention of mutation) would reduce the frequency of rare dominant and sex-linked recessive abnormalities in human populations that are kept in being by mutation. On the other hand, developing tests for heterozygosity in the female relatives of hemophilics would reduce their frequency, but it would not abolish the disease.

Haldane stated that H.J. Muller and some others who agree with him believed that there is a similar equilibrium for most human recessive genes, which, in heterozygous condition, lower longevity and fertility. However, it is certain that at some loci, several alleles remain in being because heterozygotes are biologically fitter than either homozygote. Haldane's calculations showed that gene frequencies can be stabilized in more situations than is generally realized.

Haldane believed that it is desirable to prevent the production of the most socially inefficient 5% or so of most populations. There are two quite distinct methods of negative eugenics: (a) first, persons with well-marked dominant and sex-linked recessive defects should not have children, nor should the daughters of hemophilics and others with similar sublethal sex-linked conditions. Eugenists should encourage such people to intermarry, after thorough training in the use of contraceptives or voluntary sterilization of one of the partners. Haldane wrote: "eugenic organizations could do good work by introducing pairs of people for both of whom fertile marriage was contra-indicated"; (b) screenings of whole populations for rare and harmful autosomal recessive genes. Although there is no hope of eliminating the "load" we carry for many generations, it should be possible to discourage marriages between persons heterozygous at the same locus. If, for example, 2% of all babies are homozygous for an unwanted recessive, this would only mean that 8% of intended marriages would be contraindicated. Some people have argued that such a policy would lead to increased frequency of undesirable recessive genes, which are now being eliminated because the homozygous recessives are inviable or sterile. However, the increase, which would be due to mutation, will be very slow.

Haldane agreed with H.J. Muller that as long as artificial insemination is practiced, genetically superior donors should be chosen. However, he is rather skeptical if such methods would greatly improve the genetic condition of our species. One problem is that we do not know the genetic determination of several desirable characters, nor how rapidly genetic homeostasis would prevent their realization.

Among other predictions, Haldane (1963) suggested that organ grafting holds great promise. He suggested two other possible applications of human genetics. First is *clonal reproduction*: This involves the culture of one or more cells from an adult so as to produce another individual of the same genotype. Second is *control of development*: Since genes act by initiating biochemical processes, in the absence of a gene, it should be possible to produce its effects. Haldane wrote that it is conceivable that in future our descendants may discover nongenetical methods of controlling human ontogeny other than the very crude methods of intellectual and physical training. "If a way can be found, by prenatal or postnatal treatment, to convert what would otherwise have developed into an ordinary person into a moral or intellectual genius, there may be no need for further human biological evolution."

EUGENIC VIEWS OF HALDANE AND MULLER

It is interesting to compare the views of Haldane and Herman J. Muller on the possible application of eugenics for the betterment of humanity. Their lives and careers were remarkably similar. Both made important contributions to genetics and evolution. Both were socialists who expressed admiration for Marxism, and both were disappointed in its outcome, although Haldane took longer than Muller. Both were seriously concerned about the genetic future of humanity and the need to apply scientific knowledge for its betterment. Both were atheists and materialists in their outlook. Both had a wide range of interests outside science. Both were fearless in their outspokenness in condemning social injustice. Both left their countries in disgust and disappointment (Carlson 1987).

Haldane's *Daedalus* is a man's world, reflecting the men's club mentality of the British society in the 1920s. Muller, on the other hand, emphasized the role of women in society and the need to acknowledge their contribution. Muller believed in opening new opportunities for women to follow their professions and express their genetic talents. Muller's commitment to feminism stems from his first wife, Jessie Muller, a PhD in mathematics, who was dismissed from the faculty of the University of Texas on the birth of their son. Muller was deeply affected by that incident and the irrational prejudice against women who desired to pursue both a family and a career. Muller's views on a eugenic program, which were summed up in his book,

Out of the Night, were dated at least a decade later than Haldane's *Daedalus*, and that difference may account for some of the differences in their views.

There were other significant differences between Muller and Haldane. While Haldane favored ectogenesis, Muller supported the use of cultured sperm, which was preferably kept in storage until 20 years after the death of the donor to allow a careful evaluation of the donor's worth, among other reasons. He foresaw some 250,000 inseminations from each voluntarily chosen male donor and repeating that process with each generation's best endowed individuals for intelligence and personality traits. Muller was interested in selecting not only for intelligence but also for other desirable qualities, such as compassion, cooperation, and other qualities that promote social equality and justice. Muller noted further that sexual selection could be used to prevent X-linked disorders in a relatively few generations. Muller's commitment to eugenics was much more intense than Haldane's. Muller maintained a serious interest in eugenics from his undergraduate years which lasted throughout his life. Haldane and Muller had similar eugenic goals but differed in their methods and possibilities.

Haldane (1965) returned to the subject of hypothetical biological intervention, but not systematic eugenic programs as advocated by H.J. Muller and Julian Huxley. In a paper contributed to the CIBA Foundation Symposium on the future of man in 1963, Haldane speculated on the possibilities for human evolution in the next 100,000 years. He considered the incorporation of synthesized new genes into human chromosomes, duplication of existing genes to perpetuate the advantages of heterozygosity (hybrid vigor), and intranuclear grafting to enable our descendants to incorporate many valuable capacities of other species without losing their human capacities. For instance, the disease-resisting quality of many species could be incorporated without losing human consciousness and intelligence. Haldane cited gene grafting as a means to induce various desired phenotypes suited for special tasks. One of these tasks was special adaptation for long-distance space travel. The following comment was typical of Haldane's (1949c) approach: "A regressive mutation to the condition of our ancestors in the mid-pliocene, with prehensile feet, no appreciable heels, and an ape-like pelvis, would be still better." With reference to the unlikely prospect of encountering high gravitational fields, Haldane wrote: "Presumably they should be short-legged or quadrupedal. I would back an achondroplasic against a normal man on Jupiter."

Among all the major biologists who discussed future eugenic possibilities, Haldane (1963) was unique in emphasizing the inadequacy of our technical knowledge of human genetics. With respect to the possibilities of human evolution over the next thousand years, he wrote: "It may take a thousand years or so before we have a knowledge of human genetics even as full as our present very incomplete knowledge of organic chemistry. Till then we can hardly hope to do much for evolution.... If the capacity for consciousness and control of physiological processes is prized by posterity, steps will probably be taken to make it commoner, and it may be that ten thousand years hence our descendants will differ from us not only in achievements but in capacities and aspirations, to so great an extent that it is useless to attempt to follow them further" (address to the XI International Congress of Genetics in The Hague, 1963).

SCIENCE AND ETHICS

Distinguished physicist Freeman Dyson (1995) noted that Einstein and Haldane were almost alone in early twentieth century to discuss the ethical issues that are brought on by scientific progress. At first, Haldane's discussion in *Daedalus* dealt with a whole range of ethical issues relating to the application of science to the conduct of wars in the twentieth century. Second, he raised a whole range of questions concerning the application of biological knowledge for the purpose of reproductive intervention. In what was clearly his first foray into the controversial subject of eugenic applications, Haldane appears to be mocking the kind of eugenical selection that has been (and still is) advocated by many followers of the eugenics movement, an attitude that is consistent with his later writings.

In 1923, Haldane's approach to the genetic betterment of the human species was based on hypothetical technical advances in what was later called molecular genetics. He argued that the separation of sexual intercourse and the reproductive process will facilitate greater flexibility in manipulating the human genome. Haldane was well ahead of his time in discussing these biological and ethical issues. He predicted that the term "parent" acquires a new meaning in such a society. Directed mutation and control of *in vitro* fertilization would lead to a mass production of humans with specialized skills and talents. The genetic revolution that Haldane had anticipated has begun. It is occurring a few decades later than he predicted.

A major underlying theme in *Daedalus* was the fact that new scientific discoveries are constantly changing our ethical outlook. Ethics are bound by time, knowledge, and culture. New ethical duties arise from the application of new discoveries. Haldane was especially concerned with the ethical impact of biological discoveries. There is a clear warning in *Daedalus* that progress in science must go hand in hand with progress in our ethical outlook. Haldane returned to the subject of the impact of science on ethics repeatedly in his writings.

In his book *The Inequality of Man and Other Essays*, Haldane (1932) listed five different ways in which science can impact on ethical situations:

> In the first place, by its application it creates new ethical situations. Two hundred years ago, the news of a famine in China created no duty for Englishmen.... Today the telegraph and the steam-engine have made action possible, and it becomes an ethical problem what action, if any, is right.
>
> Secondly, it may create new duties by pointing out previously unexpected consequences of our actions.... We may not all be of one mind as to whether a person likely to transmit club-foot or cataract to half his or her children should be compelled to abstain from parenthood.
>
> Thirdly, science affects our whole ethical outlook by influencing our views as to the nature of the world—in fact, by supplanting mythology.

Fourthly, anthropology...is bound to have a profound effect...by showing that any given ethical code is only one of a number practiced with equal conviction and almost equal success.

Finally, ethics may be profoundly affected by an adoption of the scientific point of view; that is to say, the attitude which men of science, in their professional capacity, adopt towards the world. This attitude includes a high (perhaps an unduly high) regard for truth, and a refusal to come to unjustifiable conclusions...agnosticism.

In his foreword to my book *Haldane's Daedalus Revisited*, Nobel laureate Joshua Lederberg (1995) drew attention to a kind of imperialism of the present over the future: the complex problem of intergenerational responsibility, a legacy of technology about whose merits they (the future generations) had no voice in deciding. Our children surely deserve adequate ethical and moral preparation to deal with new technologies.

In his commentary on *Daedalus*, physicist Freeman Dyson (1995) drew attention to the fact that both Haldane and Einstein shared a common concern in this regard. Einstein wrote: "...a positive aspiration and effort for an ethical-moral configuration of our common life is of overriding importance. Here no science can save us. I believe, indeed, that over-emphasis on the purely intellectual attitude, often directed solely to the practical and factual, in our education, has led directly to the impairment of ethical values. I am not thinking so much of the dangers with which technical progress has directly confronted mankind, as of the stifling of mutual human considerations by a 'matter-of-fact' habit of thought which has come to lie like a killing frost upon human relations."

SCIENTIFIC PREDICTIONS IN 1964

Haldane made some scientific predictions in 1964, especially what might be achieved by A.D. 2000 in an article in the *New York Times*, shortly before his death in India on December 1 of the same year. It was titled "A Scientific Revolution? Yes, Will We Be Happier? May Be." He predicted that we shall have a fair knowledge of the large-scale structure of objects within a billion light years, and in particular, "we shall be able to rule out a lot of theories which now seem possible...the theory that matter is being continuously created, so that though the galaxies move apart, matter does not thin out, will very likely be disproved. If it has *not* been disproved in its first 50 years, it will seem much more likely to be true than it does today, and may be generally accepted."

Haldane predicted that chemistry will be extremely accurate and the metachemistry of short-lived particles will be about as systematic as is classical chemistry in 1964. Nuclear fusion reactions based on the conversion of hydrogen to helium will be practicable on a laboratory scale, but not yet a source of industrial power by A.D. 2000. There will be a few people resident on the moon and Mars, but a voyage to the nearest star would not even be a possibility!

Most infectious diseases will be pretty rare, but new ones will appear often enough to warrant caution. Cancer will usually be curable, but some types will be preventable if diagnosed early. Most people will die between 80 and 100, and a few will live to be 120. Both eugenics and preventive medicine will be needed to extend life spans of most people to 100. Psychology and brain physiology will be advanced but not yet fused as physics and chemistry have done. A serious problem will be that all advanced knowledge will be expressed only in terms of advanced mathematics. This will widen the gap between different sciences and also between scientists and others. We know about processes lasting up to 4 billion years—the age of the earth, but we cannot accurately study events lasting less than about a millionth of a second, let alone the interaction between some kinds of elementary particles, which last only about one one-hundred-thousand-billionth-billionth of a second. Haldane thought we would be able to watch the rapid events that take place inside our cells and inside cellular organs such as mitochondria.

Haldane predicted further that the science of five centuries hence will be psychophysics, which will make it possible to describe any event in the language of either physics or psychology. By the year 2000, we shall be a long way from that goal and will be still further from the goal of a scientific understanding of human societies. Haldane wrote that it is the study of exceptional substances and processes—such as the alkali metals and halogens, which are not found uncombined in nature, electric conduction in metals, and radioactivity—that have led to our mastery of physics and chemistry. He wrote further that happiness is a byproduct, and if you aim at it directly, you won't achieve it. He wrote: "It *can* be achieved, but I don't think most people will have achieved it by A.D. 2000."

BIOLOGICAL POSSIBILITIES FOR THE NEXT TEN THOUSAND YEARS

In a remarkable address to the symposium on the "Biological Possibilities for the Human Species in the Next Ten Thousand Years," Haldane (1965) considered some alternative possibilities. They are as follows:

(1) Man has no future. (2) A nuclear war will cause so much biological damage that civilization will have to be rebuilt from barbarism. (3) A nuclear war with such damage will lead to a highly authoritarian world state. (4) Rational animals of the human kind cannot achieve the wisdom needed to use nuclear energy, unless they live for several centuries. And (5) a nuclear war will not occur, but some kind of world organization will gradually emerge, after a general disarmament. Haldane also considered the possibility that mankind could very probably be destroyed by processes still more lethal than nuclear reactions. He considered further the likelihood that a fringe group could get hold of enough fissile material to force their own government to precipitate an international war.

Biologically speaking, Haldane thought that qualitatively novel genetic changes such as mutations of other types are unlikely to arise than what is already known from the effects of radioactivity and cosmic particles. He blamed "imaginative writers with a superficial knowledge of biology such as Aldous Huxley and John Wyndham," who have written of new types of mutations, for having done a considerable disservice to clear thinking.

Haldane pointed out that one of the most important tasks facing mankind is a complete revision of educational methods. Teaching methods appear to be aimed at developing children of a capacity a little below the median, and very great harm is done by wrong timing and wrong methods. We may have to wait for human clonal reproduction before scientific methods can be applied.

Referring to human health situation, Haldane predicted that methods of prevention of many forms of malignant and cardiovascular diseases would soon be developed. An overwhelming desire to colonize Mars or some other planet without introducing terrestrial bacteria and viruses may well provide an incentive to render human beings completely aseptic. After eradicating many of the infectious diseases and deficiency diseases, the next stage in the struggle for health will be against congenital diseases and those that occur in later years. Haldane discussed what he called "rational geriatry." A congenitally weak organ may fail through chronic environmental stress. He wrote: "One reason why I have gone to India is to avoid chronic 'rheumatic' joint pains. I do not mind the heat, since I dress almost rationally, wearing as few clothes as decency permits. Infections such as amoebic dysentery, which are still hard to avoid, are no more trying than English respiratory infections.... Perhaps retirement may come to mean retirement to a congenial climate..."

Haldane suggested that it is far more important to discover the capacities of young people and to guide them into suitable occupations. Recognition of rare capacities is even more important. Supernormal hearing is very rare, but supernormal vision is even rarer. Supernormal smelling may be quite common. Supernormal muscular skill is highly desired, especially in sports. Aptitude tests may eliminate the worst half or three-quarters, but they do not identify that rare individual who is exceptionally talented. Once poverty becomes a memory of a distant past, there will be much less interest in acquiring material objects, and more and more interest in our bodies and minds, and those of others in whom we are interested.

ONE THOUSAND YEARS

He was inclined to be extremely cautious on the desirability of eugenic applications when many others have been rushing to apply eugenics. He predicted further that our descendants will be more interested in their own biology than us and will have far more knowledge and control of it. Haldane had hoped that, by understanding and intellectualizing their normal pleasures, our successors will convert them into servants rather than masters. One of the human goals is emotional homeostasis, which will be achieved by their integration, similar to the integration of the activities of antagonistic muscles.

CLONING

Haldane predicted that clones would be made from people aged at least 50, except for athletes or dancers, who would be cloned younger. These would be individuals who have excelled in a certain socially acceptable accomplishment. Other clones would be descendants of people with very rare capacities, such as permanent dark adaptation, lack of the pain sense, and special capacities for visceral perception and control. Centenarians, if reasonably healthy, would be useful subjects for cloning.

We have lost, in our evolutionary past, capacities that are valuable to us, such as our olfactory capacities and the capacity for healing. Hybridization with animals possessing these capacities is clearly impossible. But it is possible to introduce small fragments of the genome of one species into another. That type of intranuclear grafting might enable our descendants to incorporate many valuable capacities of other species without losing those that are specifically human. The most obvious abnormalities in extraterrestrial environments are concerned with gravitation, temperature, air pressure, air composition, and radiation. Haldane wrote: "Clearly a gibbon is better preadapted than a man for life in a low gravitational field, such as that of a space ship, an asteroid, or perhaps even the moon. A platyrhine with a prehensile tail is even more so…. The human legs and much of the pelvis are not wanted. Men who had lost their legs by accident or mutation would be specially qualified as astronauts."

Haldane wrote that a drug with an action like that of thalidomide, but on the leg rudiments only, not the arms, may be useful to prepare the crew of the first spaceship to the Alpha Centauri system, reducing their weight and their food and oxygen requirements.

One of the senses that are better developed in other species, such as birds than in ourselves, is that of time. We have largely lost this sense, mainly because of our dependency on the sun and, later, on our watches and clocks. Haldane suggested that the negative aspect of time, of which death is the most significant aspect, might not concern us so much if we could realize human life as a finite pattern in time, capable of all degrees of perfection.

Other problems that were discussed by Haldane included extremes of temperature and pressure. At air pressures below about a quarter of an atmosphere, a pressure suit is needed. It was typical of Haldane that he considered once again populations that would be easier to train, such as individuals of an Andean or Tibetan ancestry, who might be able to live at a one-fifth of an external pressure of an atmosphere, wearing a suit that allows breathing at a pressure a few millimeters above that outside. This would be both safer and more comfortable than if the difference is greater. He commented that there is no prospect, in the next 10,000 years, of adapting human beings to breathe air in which the partial pressure of oxygen is less than 1% of a terrestrial atmosphere. However, previous experience has shown that given an artificial breathing mixture, men can live quite happily at all pressures from 1/4 atmosphere to 20 atmospheres, and very likely at higher ones. Other dangers include radiation and high-speed particles. Resistance to radiation is a desirable character in astronauts, which may or may not be inherited although it occurs rarely in bacteria.

DIVISION OF HUMAN SPECIES

One idea of Haldane, which he pursued in his writings occasionally, is the real prospect of our species dividing into two or more branches, either through specialization for life on different stars or for the development of different human capacities. This could be a dangerous development as such species may fail to understand each other, leading to quarrels or even wars. Haldane predicted that, in 10,000 years, the world community will still be polytypic, but much more polymorphic than today. He wrote: "... I hope that the lowest 50 per cent of present mankind for any achievement will be represented by only 5 per cent in our descendants. I do not there will be universal racial fusion.... I do not believe in racial equality, though of course there is plenty of overlap; but I have no idea who surpasses whom in what.... When opportunities are nearly equalized some races are found to produce far more superior people at some particular skill than others." Haldane referred to the fact that men and women of African ancestry in the United States excel in sprinting. He predicted that tropical Africans include more potential biologists than potential physicists. However, the intellectual elite of the world will be of very mixed racial origins, perhaps with a median color about that of northern Indians today. The intellectual elite will be more polymorphic than the general population, partly because they will largely be products of assortative matings; i.e., individuals in the same profession tend to marry each other. The physiological and psychological polymorphisms will be far more important, leading to much more tolerance.

HALDANE'S "FIRSTS"

Some Haldane "firsts" in genetics are as follows: he discovered the first case of genetic linkage in mammals, he invented the first mapping function, he introduced the terms "morgan" and "centimorgan" as units of map distance, he estimated the first human mutation rate, he was one of the first to estimate the probability of gene fixation in a population, he calculated the selective advantage of a gene in a natural population, he estimated the cost of a gene substitution in evolution, he prepared the first human gene map, he invented the idea of partial sex-linkage, and he invented the unit "darwin" to measure the rate of evolution.

Other Haldane "firsts" in science are as follows: he was the first to "taste" oxygen at higher pressures (Case and Haldane 1941), designed methods to escape from submarines, pioneered the theory of enzyme kinetics, and proposed a new theory of the origin of life. Other ideas of Haldane promoted "cybernetics," which was developed by Norbert Wiener, and a radiation-dominated early universe (Haldane–Milne hypothesis).

BIOCHEMISTRY

During the years 1923–1933, Haldane was Sir William Dunn Reader in Biochemistry at Cambridge University. Even though he did not possess a formal qualification in science and his previous work was in respiratory physiology, not biochemistry, the Professor of Biochemistry, F. Gowland Hopkins, took some risk in appointing Haldane to the Readership. Haldane's performance as the Dunn Reader in Biochemistry more than justified the faith placed in him by Hopkins.

Haldane began his Readership with a "bombshell," an explosive lecture to the heretics at Cambridge that predicted revolutionary developments in reproductive biology and eugenics. His lecture, which was delivered on February 4, 1923, was published under the title *Daedalus or Science and the Future* by Kegan Paul in London.

Haldane discussed the impact of *in vitro* fertilization and mass production of "test-tube" babies on society. He explored the hypothetical applications of eugenics for large-scale social improvement and its ethical consequences. Haldane's predictions in molecular and reproductive biology became a reality 60 years later. However, the ethical dilemmas that he predicted still remain unresolved.

Haldane's *Daedalus* had a profound impact on other scientists and writers. His biological predictions caused an instant sensation at a time when Julian Huxley was reprimanded by Lord Reith for merely mentioning "birth control" in a talk on BBC radio! Aldous Huxley incorporated Haldane's ideas in his science-fiction, *Brave New World*, which, in turn, had a profound impact on our scientific, ethical, and moral outlook.

Haldane's Cambridge period was specially noted for his series of mathematical papers on the role of natural selection in evolution, which I have mentioned earlier.

Haldane's biochemical work at Cambridge was noted for his contribution to enzyme kinetics. In 1925, G.E. Briggs and Haldane derived a new interpretation of the enzyme kinetics law described by Victor Henri in 1903, which was different from the 1913 Michaelis–Menten equation. Michaelis and Menten assumed that enzyme (catalyst) and substrate (reactant) are in equilibrium with their complex, which then dissociates to yield product and free enzyme. The derivation of the Briggs–Haldane equation is based on the quasi-steady-state approximation; that is, the concentration of intermediate complex (or complexes) does not change. As a result, the microscopic meaning of the "Michaelis Constant" (K_m) is different. Most of the current models use the Briggs–Haldane derivation (Briggs and Haldane 1925).

Haldane instigated early research on the biochemical genetics in plants and the gene–enzyme relationship, which was later developed more extensively in *Neurospora* by Gorge Beadle and Edward Tatum (1941).

ANIMAL BEHAVIOR

Haldane was greatly interested in animal and human behavior. This was an extension of his interest in the writings of Charles Darwin. Haldane explored the origins of human behavior in evolutionary history. He was particularly interested in communication among social insects and the origin of human language. Haldane regarded the discovery of "bee language" by Karl von Frisch as one of the most important biological discoveries of the twentieth century. Haldane and Spurway presented an extensive analysis of von Frisch's observations of communication in *Apis mellifera* and summarized as follows:

> The dance conveys about 5 cybernetic units of information concerning direction, of which the average recipient received at least 2.5.... Between 100 and 3000 metres the number of turns made in a given time fall off linearly with the logarithm of the distance. At greater distances they fall off more slowly. The number of abdominal waggles made per straight run increases by 1 per 75 metres between 100 and 700 metres. It is suggested that this is the principal means by which distance is communicated.

In his classic work, *The Causes of Evolution*, Haldane (1932) discussed the consequences of selection of a gene for altruistic behavior in a colony of bees. He argued that genes causing unduly altruistic behavior in the queens would tend to be eliminated.

Other studies involved the possible inheritance of bird song, food gathering in birds, nest-building activity of wasps, and the rhythm of breathing in newts and fish. He saw unusual connections between apparently unrelated factors. He suggested that Janssens studied meiosis more fully than his contemporaries because, as a Jesuit, he believed in teleology and so had to find a purpose in the loss of half the genetic material.

In 1922, Haldane proposed a rule which later came to be known as "Haldane's rule"—"When in the *offspring* of two different animal races one *sex* is absent, rare, or sterile, that *sex* is the heterogametic *sex*." This rule has been found to be true for the offspring of several species crosses for over 80 years. It is one of the few generalizations in biology that is still considered valid today.

POPULARIZATION OF SCIENCE

Sir Arthur C. Clarke once wrote: "Professor J.B.S. Haldane was perhaps the most brilliant scientific popularizer of his generation; starting in 1923 with *Daedalus; or Science and the Future*, he must have delighted and instructed millions of readers. Unlike his equally famous contemporaries, Jeans and Eddington, he covered a vast range of subjects. Biology, astronomy, physiology, military affairs, mathematics, theology, philosophy, literature, politics—he tackled them all" (Clark 1968, Dronamraju 2009).

Haldane's popular essays covered a great number of subjects related to the impact of science and technology on human society. The topics included biological and physical sciences as well as ethics and moral questions, history and religion, and war and peace. These essays often contained passionate commentaries on a number of scientific and social problems of such fundamental character that their value has not diminished with time. They give us insights into how Haldane saw the world of science during the years 1925–1965. He wrote extensively on the problems of dissemination of scientific knowledge, emphasizing that science cannot be adequately disseminated through textbooks and lectures. He urged large-scale dissemination of scientific knowledge to millions of people. His writings were full of passion and force on this subject.

Writing on the future of the human species, Haldane argued that unless man learns to use scientific knowledge to control his evolution, disastrous consequences may follow. He wrote that only a very few species have managed to develop into something higher. A great majority of the species have degenerated and become extinct, or lost many of their functions. It is unlikely that man will evolve into something better unless he learns to control his evolution.

Haldane believed that good popular science is more valuable if it emphasizes its proper role in achieving the unity of human knowledge and endeavor at their best. He stated his belief on numerous occasions that a better-educated world will also be a happier one and urged scientists to follow his example in sharing their knowledge with the rest of humanity.

INDIA

In 1957, Haldane accepted an invitation from the Indian Statistical Institute in Calcutta, India, and moved there to continue research and teaching in genetics and biometry. The director of that institute was P.C. Mahalanobis, FRS, who was an eminent statistician and advisor to the Indian Government. Haldane was a research professor at the institute.

Haldane's conversion to life in India was complete; it included wearing Indian-style clothes, adopting a vegetarian diet, learning local languages and Hindu scriptures, as well as acquiring a new spiritual outlook. I have described these aspects in greater detail in an earlier contribution to the *Notes & Records of the Royal Society* (Dronamraju 1987). But the most important part was his interest in India's biodiversity.

The research projects undertaken by Haldane's research team in India were mostly of ecological and statistical nature. Research in biometry and genetics was pursued on a number of local plant and animal species. Some examples are selective pollination by Lepidoptera and its relevance to sympatric speciation, meristic variation in plant and animal organs, which interested Haldane's mentor William Bateson, rice grain yields in plots with mixed varieties, floral symmetry, life cycle of the tussore silk moth (*Antheraea mylitta* Drury), and nest-building activity of the solitary wasp (*Sceliphron madraspatanum* Fabr.). Projects in human genetics included inbreeding in south Indian populations, Y-chromosome linkage, color blindness, and deaf–mutism.

In 1962, Haldane resigned from the Indian Statistical Institute and moved to establish the Genetics and Biometry Laboratory supported by the State Government of Orissa in Bhubaneswar, but death came too soon because of the onset of rectal cancer. Haldane memorialized his cancer surgery by writing a poem from which I quote the following:

I wish I had the voice of Homer
To sing of rectal carcinoma,
Which kills a lot more chaps, in fact,
Than were bumped off when Troy was sacked.

I know that cancer often kills,
But so do cars and sleeping pills;
And it can hurt one till one sweats,
So can bad teeth and unpaid debts.
A spot of laughter, I am sure,
Often accelerates one's cure;
So let us patients do our bit
To help the surgeons make us fit.

PERSONAL ASPECTS

Haldane was a natural aristocrat. He seemed quite comfortable and confident with his self-assured place in society. He was a man endowed with very great intelligence and a profound memory. He enjoyed displaying his classical education during scientific meetings, reciting from classic works of Dante and Virgil with as much ease as he enjoyed quoting the Psalms from the Old Testament or the Sanskrit *slokas* from Hindu epics. Haldane's mental powers were truly amazing. He was capable of paying close attention to a lecture (as he clearly showed during the ensuing discussion) while simultaneously writing a paper on an entirely different topic. He said he taught himself to divide his attention in different directions at the same time. Haldane worked fast and enjoyed doing complex and long mathematical sums by hand. Consequently, it was not unusual to find minor (even elementary) errors in his mathematical papers. However, they seldom altered the main arguments in his papers.

Haldane generously provided advice to his students, colleagues, and others on a great number of subjects. In spite of his gruff exterior (which was mostly reserved for journalists), Haldane enjoyed scientific discussions with colleagues and was always prepared to provide a novel point of view. Because of his knowledge of multiple disciplines, he was able to transfer ideas and concepts across disciplines, such as the use of *cis* and *trans* from biochemistry to genetics.

Haldane disdained the use of any equipment for his calculations, preferring instead to perform all the sums by long hand. He encouraged us to pursue research projects that did not require much equipment. Much of our research in India was of ecological and statistical nature.

During his years in India, Haldane worked almost all the time. He relaxed in his "spare" time, either by reading science fiction or writing essays on popular science for the press. He wrote a great deal for the popular press while travelling on trains and planes. Relaxation also came occasionally when other scientists were visiting us or while visiting friends and colleagues. During his Indian period, Haldane's visitors included Sir Julian Huxley, FRS (U.K.), N.W. Pirie, FRS (U.K.), Welsh architect Sir Clough Williams-Ellis (U.K.), Sir Ronald Fisher, FRS (U.K.), Frank Fenner, FRS (Australia), Jacques Monod (France), Rene Wurmser (France), Antoine Lacassagne (France), Ernst Mayr (United States), T. Dobzhansky (United States), George Gamow (United States), Curt Stern (United States). Harlow Shapley (United States), and Joshua Lederberg (United States), among others.

Haldane died of complications arising from cancer surgery on December 1, 1964.

ENDNOTE

1. Origins of life: Haldane suggested that systems with a quasi-vital degree of complexity may have appeared independently. He was one of the first to write of the "origins" of life instead of using the more usual singular.

REFERENCES

Carlson, E.A. (1987) *Genes, Radiation and Society: The Life and Work of H.J. Muller.* Ithaca: Cornell University Press.

Briggs, G.E., and J.B.S. Haldane (1925) Note on the kinetics of enzyme action. *Biochem. J., 29*: 338–339.

Case, E.M., and J.B.S. Haldane (1941) Human physiology under high pressure. I. Effects of nitrogen, carbon dioxide, and cold. *J. Hyg., 41*: 225–249.

Clark, A.C. (1968) Haldane and space. In: K.R. Dronamraju (ed.), *Haldane and Modern Biology*. Baltimore: Johns Hopkins University Press, pp. 243–248.

Dronamraju, K.R. (ed.) (1968) *Haldane and Modern Biology*. Baltimore: Johns Hopkins University Press.

Dronamraju, K.R. (1985) *Haldane: The Life and Work of JBS Haldane with Special Reference to India*. Aberdeen: Aberdeen University Press.

Dronamraju, K.R. (1987) On some aspects of the life and work of John Burdon Sanderson Haldane, F.R.S., in India. *Notes Records R. Soc. Lond., 41*: 211–237.

Dronamraju, K.R. (1989) *Foundations of Human Genetics*. Springfield, IL: Charles C. Thomas Inc.

Dronamraju, K.R. (ed.) (1990) *Selected Genetic Papers of J.B.S. Haldane*. New York: Garland Publishing, Inc.

Dronamraju, K.R. (ed.) (1995) *Haldane's Daedalus Revisited*. Oxford: Oxford University Press.

Dronamraju, K.R. (ed.) (2009) *What I Require from Life: Writings on Science and Life from J.B.S. Haldane*. Oxford: Oxford University Press.

Dronamraju, K.R. (ed.) (2011) *Haldane, Mayr and Beanbag Genetics*. Oxford: Oxford University Press.

Dronamraju, K.R. (2017) *Popularizing Science: The Life and Work of J.B.S. Haldane*. Oxford: Oxford University Press.

Dublin, L.I., and A.J. Lotka (1925) On the true rate of natural increase. *J. Am. Stat. Assoc., 20*: 305–339.

Dyson, F. (1995) Daedalus after seventy years. In: K.R. Dronamraju (ed.), *Haldane's Daedalus Revisited*. Oxford: Oxford University Press, pp. 55–63.

Einstein, A. (1954) *Ideas and Opinions*. New York: Crown Publishers Inc.

Fisher, R.A. (1930) *The Genetical Theory of Natural Selection*. Oxford: The Clarendon Press.

Garrod, A. (1909) *Inborn Errors of Metabolism*. Oxford: Oxford University Press.

Haldane, J.B.S. (1919) The combination of linkage values, and the calculation of distances between the loci of linked factors. *Journal of Genetics, 8*: 299–309.

Haldane, J.B.S. (1922) Sex-ratio and unisexual sterility in hybrid animals. *J. Genet., 12*: 101–109.

Haldane, J.B.S. (1923) *Daedalus, or Science and the Future*. London: Kegan Paul.

Haldane, J.B.S. (1924) Mathematical theory of Natural and Artificial Selection, Part I. *Trans. Camb. Phil. Soc., 23*: 19–41.

Haldane, J.B.S. (1927) A mathematical theory of natural and artificial selection, Part V. Selection and mutation. *Proc. Cambridge Phil. Soc., 23*: 838–844.

Haldane, J.B.S. (1929) The origin of life. *The Rationalistic Annual*. Reprinted in *Science and Life* with an introduction by John Maynard Smith, London: Pemberton Publishing Co, 1968, pp. 1–11.

Haldane, J.B.S. (1931) A mathematical theory of natural selection. Part VIII. Metastable populations. *Proc. Cambridge Phil. Soc., 27*: 137–142.

Haldane, J.B.S. (1932) *The Causes of Evolution*. London: Longmans, Green.

Haldane, J.B.S. (1934) Human biology and politics. London: The British Science Guild. The Tenth Norman Lockyer Lecture, 3.

Haldane, J.B.S. (1935) The rate of spontaneous mutation of a human gene. *J. Genet., 31*: 317–326.

Haldane, J.B.S. (1937) The effect of variation on fitness. *Am. Nat., 71*: 337–349.

Haldane, J.B.S. (1942) The selective elimination of silver foxes in Eastern Canada. *J. Genet., 44*: 296–304.

Haldane, J.B.S. (1949a) The rate of mutation of human genes. *Proc. Eighth Intl. Cong. Genetics, Hereditas, 35*: 267–273.

Haldane, J.B.S. (1949b) Disease and evolution. *La Ricerca Scientifica, Supplemento, 19*: 2–11.

Haldane, J.B.S. (1949c) Human evolution: past and future. In: J.L. Jepson, G.G. Simpson, and E. Mayr (eds.), *Genetics, Paleontology and Evolution.* Princeton, NJ: Princeton University Press.

Haldane, J.B.S. (1956) The theory of selection for melanism in Lepidoptera. *Proc. R. Soc. B, 145*: 303–306.

Haldane, J.B.S. (1961) An autobiography in brief. *The Illustrated Weekly of India*, Bombay, India.

Haldane, J.B.S. (1963) Biological possibilities for the human species. In *Man and his Future*, ed. G.E.W. Wolstenholme, CIBA Foundation Symposium, London: J. & A. Churchill, pp. 337–347.

Haldane, J.B.S. (1964) A defense of beanbag genetics. *Perspect. Biol. Med., 7*: 343–359.

Haldane, J.B.S. (1965) The implications of genetics for human society. In *Genetics Today, Proceedings of the XI International Congress of Genetics*, The Hague, London: Pergamon Press, pp. xcv–xcvi.

Haldane, J.B.S., A.D. Sprunt, and N.M. Haldane (1915). Reduplication in mice. *J. Genet., 5*: 133–135.

Haldane, J.B.S., and J. Bell (1937) The linkage between the genes for colour-blindness and haemophilia in man. *Proc. R. Soc. Lond. B, 123*: 119–150.

Haldane, J.B.S. and S.D. Jayakar (1963) The solution of some equations occurring in population genetics. *Journal of Genetics, 58*: 291–317.

Huxley, A. (1932) *Brave New World.* London: Penguin Books.

Kettlewell, H.B.D. (1956) Further selection experiments on industrial melanism in the Lepidoptera. *Heredity 10*: 287–303.

Kettlewell, H.B.D. (1973) *Industrial Melanism in Biston betularia.* Oxford: Oxford University Press.

Kimura, M. (1957) Some problems of stochastic processes in Genetics. *Ann. Math. Stat., 28*: 882–901.

Lederberg, J. (1995) Foreword. In: K.R. Dronamraju (ed.), *Haldane's Daedalus Revisited.* Oxford: Oxford University Press, pp. xii–ix.

Lotka, A.J. (1922) The stability of the normal age distribution. *Proc. Nat. Acad. Sci., Wash., 8*: 339–345.

Malecot, G. (1966) *Probabilites et Heredite.* Paris: P.U.F. et I.N.E.D., Pub. No. 34, 357 pp.

Morgan, T.H., A.H. Sturtevant, C.B. Bridges, and H.J. Muller. (1915) *The Mechanism of Mendelian Heredity.* New York: Henry Holt.

Neel, J.V., and W.N. Valentine (1947) Further studies on the genetics of thalassemia. *Genetics, 32*: 38–63.

Mayr, E. (1959) Where are we? *Cold Spring Harb. Symp. Quant. Biol., 24*: 1–14.

Wright, S. (1931) Evolution in Mendelian populations. *Genetics, 16*: 97–159.

Wright, S. (1968) Contributions to genetics. In: K.R. Dronamraju (ed.), *Haldane and Modern Biology.* Baltimore: Johns Hopkins University Press.

6 Ronald Aylmer Fisher (1890–1962)

Ronald Aylmer Fisher (or R.A. Fisher) was one of the three founders of "population genetics," the other two being J.B.S. Haldane and Sewall Wright. Working independently, all three pioneers developed the mathematical theory of evolution, which examined the Darwinian theory of natural selection in the light of Mendelian genetics.

Fisher was born February 17, 1890, in East Finchley, north London, to a fine arts auctioneer named George Fisher and Katie (Heath). None of his ancestors showed any strong mathematical inclinations, with the small exception of his uncle, who graduated as a Wrangler from Cambridge.

His father's family was mostly businessmen, but an uncle, a younger brother of his father, was placed high as a Cambridge Wrangler and went into the church. His mother's father was a successful London solicitor noted for his social qualities. There was, however, an adventurous streak in the family, as his mother's only brother abandoned excellent prospects in London to collect wild animals in Africa, and one of his own brothers returned from the Argentine to serve in the First World War and was killed in 1915.

As a child, Fisher's eyesight was already poor, which inhibited his private reading. In 1904, when he was 14 years old, his mother died, and that same year, he received a scholarship in mathematics to Harrow. Because of his eyesight, he received tutoring in math under G.H.P. Mayo. In 1909, he won a mathematics scholarship to Cambridge University and graduated in 1912 with a BA in Astronomy. He was awarded a studentship for one year and studied theory of errors under F.J.M. Stratton and also more astronomy and quantum physics under J. Jeans.

Although still a student, Fisher was active in starting the Cambridge University Eugenics Society in the spring of 1911. He served on the Council and also spoke there on the theories of the day: Darwin's theory of evolution and natural selection.

EDUCATION

Fisher's special ability in mathematics showed at an early age. Before he was six, his mother read to him a popular book on astronomy, an interest that he followed eagerly through his boyhood, attending lectures by Sir Robert Ball at the age of seven or eight. Mathematics dominated his educational career. He was fortunate at Stanmore Park School in being taught by W.N. Roe, a brilliant mathematical teacher and a well-known cricketer, and at Harrow School by C.H.P. Mayo and W.N. Roseveare. His eyesight was poor as he suffered from extreme myopia and was forbidden to work by electric light. In the evenings, Roseveare would instruct him without pencil or paper or any visual aids. This gave him an exceptional ability to solve mathematical problems entirely in his mind and also a strong geometrical sense, which stood him in good stead later in the derivation of exact distributions of many well-known statistics derived from small samples.

Other statisticians, most of whom were not very skilled algebraists, consequently found his work difficult to follow and often criticized him for inadequate proofs and the use of "intuition." From Harrow, he obtained a scholarship to Gonville and Caius College, Cambridge, which he entered in 1909. He became a Wrangler in 1912, with distinction in the optical papers in Schedule B. After graduation, he spent a further year at Cambridge with a studentship in physics, studying statistical mechanics and quantum theory under James Jeans and the theory of errors under F.J.M. Stratton.

Fisher also showed an early interest in biology. As a boy, he was undecided whether to concentrate on mathematics or biology. But on a chance visit to a museum, he happened on a cod's skull with all its bones separated and labeled with their names; he decided on mathematics.

While at Cambridge, he came across Karl Pearson's Mathematical contributions to the theory of evolution. He became keenly interested in evolutionary and genetical problems, an interest that he retained throughout his life.

Fortunately, his attempts to enlist in the Army were futile because of his defective eyesight. He taught mathematics and physics, during the years 1915 to 1919, at various public schools, including Haileybury.

In 1917, Fisher married Ruth Eileen, the daughter of a physician. They had eight children, two sons and six daughters. The elder son, George, abandoned medical education to join the Royal Air Force as a fighter pilot and was killed in 1943. The younger son, Harry, also joined the Royal Air Force but was unable to serve because of motion sickness. In the meantime, Fisher had already published a paper on the fitting of frequency curves and met with Karl Pearson and also obtained the exact distribution of the correlation coefficient.

FIRST GENETICAL PAPER

In 1918, he published a monumental study on the correlation between relatives on the supposition of Mendelian inheritance. This was first submitted to the Royal Society but, on the recommendation of the referees, was withdrawn and was subsequently published by the Royal Society of Edinburgh, partly at the author's expense. This early work led to simultaneous offers in 1919 of the post of chief statistician under Karl Pearson at the Galton Laboratory and of a newly created post of statistician at Rothamsted Experimental Station under Sir John Russell. Fisher immediately accepted the Rothamsted offer, which he thought would provide considerably greater opportunities for independent research. He was also attracted by the prospect of being able to pursue his genetical studies more actively at Rothamsted.

Fisher's book *Statistical Methods for Research Workers*, which appeared in 1925, was essentially a practical handbook on these new methods. It made them generally available to biologists, who were not slow to take advantage of them. By this time, his work was becoming known to a wider circle of research workers. In addition to his own small department, an increasing number of visitors from other institutes came to work under him. He was elected a Fellow of the Royal Society in 1929.

While at Rothamsted, he pursued his research on genetics and evolution and undertook a series of breeding experiments on mice, snails, and poultry; from the last, he confirmed his theory of the evolution of dominance. Rothamsted was never officially concerned with this work, as neither genetics nor plant breeding was done there, but the Station provided land for his poultry and accommodation for his snails. The mice he kept at his own home, helped by his wife and children.

The Genetical Theory of Natural Selection, which was published in 1930, completed the reconciliation of Darwinian ideas on natural selection with Mendelian theory. In this book, he also developed his theories on the dysgenic effects on human ability of selection in civilized communities; the major factor responsible he believed to be the parallel advancement in the social scale of able and of relatively infertile individuals, the latter because of their advantages as members of small families, with consequent mating of ability with infertility. He regarded the reversal of these trends as a matter of paramount long-term importance and for a time played an active part in the affairs of the Eugenics Society. His immediate practical proposal of family allowances proportional to income was, however, so much at variance with the current thought of the time that he found few followers. His conviction of the power of genetic selection in molding organisms to fit their environment also led him to belittle the importance of reforms, such as improvements in nutrition, designed to better the environment, and this enabled progressive thinkers to dismiss his theories on human evolution, which in any event they found distasteful, without serious thought. This book became subsequently one of the foundations of population genetics, along with J.B.S. Haldane's *The Causes of Evolution* (1932) and Sewall Wright's (1931) extensive paper "Evolution in Mendelian Populations."

UNIVERSITY COLLEGE LONDON

In 1933, Fisher's genetical work led to his appointment at University College, London, as Galton Professor in succession to Karl Pearson. He did not, however, take over the whole of Karl Pearson's Department; Karl Pearson's son, E.S. Pearson, was put in charge of the statistical part, which was renamed the Department of Applied Statistics. As the accommodation was also split, the two departments lived in close proximity, sharing the same common room, which called for considerable tact by visitors with contacts with both camps, for by then, Fisher's long-standing conflict with Karl Pearson and his followers had become completely irreconcilable. The Galton Laboratory offered opportunities for the experimental breeding of animals that had not been available at Rothamsted. He brought to the laboratory the mouse colony that he had started at his home and expanded it, first to search for linkages and later for the study of modifiers in the expression of mutant genes. He continued with his snails, introduced grouse locusts, and began a co-operative experiment in dog genetics. He even brought marsupials into the department, although they were soon abandoned as impracticable for breeding experiments. Nor was the study of human genetics neglected.

He took over *The Annals of Eugenics from Karl Pearson*. This journal had originally been founded for the publication of papers on eugenics and human genetics, so as to leave *Biometrika* (which went to the Department of Applied Statistics) free for papers on statistical methodology, but under Fisher's guidance, it rapidly became a journal of importance in statistics.

In 1935, *The Design of Experiments* was published; this was the first book explicitly to be devoted to this subject and amplified and extended the somewhat cursory and elementary exposition in statistical methods. In 1938, he and Yates produced statistical tables for agricultural and medical research. During these years, he and his family continued to live in Harpenden. He kept close contact with his old friends at Rothamsted and was a familiar figure in the laboratories on Saturday mornings.

At the outbreak of war, University College was evacuated. Fisher resisted this for as long as he could, but eventually, he had to bow to the inevitable and returned to Rothamsted, where Sir John Russell found him and his department accommodation.

CAMBRIDGE UNIVERSITY

In 1943, he accepted the Arthur Balfour Chair of Genetics at Cambridge in succession to R.C. Punnett. He held this chair until his retirement in 1957 and remained in Cambridge until his successor was appointed in 1959. While in Cambridge, he continued and extended his work with mice, particularly in the study of linkage and inbreeding. Also, for the first time, he had facilities for plant genetics in the gardens of Whittingehame Lodge where the department was housed. These he used for his analysis of tristyly in *Lythrum* and *Oxalis*.

He initiated a small and short-lived, but successful, program of work with bacteria and built up a flourishing school of mathematical genetics. When he returned to Cambridge, he was reelected a Fellow of his old college, Gonville and Caius. For the most part, he resided in College, although for a time he slept at Whittingehame Lodge, at some personal inconvenience, to register his disapproval of the University regulation that required a reduction in stipend of university staff residing in College! In his later years, he became almost a legend to Caius undergraduates. His venerable figure was well known to them, for every morning he took breakfast in solitary state on High Table. He attended chapel regularly (reputedly wearing the hood of a different honorary degree each Sunday) and even on occasion preached in his own characteristic style.

PERSONAL ASPECTS

At first sight, he appeared reserved and aloof to his colleagues in College, but closer acquaintance revealed affections that could bestow warmth of heart and generosity of spirit to his friendships. In conversation, he brought not only a vast store of knowledge but also an independent mind of great rigor and penetration to bear on almost any subject. He constantly questioned conventional assumptions. He added distinction to the society of the Senior Combination Room, in which he particularly enjoyed the company of the younger Research Fellows.

He was addicted to the *Times* crossword, which he usually did alone, since he was accustomed only to fill in those letters where words crossed each other; indeed, he was engaged on such a crossword almost on his deathbed. In 1956, the Fellows did him the honor of electing him as their president; and after he had retired from his University Chair, they continued his Fellowship for life.

After leaving Cambridge, Fisher visited E.A. Cornish in Australia, head of the Division of Mathematical Statistics of the Commonwealth Scientific and Industrial Research Organization (CSIRO) in Adelaide. He found the climate and intellectual atmosphere there to his liking and was persuaded by Cornish to accept a Research Fellowship. He remained in Adelaide until his death, following an operation.

Fisher believed in suggestion rather than direction and was inclined to abandon assistants whom he found muddleheaded or lacking in initiative. Moreover, he was impatient of administrators and did not attempt to preserve good relations with them. As he said sadly after his return to Cambridge: "I used to think only University College was difficult, but now I know all officials are obstructive."

Fisher had a difficult character. He had many friends and was an intellectually stimulating man to work with and was excellent company. Generally, he had a benign and tolerant attitude and did not sit in judgment on the personal conduct of his colleagues.

His large family, in particular, reared in conditions of great financial stringency, was a personal expression of his genetic and evolutionary convictions. He was fond of children and enjoyed having them around, being completely indifferent to the disturbance they created. His respect for tradition, and his conviction that all men are not equal, inclined him politically toward conservatism and made him an outspoken and lasting opponent of Marxism.

Although he did not subscribe to the dogmas of religion, he believed that the practice of religion was a salutary and humbling human activity.

His eyesight, although a constant source of anxiety, never seriously hindered his work or his enjoyment of life. He enjoyed the company of other scientists and was a familiar figure at scientific meetings and international gatherings; the latter he attended more for the opportunity of meeting his friends than to listen to scientific communications. He was largely instrumental in setting up the Biometric Society and played a leading part in the affairs of many other societies.

He was always ready to discuss the statistical problems of others and often came up with a solution at surprising speed. He was extremely generous with his scientific ideas; many novel methods evolved by him appeared under the names of other authors.

G.I. Bliss tells a characteristic story of the origin of the maximum-likelihood method of fitting probit lines to dosage–mortality data. This occurred one Saturday in Harpenden, in 1934, when after some discussion on how to treat groups with 0% or 100% kill, which could not be dealt with by the approximate methods then current, Fisher remarked "When a biologist believes there is information in an observation, it is up to the statistician to get it out." After lunch, he obtained the maximum likelihood solution and evolved a practical arithmetical method for calculating it, which rapidly became standard for all exact work. These were the positive and very great virtues of the man.

NEGATIVE QUALITIES

On the negative side must be mentioned his notoriously contentious spirit, his quick temper, which was sometimes provoked by trivialities, and his tendency, on occasion, to be coldly rude to those whom he regarded as misguided. Nor, when his temper seized him, was he discriminating on whom his wrath should fall, or a respecter of persons; many innocent people in authority, as well as minor officials and servants, must have been amazed to find themselves blamed for faults of others or of Fisher himself. It was indeed often guilt by association, particularly in scientific matters.

The originality of his work inevitably resulted in conflicts with accepted authority, and this led to many controversies, which he entered into with vigor, but often in that indignant frame of mind that leads to a partial view of the problem and leaves unanswered objections that are obvious to the impartial observer.

On scientific matters, he was uncompromising and was intolerant of scientific pretentiousness in all its forms, especially the pretentiousness of mathematicians. He could be unforgivingly hostile to those who, in his opinion, criticized his work unjustly, particularly if he suspected they were attempting to gain credit thereby. Nevertheless, he undoubtedly enjoyed the cut and thrust of scientific controversy. His pungent verbal comments were well known; although frequently made without malice, they were nevertheless disconcerting to those of less robust temperament. However, it must be mentioned that his quarrel with Sewall Wright went beyond the academic norm and civility. They were not speaking terms for much of their lives.

SMOKING AND LUNG CANCER CONTROVERSY

Fisher's Arguments Concerning Lung Cancer

Fisher developed four lines of argument in questioning the causal relation of lung cancer to smoking. I will first list these and then briefly describe the evidence he produced in support of these arguments.

1. If A is associated with B, then not only is it possible that A causes B, but it is also possible that B is the cause of A. In other words, smoking may cause lung cancer, but it is a logical possibility that lung cancer causes smoking.
2. There may be a genetic predisposition to smoke (and that genetic predisposition is presumably also linked to lung cancer).
3. Smoking is unlikely to cause lung cancer because secular trend and other ecologic data do not support this relation.

4. Smoking does not cause lung cancer because inhalers are less likely to develop lung cancer than are noninhalers.

Fisher sees the argument that lung cancer causes smoking as an essentially unsupported speculation.

However, the scientific community generally viewed Fisher's comment on this problem with much skepticism, for two reasons: Fisher himself was addicted to smoking all his life and was clearly reluctant to condemn that habit that gave him so much pleasure; second, he was being paid by the tobacco companies as a consultant while performing his analysis.

GENETICS

Fisher's first genetical paper was published in 1918. He was greatly interested in the exciting developments in genetics that have been occurring in the early decades of the twentieth century. Mendelism had been shown to apply in all the groups of plants and animals that were investigated. Genic interaction and linkage were discovered. Morgan's group at Columbia was making a series of exciting discoveries on linkage, recombination, and gene mapping in the fruit fly *Drosophila*.

First Genetical Paper

Fisher was particularly interested in the controversy that had raged about the hereditary determination of metrical or continuous variation and the assumption of the early geneticists that their findings were in conflict with Darwin's notion of evolution by natural selection. Fisher was convinced that natural selection must be the agent of adaptation and evolution and that Mendelism and Darwinism must fit together. Darwin had seen the small differences of continuous variation as the raw material of adaptive change. Francis Galton had shown such variation to be heritable. Yule, among others, had pointed out that the basic mechanism of this continuous variation is compatible with the assumption that the expression of the character depended on the simultaneous action of many genes whose effects were additive. However, Karl Pearson and the biometricians disagreed that the correlations observed between human relatives could be interpreted successfully on this basis.

Mendel's theory of inheritance rescued Darwinism from the problem of gradual diminution of natural selection's raw material through the process of sexual reproduction. Yet due to personal and professional rivalries, many did not see in Mendelism the salvation of evolutionary theory. Pearson and the biometricians scoffed at Bateson and company's innumeracy. They also argued that Mendel's laws could not account for the wide range of continuous traits that formed the foundation of biometrics and, therefore, natural selection itself. Fisher's (1918) paper, his first genetical paper, resolved this conflict by pointing out that Mendelian inheritance is not incompatible with quantitative variation.

In his first genetical paper, "The Correlations to Be Expected between Relatives on the Supposition of Mendelian Inheritance," published in the *Transactions of the Royal Society of Edinburgh* (1918), Fisher showed clearly not only that they could be so interpreted but also that Mendelian inheritance must in fact lead to just the kind of correlations observed. He showed how the correlations could be used to partition the variation into its heritable and nonheritable fractions, how the heritable fraction could itself be broken down into further fractions relatable to additive gene action, to dominance and to genic interaction (which he generalized in a way to make it manageable by biometrical methods), and how due allowance could be made for the correlation observed between spouses. Finally, he pointed out that the excess of the sib correlation over that found between parent and offspring must follow from the Mendelian phenomena of dominance, whereas it was quite inexplicable otherwise.

Fisher's paper left little doubt that the inheritance of continuous variation provided no exception to the Mendelian rule. The first apparent conflict was resolved. In this analysis, which laid the foundation for what has now come to be called biometrical genetics, Fisher paid little attention to the experimental work being done on continuous variation with several species, especially of plants, nor did he ever discuss the biometrical consequences of linkage, which, although of course negligible in his human correlations, could have profound effects on the relationships observable in experimental investigations of continuous variation.

After dealing with the genetical mechanism underlying continuous variation, Fisher turned his attention to the relations between Darwin's principle of natural selection and Mendel's principles of heredity. He delayed publication of his consideration until 1930, when it appeared as the first chapter of his great book *The Genetical Theory of Natural Selection*. His conclusions are clear and simple: by assuming, falsely, that inheritance was blending, Darwin placed himself in a dilemma, since he then had to find some agency for replacing the variation that the blending would eliminate in large amounts in each generation. This led him to postulate a stimulating action of the environment on the production of new variation and so to destroy the whole simplicity of his notion of adaptive change resulting from natural selection.

As Fisher pointed out, the essence of Mendelian inheritance is that it conserves variation, so that far from being incompatible with Darwinism, it provides the very piece whose absence had led Darwin into his, as we could now see, unreal difficulties. Furthermore Johannsen's pure line experiments showed that new variation is not arising at the rate Darwin had been forced to postulate. He went on to show that the information already available about mutations showed them to be incapable, by virtue of both their rarity and their lack of direction, of themselves directing the course of evolution as early geneticists had supposed. Natural selection was thus displayed as the only directive agency of adaption and evolution.

Darwinism and Mendelism were indeed complementary, each supplying what the other lacked, and the need was to direct genetics toward the study of selection. Fisher's first major study of the action of natural selection was in relation to dominance. He noted that the great majority of the mutant genes observed in *Drosophila* and other species were recessive and that few, if any, fully dominant mutations had been reported. This he accounted for by the hypothesis that mutants were, in general, harmful and that selection acted to modify the phenotype of the heterozygote (itself very much more common in wild populations than the mutant homozygote) toward the wild-type, and he adduced a great deal of evidence in favor of this process of evolution of dominance. Reactions to this hypothesis were interesting: some asserted that even the heterozygote would be too rare for the forces of selection to be effective and others argued that the modification would be achieved by selection among allelic wild-type genes rather than by selection of nonallelic modifiers as Fisher had supposed. But few, if any, denied that selection and its consequences had to be taken into account.

NEO-DARWINISM

Neo-Darwinism was on its way. Fisher himself noted exceptions to his hypothesis in certain species. Most mutants in domestic poultry were not recessive but dominant, and this he sought to explain by an ingenious theory about the early domestication of poultry. To test this theory, he set up poultry breeding pens at Rothamsted and proceeded to backcross these mutants into jungle fowl, in which, in his view, their dominance should be incomplete. The results of these experiments, which he continued at Rothamsted after he had moved to the Galton Laboratory, were published in a series of papers in the mid-1930s and bore out his expectations. He also observed that polymorphism in wild populations commonly depends on dominant genes, which, contrary to his basic hypothesis, were not more common than their alleles. Now, he had already shown that if the heterozygote had a selective advantage over both the corresponding homozygotes, both alleles would be held in the population and polymorphism would result. He argued, therefore, that in cases of polymorphism, depending on dominant genes, the variant homozygotes, although not phenotypically distinct from the heterozygote, should be less fit. Nabours's data on the grouse locusts *Apotettix* and *Paratettix* provided him with the evidence he needed to show that this was indeed the case and he was able to demonstrate differences in fitness of some 10%. These differences were much larger than most geneticists were prepared to contemplate 25 or 30 ago but are in full keeping with other more recent findings.

GENETICAL THEORY OF NATURAL SELECTION

His theory of dominance was published in a series of papers between 1928 and 1932. It was also summarized in *The Genetical Theory of Natural Selection*.

The book contained a great deal more. He formulated the Malthusian parameter, as he called it, for representing the fitness or reproductive value of populations, and he developed his fundamental theorem of natural selection, that "The rate of increase in fitness of any organism at any time is equal to its genetic variance in fitness at that time"—a finding that has been "rediscovered" (using much more elaborate mathematics) at least once in the succeeding 30 years. He showed that even the smallest genetic changes can have no more than a half chance of being advantageous and that this chance falls off rapidly with the size of effect the change produces; that the environment must be constantly deteriorating from the organism's point of view and that this deterioration offsets the action of selection in raising fitness; and that any net gain in fitness is expressed as increase in size of population. More numerous species carry relatively more genetic variation so that they have a greater prospect of adaptive change, and hence of survival, than their scarcer fellow species. He argued that since mutation maintains the variation in populations, its rate of occurrence will determine the speed of evolution just as selection determines the direction.

He discussed sexual selection, developing the view that natural selection will tend to equalize the parental expenditure devoted to the two sexes rather than to equalize the sex ratio itself, and he considered the action of selection in Batesian and Mullerian mimicry.

GENE FIXATION

Fisher also developed a comprehensive mathematical theory of gene survival and spread under selection, special aspects of which he extended in a number of papers over the next 15 years or so. This mathematical study led him to conclude that the initial establishment of a favorable mutation in any population, no matter how large, would depend on random survival but that sooner or later, the recurrence of mutation would ensure the establishment of a favorable change and that once established, its fate would be governed by its consequences for selection. For a gene then to be selectively neutral, its effect on fitness must be inversely proportional to the size of the population so that, other than in special cases, it would behave as effectively neutral in selection only if its consequences for fitness were extremely small. No doubt for this reason, he would never countenance Wright's notion of gene fixation by random drift in subpopulations.

Quite independently, Haldane (1927) investigated the probability of fixation of mutant genes, showing that a dominant mutant gene having a small selective advantage k in a large random mating population has a probability of about $2k$ of ultimately becoming established in the population. The probability of fixation is much more difficult to evaluate if the advantageous mutation is completely recessive, but with remarkable insight, Haldane estimated it as of the order of k/N, where k is the selective advantage of the recessives and N is the population number.

In collaboration with E.B. Ford (1940–1947), he amply proved his point in the case of the *medionigra* gene in the moth *Panaxia dominula* by measuring the size of a colony near Oxford (using a technique of marking, release and recapture which he elaborated for the purpose) and showing that this was too large to allow the changes in gene frequency, small as they were, to be accounted for by random processes.

The last five chapters of *The Genetical Theory of Natural Selection* were devoted to man, his societies, and the decline of his civilizations. Human fertility was shown to have a heritable component, and the causes of the negative correlation between ability and fertility were analyzed in relation to the structure of society. Fisher saw the reversal, or at least abolition, of this correlation as the great problem facing our society and discussed the use of family allowances as a means to this end.

GALTON CHAIR

With his appointment to the Galton Chair in 1933, he turned more specifically to consider the promotion of human genetics, setting out his conclusions in a paper entitled "Eugenics, Academic and Practical" (1935). He believed that linkage studies could be of great value in man and published a series of papers setting out a statistical methodology for the detection and estimation of human linkage in a wide variety of circumstances. These methods have in fact had little impact, although his statistical treatment of linkage in experimental genetics, which began with the consideration of linkage estimation from data (1928) and culminated in the balanced experimental designs for detecting and estimating linkage developed 20 years later in Cambridge, is the basis of all modern practice.

It was characteristic of Fisher, too, that almost as an incidental interest, he first devised the discriminant function for the combination of multiple measurements in the comparison of skulls to replace the so-called coefficient of racial likeness, which he found in use at the Galton Laboratory and to which he objected as statistically unsound. He later developed and generalized the new technique for use in a wider range of applications. The great contribution that emerged from his promotion of human studies at the Galton Laboratory was, however, in serology.

BLOOD GROUPS

He enlisted G.L. Taylor to begin work on human blood groups, initially with the aim of finding common marker genes for use in linkage studies. The great development came, however, when Taylor, now joined by R.R. Race, turned to the Rhesus blood groups. The story as set out by Fisher in "The Rhesus Factor—a Study in Scientific Method" (1947) is fascinating. Seven alleles of the Rhesus "gene" had been recognized using four antisera, two of which gave complementary reactions with the seven genetic types. From this, he worked out the CDE structure of the gene, predicting the existence of two further antisera and one further rarc allele. These quickly came to light once the search was begun under the stimulus of his analysis. Racc and others have since shown the system to be even more complicated, but these later elaborations all fit into the trinitarian structure, which was due initially to Fisher.

The Galton Laboratory offered facilities for experimental as well as human genetics, and two lines of work were begun there, which continued to occupy his interest until his retirement from the Chair of Genetics in Cambridge. He brought with him to University College a small breeding colony of mice, which he expanded to undertake a search for linkages. The first test of seven genes yielded little except an indication of recombination values exceeding 50% between the genes for dilute pigmentation and wavy hair. To those with a background of *Drosophila* and chromosome studies, this was evidence of the nonrandom assortment of the four strands in crossing-over at successive chiasmata, but Fisher preferred to approach the problem in his own way. He devised his own mathematical approach to the theory of crossing-over and recombination, which led him to conclude that genes near the ends of long chromosome arms would characteristically show values in excess of 50% recombination.

The mouse experiments were later directed toward the adjustment of the expression of genes and their dominance properties by selection of modifiers and also toward building up inbred lines in which groups of genes were maintained segregating for the investigation of their expressions and their linkage relations. He made a thorough theoretical investigation of the progress of inbreeding under these circumstances. The results of this investigation and an account of the theory of junctions that he developed for it are set out in *The Theory of Inbreeding* (1949). In this period, he and his students were engaged in extensive theoretical investigations, one important group of which sprang from the second line of experiment begun at the Galton Laboratory. He had become interested in a complex hypothesis proposed by East to account for the inheritance of the midstyled form of flower in *Lythrum salicaria* and he set up crosses to test it. The experiments proved East's theory wrong and established that inheritance of the controlling gene was in fact tetrasomic.

Fisher's genetical experiments always started off with a precisely defined objective and were designed specifically and carefully for their purpose. Some, like the poultry experiments, ceased when the objective was gained. Others, like those with *Lythrum*, went on to explore further the consequences of the situation as it had been revealed. Even then, however, there was no departure from the basic pattern, for theoretical analysis was always sufficiently ahead of experimental investigation to ensure that new objectives would be successively defined.

He had a real feeling for the animals and plants, and despite his extreme myopia, he had an eye for their variation and classification that was a lesson to all who worked with him. In a sense, his interests were always more in working out consequences, especially selective consequences, than in exploring the heritable materials and their mechanisms.

Fisher never showed much interest in chromosomes or in genetic systems. Yet he was far from blind to novel developments. This was clear from his initiation (in Cambridge) of experimental work with bacteria in the early days of bacterial genetics—work that, in the hands of Luca L. Cavalli-Sforza, was fruitful despite the physical difficulties of the accommodation available for it, but which unfortunately came to an early end.

His study of Mendel's experiments (1936) was a delightful example of statistical analysis applied to the better understanding of an important chapter in the history of science. He brought to the belief that the mathematician's approach and imagination were complementary to the biologist's, that each stood only to gain from learning something of the other's world, and he handed on this belief to his students. His own work attested its truth; few could bring the same power to bear on their problems.

Fisher's contributions to statistics were far more important than those to genetics. Nevertheless, his position in genetics was unique.

EUGENICS

As a Darwinian, Fisher believed that natural selection worked in contexts where genotypes, once adaptive, might become maladaptive given specific environmental changes. He wrote: "Although the progressive strengthening of the reproductive instincts, which may be regarded as the principal effect of reproductive selection, will be steadily pursued, a variety of psychological modifications, together with sociological changes consequent upon these, may all be, in different circumstances, effective means to that end" (Fisher, 207, 1930).

SELECTED PUBLICATIONS BY R.A. FISHER

Books

The Genetical Theory of Natural Selection. Oxford: University Press, 1930; New York: Dover Pubns, 1958.

The Theory of Inbreeding. Edinburgh: Oliver & Boyd, 1949.

Statistical Methods for Research Workers. Edinburgh: Oliver & Boyd, 1958.

Statistical Tables for Biological, Agricultural and Medical Research (with F. Yates). Edinburgh: Oliver & Boyd, 1963.
Statistical Methods and Scientific Inference. Edinburgh: Oliver & Boyd, 1959.
Smoking—The Cancer Controversy. Edinburgh: Oliver and Boyd, 1960.

Other Publications

The correlation between relatives on the supposition of Mendelian inheritance. *Trans. R. Soc. Edinb.*, *52*: 399–433. 1919.
The genesis of twins. *Genetics*, *4*: 489–499. 1922.
On the dominance ratio. *Proc. R. Soc. Edinb.*, *42*: 321–341. 1922.
The systematic location of genes by means of cross-over ratios. *Am. Nat.*, *56*: 406–411. 1925.
The estimation of linkage from the offspring of selfed heterozygotes. *Genet.*, *20*: 79–92. 1928.
Two further notes on the origin of dominance. *Am. Nat.*, *62*: 571–574. 1928.
(With E. B. Ford.) The variability of species in the Lepidoptera with reference to abundance and sex. *Trans. R. Ent. Soc. Lond.*, *76*: 367–379. 1929.
The evolution of dominance; reply to Prof. Sewall Wright. *Am. Nat.*, *63*: 553–556. 1930.
Mortality among plants and its bearing on natural selection. *Nature, Lond.*, *125*: 972–973. 1930.
The distribution of gene ratios for rare mutations. *Proc. R. Soc. Edinb.*, *50*: 205–220. 1930.
Biometry and evolution. *Nature, Lond.*, *126*: 246–247. 1930.
The evolution of dominance in certain polymorphic species. *Am. Nat.*, *64*: 385–406. 1930.
Note on a tri-colour (mosaic) mouse. *J. Genet.*, *23*: 77–81. 1930.
Genetics, mathematics and natural selection. *Nature, Lond.*, *126*: 805–806. 1931.
The evolution of dominance. *Biol. Rev.*, *6*: 345–368. 1932.
A new series of allelomorphs in mice. *Nature, Lond.*, *129*: 130. 1932.
(With F. R. Immer and O. Tedin.) The genetical interpretation of statistics of the third degree in the study of quantitative inheritance. *Genetics*, *17*: 107–124. 1932.
The bearing of genetics on theories of evolution. *Sci. Progr. Twent. Cent.*, *26*: 273–287. 1932.
Inheritance of acquired characters. *Nature, Lond.*, *130*: 579. 1932.
The evolutionary modification of genetic phenomena. *Proc. 6th Int. Cong. Genetics*, *1*: 165–172. 1933.
Number of Mendelian factors in quantitative inheritance. *Nature, Lond.*, *131*: 400–401. 1933.
Selection in the production of the ever-sporting stocks. *Ann. Bot.*, *47*: 727–733. 1934.
Indeterminism and natural selection. *Philos. Sci.*, *1*: 99–117. 1934.
Adaptation and mutations. *School Sci. Rev.*, *59*: 294–301. 1934.
(With C. Diver.) Crossing-over in the land snail *Cepaea nemoralis* L. *Nature, Lond.*, *133*: 834–835. 1934.
Prof. Wright on the theory of dominance. *Am. Nat.*, *68*: 370–374. 1934.
Crest and hernia in fowls due to a single gene without dominance. *Science*, *80*: 288–289. 1934.
The effect of methods of ascertainment upon the estimation of frequencies. *Ann. Eugen. Lond.*, *6*: 13–25. 1934.
The amount of information supplied by records of families as a function of the linkage in the population sampled. *Ann. Eugen. Lond.*, *6*: 66–70. 1934.
The use of simultaneous estimation in the evaluation of linkage. *Ann. Eugen. Lond.*, *6*: 71–76. 1934.
Some results of an experiment on dominance in poultry, with special reference to polydactyly. *Proc. Linn. Soc. Lond.*, *147*: 71–88. 1934.
The detection of linkage with 'dominant' abnormalities. *Ann. Eugen. Lond.*, *6*: 187–201. 1935.
Dominance in poultry. *Phil. Trans. B*, *225*: 195–226. 1935.

The sheltering of lethals. *Am. Nat., 69*: 446–455. 1935.

Linkage studies and the prognosis of hereditary ailments. *Proc. Int. Cong. Life Assurance Medicine*. 1935.

The detection of linkage with recessive abnormalities. *Ann. Eugen. Lond., 6*: 339–351. 1936.

(With K. Mather.) Verification in mice of the possibility of more than fifty per cent recombination. *Nature, Lond., 137*: 362–363. 1936.

Has Mendel's work been rediscovered? *Ann. Sci., 1*: 115–137. 1936.

Heterogeneity of linkage data for Friedreich's ataxia and the spontaneous antigens. *Ann. Eugen. Lond., 7*: 17–21. 1936.

The measurement of selective intensity. *Proc. R. Soc. B, 121*: 58–62. 1936.

Tests of significance applied to Haldane's data on partial sex linkage. *Ann. Eugen. Lond., 7*: 87–104. 1936.

(With K. Mather.) A linkage test with mice. *Ann. Eugen. Lond., 7*: 265–280.

The wave of advance of advantageous genes. *Eugen. Lond., 7*: 355–369. 1937.

(With H. Gray.) Inheritance in man; Boas' data studied by the method of analysis of variance. *Ann. Eugen. Lond., 8*: 74–93. 1938.

Dominance in poultry; feathered feet, rose comb, internal pigment and pile. *Proc. R. Soc. B, 125*: 25–48. 1938.

The precision of the product formula for the estimation of linkage. *Ann. Eugen. Lond., 9*: 50–54. 1939.

Selective forces in wild populations of *Paratettix texanus*. *Ann. Eugen. Lond., 9*: 109–122. 1939.

(With E.B. Ford and J. Huxley.) Taste-testing the anthropoid apes. *Nature, Lond., 144*: 750. 1939.

(With J. Vaughan.) Surnames and blood-groups. *Nature, Lond., 144*: 1047–1048. 1939.

Stage of development as a factor influencing the variance in the number of offspring, frequency of mutants and related quantities. *Ann. Eugen. Lond., 9*: 406–408. 1939.

(With G.L. Taylor.) Blood groups in Great Britain. *Br. Med. J., 2*: 826. 1940.

(With G.L. Taylor.) Scandinavian influence in Scottish ethnology. *Nature, Lond., 145*: 590. 1940.

The estimation of the proportion of recessives from tests carried out on a sample not wholly unrelated. *Ann. Eugen. Lond., 10*: 160–170. 1940.

(With K. Mather.) Non-lethality of the Mid factor in *Lythrum salicaria*. *Nature, Lond., 146*: 521. 1941.

The theoretical consequences of polyploid inheritance for the Mid style form of Lythrum salicaria. *Ann. Eugen. Lond., 11*: 31–38. 1941.

Average excess and average effect of a gene substitution. *Ann. Eugen. Lond., 11*: 53–63. 1942.

The polygene concept. *Nature, Lond., 150*: 154. 1942.

(With J.A. Fraser Roberts.) A sex-difference in blood-group frequencies. *Nature, Lond., 151*: 640–641. 1944.

(With R.R. Race and G.L. Taylor.) Mutation and Rhesus reaction. *Nature, Lond., 153*: 106. 1944.

(With S.B. Holt.) The experimental modification of dominance in Danforth's shorttailed mutant mice. *Ann. Eugen. Lond., 12*: 102–120. 1944.

Anti-Hr serum of Levine. *Nature, Lond., 155*: 543. 1945.

The hereditary and familial aspects of exophthalmic goitre and nodular goitre. Note on a paper by L. Martin. *Quart. J. Med. (N.S.), 14*: 207–219. 1946.

(With R.R. Race.) Rh gene frequencies in Britain. *Nature, Lond., 157*: 48–49. 1946.

A system of scoring linkage data, with special reference to the pied factors in mice. *Am. Nat., 80*: 568–578. 1946–1947.

The fitting of gene frequencies to data on *Rhesus* reactions; with addendum. Note on the calculation of the frequencies of *Rhesus* allelomorphs. *Ann. Eugen. Lond., 13*: 150–155, 223–224. 1947.

The Rhesus factor—a study in scientific method. *Am. Sci., 35*: 95–103. 1947.
The theory of linkage in polysomic inheritance. *Trans. B, 233*: 55–87. 1947.
(With E.B. Ford.) The spread of a gene in natural conditions in a colony of the moth *Panaxia dominula* L. *Heredity, 1*: 143–174. 1947.
Number of self-sterility alleles. *Nature, Lond., 160*: 797–798. 1947.
(With M.F. Lyon and A.R.G. Owen.) The sex chromosome in the house mouse. *Heredity, 1*: 355–365. 1948.
A quantitative theory of genetic recombination and chiasma formation. *Biometrics, 4*: 1–13. 1948.
(With G.D. Snell.) A twelfth linkage group of the house mouse. *Heredity, 2*: 271–273. 1949.
Papers on the Soviet genetics controversy. in Science, Occasional Pamphlet No. 9. 1949.
The linkage problem in a tetrasomic wild plant, *Lythrum salicaria. Proc. 8th Int. Cong. Genet* (suppl. to Hereditas, Lund.): 225–233. 1949.
(With P.H. Andresen, S.T. Callender, R. Grubb, W.T.J. Morgan, A.E. Mourant, M.M. Pickles, and R.R. Race.) A notation for the Lewis and Lutheran blood-group systems. *Nature, Lond., 163*: 580–581. 1949.
(With W.H. Dowdeswell and E.B. Ford.) The quantitative study of populations in the Lepidoptera. 2. *Maniola jurtina* L. *Heredity, 3*: 67–84. 1949.
Note on the test of significance for differential viability in frequency data from a complete three-point test. *Heredity, 3*: 215–219. 1949.
A preliminary linkage test with agouti and undulated mice. 1. The fifth linkage group. *Heredity, 3*: 229–241. 1949.
A class of enumerations of importance in genetics. *Proc. R. Soc. B, 136*: 509–520. 1950.
Polydactyly in mice. *Nature, Lond., 165*: 407. 1950.
(With E.B. Ford.) The 'Sewell Wright effect'. *Heredity, 4*: 117–119. 1950.
Gene frequencies in a cline determined by selection and diffusion. *Biometrics, 6*: 353–361. 1951.
A combinatorial formulation of multiple linkage tests. *Nature, Lond., 167*: 520. 1951.
(With L. Martin.) Standard calculations for evaluating a blood-group system. *Heredity, 5*: 95–102. 1951.
Statistical methods in genetics. The Bateson Lecture, 1951. *Heredity, 6*: 1–12. 1953.
The variation in strength of the human blood-group P. *Heredity, 7*: 81–89. 1953.
The linkage of polydactyly with leaden in the house mouse. *Heredity, 7*: 91–95. 1953.
(With W. Landauer.) Sex differences of crossing-over in close linkage. *Am. Nat., 87*: 116. 1953.
Population genetics. The Croonian Lecture. *Proc. R. Soc. B, 141*: 510–523. 1954.
A fuller theory of 'junctions' in inbreeding. *Heredity, 8*: 187–197. 1954.
The experimental study of multiple crossing over. *Proc. 9th Int. Congr. Genet (suppl. to Caryologia), 6*: 227–231. 1955
Blood groups and population genetics. *Acta Genet., 6*: 507–509. 1957.
Methods in human genetics. *Acta Genet., 7*: 7–10. 1958. Polymorphism and natural selection. *J. Ecol., 46*: 289–293. 1958.
Retrospect of the criticisms of the theory of natural selection. In J.S. Huxley (ed.), *Evolution as a Process*. London: Allen and Unwin, pp. 84–98. 1958.
An algebraically exact examination of junction formation and transmission in parent offspring inbreeding. *Heredity, 13*: 179–186, 1959.
A model for the generation of self-sterility alleles. *J. Theoret. Biol., 1*: 411–414. 1962.
Enumeration and classification in polysomic inheritance. *J. Theoret. Biol., 2*: 309–311. 1962.
The detection of a sex difference in recombination values using double heterozygotes. *J. Theoret. Biol., 3*: 509–513. 128.
Some hopes of a eugenist. *Eugen. Rev., 5*: 309–315. 1915.
(With G.S. Stock.) Cuenot on preadaptation: a criticism. *Eugen. Rev., 7*: 46–61. 1915.

The evolution of sexual preference. *Eugen. Rev., 7*: 184–192. 1917.
Positive eugenics. *Eugen. Rev., 9*: 206–212. 1919.
The causes of human variability. *Eugen. Rev., 10*: 213–220. 1922.
Darwinian evolution by mutation. *Eugen. Rev., 14:* 31–34. 1922.
New data on the genesis of twins. *Eugen. Rev., 14*: 115–117. 1922.
The elimination of mental defect. *Eugen. Rev., 16*: 114–116. 1924.
The biometrical study of heredity. *Eugen. Rev., 16*: 189–210. 1926.
Eugenics—can it solve the problem of decay of civilizations? *Eugen. Rev., 18*: 120–136. 1927.
The effect of family allowances on population. Report of the Public Conference on Family Allowances, 7–11. London: Family Endowment Society. 1928.
Income-tax rebates; the birth-rate and our future policy. *Eugen. Rev., 20*: 79–81. 1928.
The differential birth-rate: new light on causes from American figures. *Eugen. Rev., 20*: 183–184. 1931.
The biological effects of family allowances. *Family Endowment Chronicle, 1*: 21–25. 1932.
The social selection of human fertility. The Herbert Spencer Lecture, 1932. Oxford: Clarendon Press. 1932.
Family allowances in the contemporary economic situation. *Eugen. Rev., 24*: 87–95. 1934.
The children of mental defectives. Departmental Committee on Sterilisation, Report, 60–74. 1935.
Eugenics, academic and practical. *Eugen. Rev., 27*: 95–100. 1936.
Income-tax and birth-rates: family allowances. *London Times.* April 30,1943.
'Student' (obituary). *Ann. Eugen. Lond., 9*: 1–9. 1945.
G.L. Taylor, M.D., Ph.D., F.R.C.P. (obituary). *Br. Med. J., 1*: 463. 1945.
The Indian Statistical Institute. *Nature, Lond., 156*: 722. 1947.
Thomas Hunt Morgan, 1866–1945 (obituary). *Obit. Not. R. Soc., 5*: 451–454. 1948.
Biometry. *Biometrics, 4*: 217–219. 1949.
The report of the Royal Commission on Population. *Cambridge J., 3*: 32–39. 1949.
The Sub-Commission on Statistical Sampling of the United Nations. *Bull. Inst. Int. Stat., 32*: 207–209. 1950.
Creative aspects of natural law. The Eddington Memorial Lecture, 1950. Cambridge Univ. Press. 1953.
The expansion of statistics. *J. R. Stat. Soc. A, 116*: 1–9. 1955.
(With W.H. McCrea.) Space travel and ageing. *Philos. of Sci., 18*: 56–58, 174–175. 1957.
Dangers of cigarette smoking. *Br. Med. J., 2*: 43, 297–298. 1958.
Cigarettes, cancer and statistics. *Centennial Rev., 2*: 151–166. 1958.
Lung cancer and cigarettes? *Nature, Lond., 182*: 108. 1958.
Cancer and smoking. *Nature, Lond., 182*: 596. 1959.
Comparison between balanced and random arrangements of field plots. *Biometrika, 29*: 363–379. 1937.

Publications by Others

Haldane, J.B.S. (1927) A mathematical theory of natural and artificial selection. Part V. Selection and mutation. *Proc. Camb. Philos. Soc., 23*: 838–844.
Tedin, O. (1931) The influence of systematic plot arrangement upon the estimate of error in field experiments. *J. Agric. Sci., 21*: 191–208.

7 Sewall Wright (1889–1988)

Sewall Wright was one of the great trio of the founders of population genetics, the other two being J.B.S. Haldane and R.A. Fisher (Dronamraju 1990).

Sewall Green Wright (he later dropped the middle name) was born in Melrose, Massachusetts, on December 21, 1889. His father, Philip Green Wright, was an economist who moved with his family to Galesburg, Illinois, in 1892 to join the faculty at Lombard College. There, he taught a great variety of subjects, including economics, mathematics, astronomy, surveying, and English composition, and also acted as director of the gymnasium. He also printed the Lombard College bulletin on his own printing press. Later, he conducted research at the Brookings Institution and published several books. One of these, *The Tariff on Animal and Vegetable Oils*, included a statistical appendix by his son Sewall Wright.

Sewall had two distinguished brothers: Quincy in international law and Theodore in aeronautical engineering. Quincy and Sewall regularly operated their father's printing press and were the first to publish the poetry of Carl Sandburg, who was then studying writing with their father at Lombard College. Philip Wright was indeed a polymath. Carl Sandburg called him the "Illinois Prairie Leonardo."

CHILDHOOD AND EDUCATION

Sewall was a precocious child. He wrote a pamphlet, at the age of seven, on natural history, with chapters on marmosets, ants, dinosaurs, and astronomy, and a wren that could not be discouraged from nesting in the family mailbox. He learned to extract cube roots by reading his father's books before entering school, a skill that he said brought him instant, lasting unpopularity with the other students.

Later, he became interested in analytical geometry and invented for himself a way of determining areas, somewhat like the integral calculus that he would learn later from his father at Lombard. He was clearly interested in science, not Greek and poetry, which his father enjoyed.

Sewall found grade school a disappointment. In high school, he pursued his interests in natural history and took what science courses were offered, but he did most of his learning outside. In his senior year, he read Charles Darwin's *Origin of Species*.

Entering Lombard College, Wright started to major in chemistry but found much of analytical chemistry not to his liking. He took math courses from his father, going as far as differential and integral calculus. His later theoretical work in population genetics depended on methods that Sewall learned on his own.

Philip Wright also taught a course in surveying, which led to Sewall's employment with a railroad company between his junior and senior years. He also used his mathematical skills to calculate the rail curvature. The year was a rich experience in the old west tradition. In his nineties, Wright still remembered words from the Sioux language. These were the same local tribes that had destroyed General Custer's troops at the Little Big Horn 33 years earlier (Jim Crow, personal communication, 1985).

During a brief illness, Sewall lived in a caboose and read about quaternions. It is interesting that another founder of population genetics, J.B.S. Haldane, also read the same book (Tait's *Elementary Treatise on Quaternions*) while convalescing from war injuries in Iraq during the First World War.

It was also the year of Halley's Comet, and Wright saw it from the roof of his caboose. Unfortunately, his failing eyesight prevented his seeing it again in the 1980s. As a result of a lung infection, Wright was refused standard life insurance, a fact that he found increasingly ironic as he continued to live into his late nineties.

GENETICS

Returning to Lombard for his senior year, Wright took a biology course for the first time. Wilhelmine Entemann Key, one of the first women to receive a PhD from the University of Chicago, was his inspiring teacher and graduate advisor. Wright learned his first genetics by reading Reginald Punnett's article in the 11th edition of the *Encyclopedia Britannica*. His professional interests were now clear. He obtained a $250 scholarship to the University of Illinois. Noted Harvard University geneticist, William E. Castle, visited the University of Illinois during this year and, on meeting Wright, offered him a Harvard assistantship on the spot.

Castle was then the nation's leading mammalian geneticist. Each student had a species to study. C.C. Little worked on mice and later founded the Jackson Laboratory. E.C. MacDowell studied rabbits and Wright took over the guinea pig work, completing his PhD at Harvard in 1915.

At the time, Castle was selecting hooded rats for greater and lesser amounts of white. Wright played a crucial role by suggesting the experiments to distinguish between the view, wrongly held by Castle, that the color changes were in the major gene itself, and the opposing (and correct) one, that there were many segregating modifiers. Wright continued his guinea pig breeding and research with coat-color experiments for more than 40 years.

U.S. DEPARTMENT OF AGRICULTURE

Upon receiving his doctorate from Harvard in 1915, Wright moved to Washington, where he became senior animal husbandry man in the U.S. Department of Agriculture (USDA). There, he took over the analysis of a colony of guinea pigs, some of which had been sib-mated for many generations. Wright continued to analyze the effects of inbreeding and hybridization and at the same time pursued his studies of coat-color inheritance.

This was the period when Wright began to make major theoretical advances in genetics and established himself as a founder of population genetics. He worked out the consequences of various mating systems, and his studies on quantitative inheritance, along with those of R.A. Fisher, became the foundation for scientific animal breeding. Simultaneously, Wright also developed what he later called the "shifting balance theory."

UNIVERSITY OF CHICAGO

In 1926, he moved to the University of Chicago, where he continued his theoretical work as well as his experiments with guinea pigs. He also took up the standard academic duties, teaching several courses and supervising graduate students. This continued until 1955, when he retired from Chicago at age 65 and moved to Wisconsin, which had a retirement age of 70. Wright was not paid a full salary, only a supplement to his Chicago retirement annuity. This lasted for five years, after which Wright continued to work an additional quarter century.

After his second retirement, Wright completed the monumental set of four volumes, *Evolution and the Genetics of Populations* (1968, 1969, 1977, and 1978), in which he not only summarized his own work but also reviewed and analyzed an enormous body of experimental and theoretical literature.

In his nineties, Wright's eyesight became so poor that he could read only with the aid of an enlarging machine. He gradually gave up active research and scientific reading. Yet he continued to write. His last paper was published in 1988. Wright was in excellent health until the end. It was on one of his customary long walks that he slipped on an icy spot. He died suddenly and unexpectedly a few days later, March 3, 1988, from a pulmonary embolism, the consequence of a pelvic fracture. He was 98.

WRIGHT'S SCIENTIFIC WORK

Wright's first scientific paper was published in 1912. It was a morphological study of a fish parasite, a trematode, done while he was at the University of Illinois. His first genetic paper (1914) was a suggestion that one could make a distinction between auto- and allo-polyploidy by the frequency of homozygosis for recessive genes.

Three of Wright's major areas of interest were apparent in the next few years, at Harvard and USDA. These were correlation analysis, animal breeding, and mammalian physiological genetics. His evolutionary ideas followed soon after.

WRIGHT'S CONTRIBUTIONS TO GENETICS

Working independently, J.B.S. Haldane, R.A. Fisher, and Sewall Wright founded mathematical population genetics in the 1920s and 1930s. They dominated this field until the 1960s, when both Haldane and Fisher died. Wright continued alone until his death in 1988. Wright's early interest in genetics was concerned with coat color inheritance in mammals. In fact, he published a series of papers on comparative coat color inheritance in guinea pigs, mice, rats, and several other mammalian species. His studies of coat color inheritance in guinea pigs eventually led to his views on evolution.

Wright's contribution to the theory of evolution by natural selection, shifting balance theory, was a direct outgrowth of his experience of breeding domestic livestock as well as his experiments with coat color in guinea pigs. By devising a number of combinations of crosses involving various color genes, Wright was able to study and analyze the complex interactions. He viewed evolution as a process of moving from one harmonious combination of gene frequencies to another that is better adapted to the environment. However, Wright emphasized that the transition from one favorable combination to a better adapted one involves passage through a number of inferior combinations. In order to explain this passage through a valley between two peaks, Wright emphasized, at least in some cases, the role of random factors.

Wright (1931) wrote: "Finally in a large population, divided and subdivided into partially isolated local races of small size, there is a continually shifting differentiation among the latter (intensified by local differences in selection but occurring under uniform and static conditions) which inevitably brings about an indefinitely continuing, irreversible, adaptive, and much more rapid evolution of the species." Wright stated that complete isolation in this case originates new species.

Wright's model was not unlike that suggested by Haldane (1931), who anticipated Wright in his remarkable paper dealing with the evolution of genes in "metastable populations."

Wright (1931) discussed his model in three phases. The first phase is stochastic variability of all gene frequencies in each local population about its set of equilibrium frequencies. The underlying assumption for Wright's model is as follows. The sort of selection involved in influencing their equilibrium values is assumed to result in the same intermediate grade for quantitative characters under multifactorial heredity. Consequently, the result may lead one to believe (with Wright 1968) that "selection toward such a grade implies the existence of many selective peaks in the 'surface' of selective values, separated by saddles, and at various heights because of secondary effects on other characters (pleiotropy)."

The occasional crossing of a saddle between a lower and a higher peak as a result of extreme stochastic variation leads to the second phase in which selection favors the set of equilibrium frequencies of the higher peak. The third (and final) phase is concerned with the successful establishment of this "superior" system by means of excess population growth and emigration to neighboring populations "and ultimately, perhaps, throughout the species as a whole." It was recognized further that local selection in a different direction may also lead to superior general adaptation and replace the first two phases. Wright emphasized that such "random" drift toward favorable combinations of gene frequencies is likely to occur more frequently in small, partially isolated populations.

HALDANE AND WRIGHT'S THIRD PHASE

We must also consider what Haldane had called "the weakest point in Wright's argument." Haldane (1959) wrote: "Wright considered a species divided up into a number of small endogamous tribes. Such a tribe might easily, by chance, become homozygous for several slightly unfavorable genes. If it happened to hit upon a favorable combination, its numbers would increase. To my mind the weakest point in Wright's argument is that he has not adequately considered what happens when this tribe starts hybridizing with others." Crow et al. (1990) examined this aspect of Wright's work. In particular, they considered the following questions: (a) Does recombination break up favorable combinations before they can spread through the population? And (b) How much migration is required to prevent this and carry out the third phase? Crow et al. (1990) concluded that remarkably little migration proves to be sufficient to bring about the final phase of Wright's shifting balance process. They wrote further that the general importance of Wright's shifting balance theory of evolution remains uncertain. Perhaps, new data on physiological aspects as well as population structure could throw light on this problem. If there are widespread circumstances that can hinder major evolutionary advances by means of small selectively advantageous steps, then Wright's theory or something similar is required. If the situation were different, however, Fisher's approach would be adequate to explain the general process of evolution.

WRIGHT'S LAST PAPER

In a defense of his "shifting balance" theory, Wright (1988) wrote a paper—his last paper, written shortly before his death—entitled "Surfaces of Selective Value Revisited." With reference to Provine's (1986) criticism of the concept of "surfaces of selective value," Wright (1988) commented that Provine was "looking for something more mathematical than was intended." It is of great interest that Wright's (1931) essential concept was largely based on the multifactorial viewpoint of genetics. As he put it: "The early viewpoint changed with the demonstration by Nilsson-Ehle (1909) and East (1910) that quantitative variation usually depends on the total effect of multiple minor factors." He emphasized once again that this implied the occurrence of "multiple selective peaks" because numerous superior combinations could exert more or less similar effects and that the selective value of any particular gene depends on the rest of the genome. Wright's 1931 paper dealt with a mathematical analysis of the allelic frequencies for pairs of alleles at a single locus under various conditions. The evolutionary implications of multiple loci, each with multiple alleles, were first discussed verbally (in terms of what Wright later called his "shifting balance" theory), and were first published in an abstract in 1929. The pictorial representation of the basic evolutionary process consisted of six diagrams, each under a different set of conditions: (a) increased mutation or reduced selection, (b) increased selection or reduced mutation, (c) qualitative change of environment, (d) close inbreeding, (e) slight inbreeding, and (f) division into local races. Wright (1988) stated that the most favorable circumstance for the operation of the shifting balance process would be a sparse population of individuals that are neither excessively mobile nor immobile over a widely extended range.

In Wright's model, random gene fluctuations are regarded not as impediments to the process of natural selection but as the raw material on which evolutionary progress is based. On the other hand, Fisher believed that random gene frequency fluctuations were relatively unimportant to the main process of evolution by natural selection. Wright's views have been held consistently throughout his life's work, during 1920–1988. Contrary to Provine's (1986) evaluation of Wright's work,

Wright (1980, 1988) himself stated that his views on the evolutionary process had *not* become more selectionist after 1932. With respect to Kimura's (1968) neutral theory, Wright (1988) wrote that his own chief interest had not been evolution in general, but *adaptive* evolution. Kimura, on the other hand, developed his neutral theory on the basis of slowly accumulating biochemical changes in the genome along long successions of species.

MAMMALIAN GENETICS

Wright's work on physiological and developmental genetics is much less well known than his work on animal breeding and evolution. Yet for many years, Wright devoted the major share of his research time to guinea pig studies. He did his own mating and record keeping; the guinea pig colony was often the best place to find him. He continued this work throughout the Chicago years and stopped on moving to Madison only because the University of Wisconsin could not furnish guinea pig facilities. I believe this was fortunate, for it gave Wright the chance to complete his long-contemplated project, writing his four-volume monument (1968, 1969, 1977, 1978). As it was, he spent his first five years at Wisconsin writing up his guinea pig studies, some done years before.

Early in his Washington years, Wright wrote a series of 11 papers on color inheritance in various mammals (1917, 1918). These papers are noteworthy in two regards. First, Wright interpreted the color interactions in terms of the latest knowledge of pigment chemistry and enzyme kinetics. Second, he discovered extensive similarities among the mammals and inferred that the causative genes had a common ancestry, facts that are now being definitively confirmed by DNA similarity.

Throughout his guinea pig studies, Wright went as far toward a chemical explanation as knowledge of the time would permit; he wanted to explain dominance and epistasis in chemical terms. His quantitative bent led him to formulate the relationships in path diagrams and to express the kinetics as differential equations, assuming flux equilibrium kinetics. Wright's major analyses (1941) appeared the same year as the work of George Beadle and Edward Tatum on biochemical mutants in *Neurospora*. This started a new direction in genetic research, and molecular biology and microorganisms took over. Wright continued his guinea pig studies for another 15 years.

In some respects, Wright's early work was ahead of its time. One of these was in the correlation of size of various body parts (Wright 1918). He analyzed the phenotypic variance into components associated with general size, limb-specific factors, fore- and hind-limb specific factors, and factors associated with the upper and lower limb (whether fore or hind). This kind of work has had a recent resurgence of interest.

STATISTICS

Wright's first statistical paper (1917) corrected Raymond Pearl on the use of probable error to test Mendelian ratios. In the same year (1917), he used the additivity of variances and covariances to separate guinea pig weights into within- and between-strain components. This was actually analysis of covariance, although he was unaware of Fisher's work and the terminology came later. Wright (1920, 1926) also found a transformation to linearize cumulative percentage data, now called the probit transformation.

Wright's most important contribution to statistics is his method of path analysis (1921, 1934, 1983, 1984). He always wanted to use statistics interpretatively rather than for description and prediction. Although the mathematics are those of partial regression, the point of view is original. A simple and useful Wrightian device is to diagram causal sequences so that paths of direct causation are indicated by arrows, while correlations between anterior, unanalyzed causes are represented by double-headed arrows. Each causal step is associated with a path coefficient, a partial regression coefficient standardized by being measured in standard deviation units. These coefficients measure the relative importance of the different paths. From such a diagram, Wright found simple rules by which one can easily write all the appropriate equations.

The method has the virtue of making immediately obvious whether there are enough data and relationships to permit a solution. In addition to using the method for genetic problems, Wright applied it to such diverse situations as growth and transpiration of plants, respiratory physiology, prey–predator relations, and the relative importance of heredity and environment in human IQ. The most impressive analysis is that of the production and prices of hogs and corn. Wright had 510 correlations and did the calculations himself, a time-consuming job in those days before computers. He was able to account for 80% of the variance of hog production and prices by fluctuations in the corn crop, various intercorrelations, and cleverly adjusted time lags. However, this paper (1925) was not published immediately as it was deemed improper for an animal husbandman to write a paper in economics. It required the help of Henry Wallace, who prevailed on his father, then secretary of agriculture, to intervene and see that the paper was published.

From 1920 to 1960, the method was seldom used outside of animal breeding circles. Scientists in general and biologists in particular made almost no use of it. Why? One reason is that the method cannot be applied routinely; it doesn't lend itself to "canned" programs. The user must have a hypothesis and diagram it. Biologists have made a great deal of use of correlation and regression analysis, but the emphasis has been on prediction and significance tests, for which Fisherian methods are more appropriate. At the same time, psychologists preferred to use factor analysis, which uses much of the same algebra but has a different conceptual basis.

Recently, however, path analysis has become popular in the social sciences. New methods of formulation, and particularly the use of computers, have greatly increased the power of Wright's methods. Yet, he was not always pleased with the uses, or with mathematical criticisms of it. One of his last papers (1983) was a spirited defense of his methods.

WRIGHT'S INFLUENCE

Wright's move from Chicago to Madison brought him into daily contact with James F. Crow, a well-known population geneticist and a distinguished member of the University of Wisconsin's Genetics Department in Madison. Dr. Crow had a number of highly gifted students, among them were Motoo Kimura and Newton Morton.

Kimura's PhD work under Crow began only a few months before Wright's arrival. This chain of events had a very significant impact on the development of population genetics. Kimura's work was deeply influenced by Sewall Wright. Kimura (1985) wrote: "Through reading Wright's papers, I became more and more absorbed in problems of random genetic drift in finite populations. …I was also attracted by Wright's idea that individually deleterious but jointly advantageous mutants can be utilized in evolution when the population is subdivided; the advantageous gene combination will become fixed in some of the local populations due to sampling drift."

During his Chicago period, Wright had several graduate students, who later became well-known geneticists in their own right. They include W.L. Russell, E.S. Russell, L.B. Russell, J.P. Scott, Herman Slatis, and Janice Spofford.

After moving to the University of Wisconsin, Wright deeply influenced the ideas of Motoo Kimura (who had just started his graduate work under James F. Crow in the Genetics Department), especially the problems of random genetic drift in finite populations, and the neutral theory of evolution. Indirectly, he influenced the careers of many geneticists worldwide.

ANIMAL BREEDING

Wright's (1922) studies on inbreeding and crossbreeding of guinea pigs, utilizing the accumulated USDA records and data of his own, were masterful. The meticulously kept records included not only pedigree information but also many kinds of measurements—litter size, individual and litter weight at various stages, and viability. The husbandry conditions were often miserable, including wartime shortages and the Washington summer heat. It is a testimony to Wright's analytical skills and hard work that he could extract so much consistent and useful information. He documented the usual, but not invariable, decline on inbreeding; the recovery on crossbreeding; and the *quantitative* predictability of decline when these hybrids were inbred. He showed that all this was entirely consistent with Mendelian inheritance and dominance.

At the same time, Wright developed his widely used algorithm for computing the inbreeding coefficient for any pedigree, however complex (1922), and wrote a series of papers on the consequences of different mating systems (1921). He later (1925, 1926, 1943) showed how to separate the effects of nonrandom mating from those of reduction in population size, and showed that in short-horn cattle, the small size of the breeding population was by far the most important.

For many years, animal breeding was dominated by a single figure, Jay L. Lush, of Iowa State University. A Wright disciple, he carried the gospel. He wrote a book that became the standard, and his numerous students came from all over the globe. As a result, Wright's path analysis, inbreeding theory, and prediction formula for selection of quantitative traits spread widely and rapidly. Animal breeding changed from an art to a quantitative science. In recent years, with computerized records and artificial insemination, the methods have become very sophisticated. The steady improvement of milk production testifies to the effectiveness of a well-organized, cooperative selection program. The current methods superficially look quite different from path analysis, but they trace back to the Wright-Lush influence.

WRIGHT AND HALDANE

Sewall Wright was a mild mannered and quiet gentleman. He was known to give his time freely to anyone who sought his help in genetic research. Like his contemporary, J.B.S. Haldane, Wright was totally dedicated to his work in genetics. Neither had any other distractions. I was fortunate to have had opportunities to enjoy the friendship of both Haldane and Wright. Wright participated in the XII International Congress of Genetics in Tokyo in 1968, lecturing on Haldane's contributions to genetics in a symposium that I had organized (Wright 1969). During an informal luncheon in Madison shortly before his death, in 1988, Wright stated that he first met Haldane during the International Congress of Genetics in Ithaca, New York, in 1932. There were some important differences between Haldane and Wright. Haldane was a highly skilled and prolific popularizer of science, whereas Wright's writings were almost entirely of a technical nature. Wright's scientific work was entirely in genetics, whereas Haldane's publications covered physiology, biochemistry, biometry, cosmology, and several other branches of science including genetics.

Wright was working and publishing papers in genetics almost until death in his 99th year, when he slipped on a patch of ice during one of his frequent walks in Madison. His last years were spent in close friendship with his friends, Dr. and Mrs. James F. Crow.

WRIGHT AND FISHER

Fisher and Wright disagreed on a number of aspects of population genetics. One question concerned the importance of random gene frequency drift and its role in Wright's shifting-balance theory of evolution. Wright thought that a structured population with many partially isolated subpopulations, within which there was random drift and among which there was an appropriate amount of migration, offered the greatest chance for evolutionary novelty and could greatly increase the speed of evolution. Fisher thought that a large panmictic population offered the best chance for advantageous genes and gene combinations to spread through the population, unimpeded by random processes.

Another area of disagreement is the origin of dominance. Fisher believed that it evolved by selection of dominance modifiers and Wright that it was a consequence of the nature of gene action.

However, now that there is a general quantitative disagreement with Fisher's explanation of modifiers, other mechanisms (e.g., selection for more active alleles) have to some extent replaced it. Wright's theory remains popular and has been generalized and extended.

CONFLICT BETWEEN WRIGHT AND FISHER

Dr. James Crow has written a great deal on the founders of population genetics. The following account benefitted from my discussions with him over many years.

In the 1920s, Fisher and Wright had a friendly relationship. Their methods in population genetics were quite different, but they exchanged letters and compared results. Sometimes, one would correct the other's errors, and the corrections were gratefully accepted. Their differences were in interpretation, not in the mathematics. Although their techniques were different, for any particular problem, their theoretical conclusions were the same.

However, this happy situation did not prevail after 1931. According to Dr. James Crow, something seems to have gone wrong at that point, which involved Wright's famous paper of 1931. It was largely written several years before publication and Wright sent a manuscript copy to Fisher, who was then working on his book *The Genetical Theory of Natural Selection* (1930). Wright's paper had an error, later corrected, but this was not pointed out by Fisher. Furthermore, some of Wright's formulas appeared later in Fisher's book. Wright thought that Fisher deliberately did not tell him of the error and that he appropriated some of Wright's results in his book, cited in Provine (1986, pp. 259–260). This version agrees with what Wright told Crow. It is possible that, since their notation and methods were so different, it was hard to translate from one to the other, and since he knew how to solve the problems, Fisher simply did not realize that Wright had already arrived at some of his results. In any case, this had a profound effect on Wright, and they no longer were on friendly terms. Correspondence ceased and Wright thereafter never sent manuscripts to colleagues prior to submission.

Fisher died in 1962, however, Wright lived until 1988.

Wright continued to write, repeatedly supporting his views and especially his shifting-balance theory. Counterarguments against Wright's model were made by others, stating that his preferred population structure with subdivisions and appropriate balance of selection, mutation, random drift, and migration is very rare and that while his mechanism is at work, the population fitness is substantially reduced. A model (or process) that produces a loss of fitness is generally not acceptable. Recently, the emphasis has become more on obtaining and examining the data rather than supporting a particular theory.

Toward the end of his long life, Wright (1988) became more conciliatory. As to the different approaches of Haldane, Fisher, Kimura, and himself, Wright wrote "All four are valid."

Random Genetic Drift and Inbreeding

Another difference between Wright and Fisher had created much less rancor, although it was a significant source of disagreement. The issue was inbreeding vs. random gene frequency drift. Random drift is especially important in small populations where random mates may be related and share a fraction of their genes. In many ways, the two processes are similar; in particular, both lead to an increase in homozygosity and ultimate fixation.

Fisher regarded them as quite different. He clearly thought that consanguineous mating within a large population, whether systematic or not, was quite different from increased fixation due to small population size.

Wright's view was different. He is justly famous for inventing an inbreeding coefficient and for producing a simple algorithm, well known to generations of genetics students, for computing this coefficient in a pedigree of any degree of complexity (Wright 1922). Wright was particularly pleased that his F statistics could be used to measure random drift as well as consanguineous mating. In his words, it is important to note that the same coefficient, F, that measures the degree of approach toward fixation (in a finite population) is also the Galtonian correlation coefficient r_{es} [F] for the alleles that come together at fertilization.

EFFECTIVE POPULATION NUMBER

Wright (1931) introduced the popular and very useful concept of Effective Population Number, N_e. This is the size of an *idealized* population with the same gene frequency drift or inbreeding as the observed population. An idealized population is panmictic, with each parent having an equal expectation of progeny. Note that this does not mean that each parent has the same number. Rather, the number of progeny has a Poisson or binomial distribution. In this idealized case, N_e is simply the number of breeding adults. This concept of effective population has been widely used; N_e is regularly employed as a surrogate for the actual number in countless formulas.

WHO WAS RIGHT?

The fact that the two effective numbers are different would seem to resolve the question in favor of Fisher's view: inbreeding and random drift are not the same thing. A single formula does not describe both.

But consider a special situation. If only populations of constant size are considered, Wright's view makes sense.

But of course, a general theory should not be so constricted. Remarkably, the only cases that Wright ever wrote about were those in which the two effective numbers were the same, so he naturally regarded this as additional evidence for his view that inbreeding and random drift can be dealt with the same way. As far as I know, Wright never made a mistake in dealing with this problem, for he never studied cases where the two effective numbers differed.

Later, Wright (1969) discussed them in some detail. I do not think this caused him to change his view. His discussions almost always dealt with populations in which the two effective numbers are the same.

WRIGHT'S IMPACT

Wright made lasting contributions in statistics, mammalian genetics, animal breeding, population genetics, and the theory of evolution. He would rank as an important contributor in any of these areas. Collectively, they place him among the greatest of twentieth century biologists. *The 1988 Science Citation Index* lists some 500 articles that refer to his papers.

PERSONAL ASPECTS

Socially, Wright was shy and retiring. He was not inclined to engage in small talk. But, paradoxically, when he did start to talk about something of interest—his childhood, his experience on the railroad surveying team, his ancestors, guinea pigs, evolution, genetics, politics—he could, and would, talk at length. His lectures invariably ran far over the allotted time. He was always gentle, yet he defended his views forcefully and he stated them fully.

Wright spent an inordinate amount of time helping others with their papers and data analysis, and often, this involved extensive calculations. Likewise, he was an extremely careful reviewer of manuscripts, often providing the author with substantial improvements. He was a conscientious teacher and spent many hours in the classroom and in the laboratory, which he ran himself. He published 211 scientific papers, most of them alone.

ACKNOWLEDGMENT

I am grateful to Dr. James Crow for advice in writing this chapter.

REFERENCES

Crow, J.F., Engels, W.R., and Denniston, C. (1990) Phase three of Wright's shifting balance theory. *Evolution., 44*: 233–247.

Dronamraju, K.R. (1990) Sewall Wright. *Jap. J. Genet., 65*: 25–31.

East, E.M. (1910) A Mendelian interpretation of variation that is apparently continuous. *Am. Nat, 44:* 65–82.

Haldane, J.B.S. (1931) A mathematical theory of natural selection. Part VIII. Metastable populations. *Proceedings of the Cambridge Philosophical Society, 27*: 137–142.

Kimura, M. (1968) Evolutionary rate at the molecular level. *Nature, 217*: 624–626.

Kimura, M. (1985) Diffusion models in population genetics with special reference to fixation time of molecular mutants under mutational pressure. In: T. Ohta and K. Aoki (eds.), *Population Genetics and Molecular Evolution*. Berlin: Springer-Verlag, pp. 19–39.

Nilsson-Ehle, H. (1909) Kreuzunguntersuchungen an Hafer und Weizen. Lunds Univ. Åarskr. *5(2)*: 1–122.

Provine, W. (1986) *Sewall Wright and Evolutionary Biology*. Chicago: University of Chicago Press.

WRIGHT'S PUBLICATIONS

Notes on the anatomy of the trematode, *Microphallus opacus. Trans. Am. Microscop. Soc., 31*: 167–175. 1912

Duplicate genes. *Am. Nat., 48*: 638–639. 1914.

The albino series of allelomorphs in guinea pigs. *Am. Nat., 49*: 140–148. 1915.

(With W.E. Castle.) Two color mutations of rats which show partial coupling. *Science, 42*: 193–195. 1915.

An intensive study of the inheritance of color and of other coat characters in guinea pigs with especial reference to graded variation. *Carnegie Institute Washington Publ., 241*: 59–160. 1916.

On the probable error of Mendelian class frequencies. *Am. Nat., 51*: 373–375. 1917.

The average correlation within subgroups of a population. *J. Wash. Acad. Sci., 7*: 532–535. 1917.

Color inheritance in mammals. I. *J. Hered., 8*: 224–235. 1917.

Color inheritance in mammals. II. The mouse. *J. Hered., 8*: 373–378. 1917.

Color inheritance in mammals. III. The rat. *J. Hered., 8*: 426–430. 1917.

Color inheritance in mammals. IV. The rabbit. *J. Hered., 8*: 473–475. 1917.

Color inheritance in mammals. V. The guinea pig. *J. Hered., 8*: 476–480. 1917.

Color inheritance in mammals. VI. Cattle. *J. Hered., 8*: 521–527. 1917.

Color inheritance in mammals. VII. The horse. *J. Hered., 8*: 561–564. 1917.
Color inheritance in mammals. VIII. Swine. *J. Hered., 9*: 33–38. 1918.
Color inheritance in mammals. IX. The dog. *J. Hered., 9*: 87–90. 1918.
Color inheritance in mammals. X. The cat. *J. Hered., 9*: 139–144. 1918.
Color inheritance in mammals. XI. Man. *J. Hered., 9*: 227–240. 1918.
(With H.R. Hunt.) Pigmentation in guinea pig hair. *J. Hered., 9*: 178–181. 1918.
On the nature of size factors. *Genetics, 3*: 367–374. 1918.
The relative importance of heredity and environment in determining the piebald pattern of guinea pigs. *Proc. Natl. Acad. Sci. USA, 6*: 320–332. 1920.
Principles of livestock breeding. *U.S. Dept. Agric. Bull., 905*. 1920.
Correlation and causation. *J. Agric. Res., 20*: 557–585. 1921.
(With P.A. Lewis.) Factors in the resistance of guinea pigs to tuberculosis with special regard to inbreeding and heredity. *Am. Nat., 55*: 20–50. 1921.
Systems of mating. I. The biometric relation between parent and offspring. *Genetics, 6*: 111–123. 1921.
Systems of mating. II. The effects of inbreeding on the genetic composition of a population. *Genetics, 6*: 124–143. 1921.
Systems of mating. III. Assortative mating based on somatic resemblance. *Genetics, 6*: 144–161. 1921.
Systems of mating. IV. The effects of selection. *Genetics, 6*: 162–166. 1921.
Systems of mating. V. General considerations. *Genetics, 6*: 167–178. 1921.
Coefficients on inbreeding and relationship. *Am. Nat., 56*: 330–338. 1922.
The effects of inbreeding and crossbreeding on guinea pigs. I. Decline in vigor. *U.S. Dept. Agric. Bull., 1090*: 1–36. 1922.
The effects of inbreeding and crossbreeding on guinea pigs. II. Differentiation among inbred families. *U.S. Dept. Agric. Bull., 1090*: 37–63. 1922.
The effects of inbreeding and crossbreeding on guinea pigs. III. Crosses between highly inbred families. *U.S. Dept. Agric. Bull., 1121*: 1–59. 1922.
Two new color factors of the guinea pig. *Am. Nat., 57*: 42–51. 1923.
The theory of path coefficients: a reply to Niles' criticism. *Genetics, 8*: 239–255. 1923.
Mendelian analysis of the pure breeds of livestock. I. The measurement of inbreeding and relationship. *J. Hered., 14*: 339–348. 1923.
Mendelian analysis of the pure breeds of livestock. II. The Duchess family of short-horns as bred by Thomas Bates. *J. Hered., 14*: 405–422. 1923.
The relation between piebald and tortoise shell color pattern in guinea pigs. *Anat. Rec., 23*: 393. 1923.
(With O.N. Eaton.) Factors which determine otocephaly in guinea pigs. *J. Agric. Res., 26*: 161–182. 1923.
Corn and hog correlations. *U.S. Dept. Agric. Bull., 1300*: 1–60. 1925.
The factors of the albino series of guinea pigs and their effects on black and yellow pigmentation. *Genetics, 10*: 223–260. 1925.
(With H.C. McPhee.) Mendelian analysis of the pure breeds of livestock. III. The short-horns. *J. Hered., 16*: 205–215. 1925.
An approximate method of calculating coefficients of inbreeding and relationship from livestock pedigrees. *J. Agric. Res., 31*: 377–383. 1925.
A frequency curve adapted to variation in percentage occurrence. *J. Am. Stat. Assoc., 21*: 162–178. 1926.
(With O.N. Eaton.) Mutational mosaic coat patterns of the guinea pig. *Genetics, 11*: 333–351. 1926.
Effects of age of parents on characteristics of the guinea pig. *Am. Nat., 60*: 552–559. 1926.
(With H.C. McPhee.) Mendelian analysis of the pure breeds of livestock. IV. The British dairy short-horns. *J. Hered., 17*: 397–401. 1926.

(With L. Loeb.) Transplantation and individuality differentials in inbred families of guinea pigs. *Am. J. Pathol., 3*: 251–285. 1927.

The effects in combination of the major color-factors of the guinea pig. *Genetics, 12*: 530–569. 1927.

An eight-factor cross in the guinea pig. *Genetics, 13*: 508–531. 1928.

(With O.N. Eaton.) The persistence of differentiation among inbred families of guinea pigs. *U.S. Dept. Agric. Tech. Bull., 103*: 1–45. 1929.

Fisher's theory of dominance. *Am. Nat., 63*: 274–279. 1929.

The dominance of bar over intra bar in *Drosophila*. *Am. Nat., 63*: 479–480. 1929.

The evolution of dominance. *Am. Nat., 63*: 556–561. 1929.

The genetical theory of natural selection (by R.A. Fisher). A review. *J. Hered., 21*: 349–356. 1930.

Statistical methods in biology. *J. Am. Stat. Assoc., 26(suppl.)*: 155–163. 1931.

Statistical theory of evolution. *J. Am. Stat. Assoc., 26(suppl.)*: 201–208. 1931.

Evolution in Mendelian populations. *Genetics, 16*: 97–159. 1931.

Complementary factors for eye color in *Drosophila*. *Am. Nat., 66*: 282–283. 1932.

General, group, and special size factors. *Genetics, 17*: 603–619. 1932.

The roles of mutation, inbreeding, crossbreeding, and selection in evolution. *Proc. 6th Intl. Congr. Genet., 1*: 356–366. 1932.

Hereditary variations of the guinea pig. *Proc. 6th Intl. Congr. Genet., 2*: 247–249. 1932.

Inbreeding and homozygosis. *Proc. Natl. Acad. Sci. USA, 19*: 411–420. 1933.

Inbreeding and recombination. *Proc. Natl. Acad. Sci. USA, 19*: 420–433. 1933.

Physiological and evolutionary theories of dominance. *Am. Nat., 68*: 25–53. 1934.

The genetics of growth. *Proc. Am. Soc. Anim. Prod., 1933*: 233–237. 1934.

(With K. Wagner.) Types of subnormal development of the head from inbred strains of guinea pigs and their bearing on the classification and interpretation of vertebrate monsters. *Am. J. Anat., 54*: 383–447. 1934.

On genetics of subnormal development of the head (otocephaly) in the guinea pig. *Genetics, 19*: 471–505. 1934.

Polydactylous guinea pigs. Two types respectively heterozygous and homozygous in the same mutant gene. *J. Hered., 25*: 359–362. 1934.

An analysis of variability in number of digits and in an inbred strain of guinea pigs. *Genetics, 19*: 506–536. 1934.

The results of crosses between inbred strains of guinea pigs differing in number of digits. *Genetics, 19*: 537–551. 1934.

Genetics of abnormal growth in the guinea pig. *Cold Spring Harbor Symp. Quant. Biol., 2*: 137–147. 1934.

Professor Fisher on the theory of dominance. *Am. Nat., 68*: 562–565. 1934.

The method of path coefficients. *Ann. Math. Stat., 5*: 161–215. 1934.

A mutation of the guinea pig, tending to restore the pentadactyl foot when heterozygous, producing a monstrosity when homozygous. *Genetics, 20*: 84–107. 1935.

The analysis of variance and the correlation between relatives with respect to deviations from an optimum. *J. Genet., 30*: 243–256. 1935.

Evolution in populations in approximate equilibrium. *J. Genet., 30*: 257–266. 1935.

On the genetics of rosette pattern in guinea pigs. *Genetics, 17*: 547–560. 1935.

(With H.B. Chase.) On the genetics of the spotted pattern of the guinea pig. *Genetics, 21*: 758–787. 1936.

The distribution of gene frequencies in populations. *Proc. Natl. Acad. Sci. USA, 23*: 307–320. 1937.

Size of population and breeding structure in relation to evolution. *Science, 87*: 430–331. 1938.

The distribution of gene frequencies under irreversible mutation. *Proc. Natl. Acad. Sci. USA, 24*: 253–259. 1938.

The distribution of gene frequencies in populations of polyploids. *Proc. Natl. Acad. Sci. USA, 24*: 372–377. 1938.
The distribution of self-sterility alleles in populations. *Genetics, 24*: 538–552. 1939.
Genetic principles governing the rate of progress of livestock breeding. *Proc. Am. Soc. Anim. Prod., 1939*: 18–26. 1939.
Statistical genetics in relation to evolution. Actualités scientific et industrielles. *Exposés de Biometrie et de la Statistique Biologique* XIII. Paris: Hermann & Cie. 1939.
Breeding structure of populations in relation to speciation. *Am. Nat., 74*: 232–248. 1940.
The statistical consequences of Mendelian heredity in relation to speciation. In J.S. Huxley (ed.), *The New Systematics*. London: Oxford at the Clarendon Press, pp. 161–183. 1940.
A quantitative study of the interactions of the major colour factors of the guinea pig. *Proc. 7th Int. Genet. Congr., 1939*: 319–329. 1941.
(With Th. Dobzhansky.) Genetics of natural populations. V. Relations between mutation rate and accumulation of lethals in populations of *Drosophila pseudoobscura. Genetics, 26*: 23–51. 1941.
The physiology of the gene. *Physiol. Rev., 21*: 487–527. 1941.
Tests for linkage in the guinea pig. *Genetics, 26*: 650–669. 1941.
On the probability of fixation of reciprocal translocations. *Am. Nat., 75*: 513–522. 1941.
(With Th. Dobzhansky and W. Hovanitz.) Genetics of natural populations. VII. The allelism of lethals in the third chromosome of *Drosophila pseudoobscura. Genetics, 27*: 363–394. 1942.
The physiological genetics of coat color of the guinea pig. *Biol. Symp., 6*: 337–355. 1942.
Statistical genetics and evolution. *Bull. Am. Math. Soc., 48*: 223–246. 1942.
Isolation by distance. *Genetics, 28*: 114–138. 1943.
The analysis of local variability of flower color in *Linanthus parryae. Genetics, 28*: 139–156. 1943.
(With Th. Dobzhansky.) Genetics of natural populations. X. Dispersion rates in *Drosophila pseudoobscura. Genetics, 28*: 304–340. 1943.
Physiological aspects of genetics. *Annu. Rev. Physiol., 5*: 75–106. 1945.
Genes as physiological agents. General considerations. *Am. Nat., 79*: 289–303. 1945.
Tempo and mode in evolution: A critical review. *Ecology, 26*: 415–419. 1945.
The differential equation of the distribution of gene frequencies. *Proc. Natl. Acad. Sci. USA, 31*: 383–389. 1945.
Isolation by distance under diverse systems of mating. *Genetics, 31*: 39–59. 1946.
(With Th. Dobzhansky.) Genetics of natural populations. XII. Experimental reproduction of some of the changes caused by natural selection in certain populations of *Drosophila pseudoobscura. Genetics, 31*: 125–156. 1946.
On the genetics of several types of silvering in the guinea pig. *Genetics, 32*: 115–141. 1947.
(With Th. Dobzhansky.) Genetics of natural populations. XV. Rate of diffusion of a mutant gene through a population of *Drosophila pseudoobscura. Genetics, 32*: 303–324. 1947.
On the roles of directed and random changes in gene frequency in the genetics of populations. *Evolution, 2*: 279–294. 1948.
Evolution, organic. *Encyclopedia Britannica, 14th ed., 8*: 915–929. 1948.
Genetics of populations. *Encyclopedia Britannica, 14th ed., 10*: 111–112. 1948.
Adaptation and selection. In: G.L. Jepson, G.G. Simpson, and E. Mayr (eds.), *Genetics, Paleontology, and Evolution*. Princeton: Princeton University Press, pp. 365–389. 1949.
(With Z.I. Braddock.) Colorimetric determination of the amounts ofmelanin in the hair of diverse genotypes of the guinea pig. *Genetics, 34*: 223–224. 1949.
Estimates of the amounts of melanin in the hair of diverse genotypes of guinea pig from transformation of empirical grades. *Genetics, 34*: 245–271. 1949.
Population structure in evolution. *Proc. Am. Philos. Soc., 93*: 471–478. 1949.

On the genetics of hair direction in the guinea pig. I. Variability in the patterns found in combinations of the *R* and M loci. *J. Exp. Zool., 112*: 303–324. 1949.
On the genetics of hair direction in the guinea pig. II. Evidence for a new dominant gene, Star, and tests for linkage with eleven other loci. *J. Exp. Zool., 112*: 325–340. 1949.
On the genetics of hair direction in the guinea pig. III. Interaction between the processes due to loci *R* and *St*. *J. Exp. Zool., 113*: 33–64. 1950.
Discussion on population genetics and radiation. *J. Cell Comp. Physiol., 35*: 187–210. 1950.
Genetical structure of populations. *Nature, 166*: 247–253. 1950.
The genetical structure of populations. *Ann. Eugen., 15*: 323–254. 1951.
Fisher and Ford on the Sewall Wright effect. *Am. Sci., 39*: 452–458. 1951.
The Genetics of Quantitative Variability. London: Her Majesty's Stationery Office. 1952.
The theoretical variance within and among subdivisions of a population that is in a steady state. *Genetics, 27*: 312–321. 1952.
Gene and organism. *Am. Nat., 87*: 5–18. 1953.
The interpretation of multivariate systems. In: O. Kempthorne, T.A. Bancroft, J.W. Gowen, and J.L. Lush (eds.), *Statistics and Mathematics in Biology.* Ames: Iowa State College Press, pp. 11–33. 1953.
(With W.E. Kerr.) Experimental studies of the distribution of gene frequencies in very small populations of *Drosophila melanogaster.* I. Forked. *Evolution, 8*: 172–177. 1954.
(With W.E. Kerr.) Experimental studies of the distribution of gene frequencies in very small populations of *Drosophila melanogaster.* II. Bar. *Evolution, 8*: 225–240. 1954.
(With W.E. Kerr.) Experimental studies of the distribution of gene frequencies in very small populations of *Drosophila melanogaster.* II. Aristapedia and spineless. *Evolution, 8*: 293–302. 1954.
Summary of patterns of mammalian gene action. *J. Natl. Cancer Inst., 15*: 837–851. 1954.
Classification of the factors of evolution. *Cold Springs Harbor Symp. Quant. Biol., 20*: 16–24. 1956.
Modes of selection. *Am. Nat., 90*: 5–24. 1956.
Genetics, the gene, and the hierarchy of biological sciences. *Proc. 10th Int. Congr. Genet., 1*: 475–489. 1958.
On the genetics of silvering in the guinea pig with special reference to interaction and linkage. *Genetics, 44*: 387–405. 1959.
Silvering (si) and diminution (dm) of coat color of the guinea pig and male sterility of the white or near-white combination of these. *Genetics, 44*: 563–590. 1959.
A quantitative study of variations in intensity of genotypes of the guinea pig at birth. *Genetics, 44*: 1001–1026. 1959.
Physiological genetics, ecology of populations, and natural selection. *Persp. Biol. Med., 3*: 107–151. 1959.
Qualitative differences among colors of the guinea pig due to diverse genotypes. *J. Exp. Zool., 142*: 75–114. 1959.
On the number of self-incompatibility alleles maintained in equilibrium by a given mutation rate in a population of given size: a reexamination. *Biometrics, 16*: 61–85. 1960.
Path coefficients and path regressions: alternative or complementary concepts. *Biometrics, 16*: 189–202. 1960.
The treatment of reciprocal interaction, with or without lag, in path analysis. *Biometrics, 16*: 423–445. 1960.
Residual variability in intensity of coat color in the guinea pig. *Genetics, 45*: 583–612. 1960.
Postnatal changes in the intensity of coat color in diverse genotypes of the guinea pig. *Genetics, 45*: 1503–1529. 1960.
The genetics of vital characters of the guinea pig. *J. Cell Comp. Physiol., 56*: 123–151. 1960.

Plant and animal improvement in the presence of multiple selection peaks. In: W.D. Hanson and H.F. Robinson (eds.), *Statistical Genetics in Plant Breeding*. Washington, DC: National Academy Press, pp. 116–122. 1963.

Gene interaction. In: W.J. Burdette (ed.), *Methodology in Mammalian Genetics*. San Francisco: Holden-Day, pp. 159–192. 1963.

Biology and the philosophy of science. *The Monist, 48*: 265–290. 1964.

Pleiotropy in the evolution of structural reduction and of dominance. *Am. Nat., 98*: 65–69. 1964.

Stochastic processes in evolution. In J. Gurland (ed.), *Symposium on Stochastic Models in Medicine and Biology*. Madison: University of Wisconsin Press, pp. 199–244. 1964.

Factor interaction and linkage in evolution. *Proc. R. Soc. B, 162*: 80–104. 1965.

The distribution of self-incompatibility alleles in populations. *Evolution, 18*: 609–619. 1965.

The interpretation of population structure by F-statistics with special regard to systems of mating. *Evolution, 19*: 395–420. 1965.

Polyallelic random drift in relation to evolution. *Proc. Natl. Acad. Sci. USA, 55*: 1074–1081. 1966.

"Surfaces" of selective value. *Proc. Natl. Acad. Sci. USA, 58*: 165–172. 1967.

1968 Dispersion of *Drosophila pseudoobscura*. *Am. Nat., 102*: 81–84. 1968.

Evolution and the Genetics of Populations, vol. 1. *Genetic and Biometric Foundations*. Chicago: University of Chicago Press. 1968.

Deviations from random combination in the optimum model. *Jap. J. Genet., 44(suppl 1)*: 152–159. 1969.

Evolution and the Genetics of Populations, vol. 2. *The Theory of Gene Frequencies*. Chicago: University of Chicago Press. 1969.

The theoretical course of directional selection. *Am. Nat., 103*: 561–574. 1969.

Random drift and the shifting balance theory of evolution. In K. Kojima (ed.), *Mathematical Topics in Population Genetics*. Heidelberg: Springer-Verlag, pp. 1–31. 1970

Panpsychism and science. In J.E. Cobb and D.R. Griffen (eds.), *Mind and Nature*. Washington, DC: University Press of America, pp. 79–88. 1975.

Evolution and the Genetics of Populations, vol. 3. *Experimental Results and Evolutionary Deductions*. Chicago: University of Chicago Press. 1977.

Evolution and the Genetics of Populations, vol. 4. *Variability Within and Among Natural Populations*. Chicago: University of Chicago Press. 1978.

The relation of livestock breeding to theories of evolution. *J. Anim. Sci., 46*: 1192–1200. 1978.

Genic and organismic selection. *Evolution, 34*: 825–843. 1980.

Character change, speciation, and the higher taxa. *Evolution, 36*: 427–443. 1982.

The shifting balance theory and macroevolution. *Annu. Rev. Genet., 16*: 1–19. 1982.

On "Path analysis in genetic epidemiology: a critique." *Am. J. Hum. Genet., 35*: 757–768. 1983.

The first Meckel oration: on the causes of morphological differences in a population of guinea pigs. *Am. J. Med. Genet., 18*: 591–616. 1984.

Diverse uses of path analysis. In A. Chakravarti (ed.), *Human Population Genetics: The Pittsburgh Symposium*. New York: Van Nostrand Reinhold. 1984.

Evolution, Selected Papers, W.B. Provine (ed.). Chicago: University of Chicago Press. 1986.

Surfaces of selective value revisited. *Am. Nat., 131*: 115–123. 1988.

8 Motoo Kimura (1924–1994)

For four decades, 1920s–1950s, the field of mathematical population genetics and evolutionary theory was dominated by the three pioneers, J.B.S. Haldane, R.A. Fisher, and Sewall Wright. The leading successor to this great heritage was Motoo Kimura. Although best known for his daring neutral theory of molecular evolution, a concept of great interest and equally great controversy, he is admired by population geneticists even more for his extensive contributions to the mathematical theory.

Motoo Kimura was born November 13, 1924, in Okazaki, Japan. He died November 13, 1994, at the age of 70 years. For some time, he had been a victim of amyotrophic lateral sclerosis (ALS) and was progressively weakening. Nevertheless, his death was accidental. He fell, hitting his head, and never regained consciousness. His death came as a blessing since his disease was worsening with age.

Kimura was interested in plants throughout his life. His father was a businessman who loved flowers and raised ornamentals in the home. Young Kimura was fascinated by their beauty and development. When his father bought him a microscope, he spent hours with it. In school, he developed an interest in botany and aspired to become a systematic botanist. However, he also enjoyed mathematics, especially Euclid. Because of his precocious talent for mathematics, his teacher advised him to become a mathematician, which he ignored. He continued to retain an interest in botany throughout his life and later became a keen orchid breeder. Many of his creations were prize-winners. Kimura used his artistic talent to paint pictures of his favorite flowers, usually on chinaware.

According to his collaborator in population genetics, Dr. James Crow, Kimura used the royalties from their book to build a small greenhouse to grow orchids. Every Sunday was spent looking after his orchids. He was encouraged to study chromosome morphology by his high school teacher. Kimura decided to become a plant cytogeneticist.

Following the tradition of botanical interest led by the Japanese Royal family, Kimura developed his botanical interest and hoped to specialize in cytogenetics. During this period, he was also fascinated by a physics course given by Hideki Yukawa, later to win the Nobel Prize for predicting the meson. Kimura began to take an interest in mathematics as the language of science. Japan was then in the midst of World War II, and the normal high school period was shortened from three to two and a half years. In 1944, he was admitted to Kyoto Imperial University.

Hitoshi Kihara was Japan's foremost geneticist, a world leader in the cytogenetics of wheat, and Kimura might have been expected to study with him. Nevertheless, Kihara advised him to enroll in botany. There was a reason. At that time, students in botany were exempt from military service until graduation. Curiously, students of agriculture—Kihara's area—did not enjoy this privilege. In this way, Kimura escaped military service, but life was far from easy. There was the irritation of regular military drill, and there was never enough good food. The atomic bombing on Hiroshima came before his first university year was finished, but conditions immediately after the war were even worse. Food was even harder to obtain and Kimura made regular Sunday visits to a cousin for a good meal. The cousin was a quantum physicist, so these Sundays were also occasions for scientific talk.

Although still a student of cytology, Kimura became increasingly interested in mathematical questions. He was first attracted to mapping functions and through this he came to know the work of J.B.S. Haldane. Reading Dobzhansky's *Genetics and the Origin of Species* led him to the work of Sewall Wright. By this time, the war was over and he had moved to Kihara's department. Kihara was doing backcrosses in wheat to introduce parts of a genome into a different cytoplasm, and Kimura helped him by deriving the frequency distribution of introduced chromosomes in successive backcrossed generations. This led to his first published scientific paper (Kimura 1950).

Kihara left young Kimura alone, giving him plenty of leisure time to pursue his interests. This was fortunate because Kimura found time to read Sewall Wright's papers. There was only a single copy and of course no duplicating facilities, so he copied the papers by hand. In 1949, Kimura joined the research staff of the National Institute of Genetics in Mishima, a position he retained for the rest of his life. That laboratory was a crude wooden building. It was used as an airplane factory during the war. Mishima was small and provincial, in sharp contrast to the sophisticated intellectual and cultural activities of Kyoto. However, on clear days, it provided an impressive view of Mount Fuji.

Kimura continued to spend much of his time in Kihara's laboratory in Kyoto, where there was a much better library. Studying probability textbooks, he discovered that the Fokker–Planck equation that Wright had used was only one of two Kolmogorov equations, the forward one. Later, Kimura proved to be especially skillful in using the backward equation. By this time, Kimura was spending almost all his time in mathematical genetics. It was typical of him that he learned the subject by himself. His formal training in mathematics was quite limited; he simply learned what he had to learn to solve the problem at hand. He was fortunate, of course, in having an exceptionally gifted mind.

During this time he proposed the "stepping stone" migration model (Kimura 1953). Wright's island model, the standard of the time, assumed that immigrants come at random from a larger population. Kimura introduced the more realistic model that immigrants come from a near neighbor. By adding long-range migrants, he could include the island model as a special case. This was a lonely period for Kimura, for none of his associates understood his work or thought it to be of any interest. The exception was Taku Komai, also in Kyoto. Komai had studied with T.H. Morgan in the United States, and although he didn't understand Wright's mathematics, he encouraged Kimura to study his papers. That Kimura persisted in working alone in an indifferent, if not hostile, environment was characteristic. Then, as later, he was confident of his own abilities and knew what he wanted to do.

EARLY CAREER

According to the distinguished population geneticist James F. Crow (who was Kimura's Ph.D. supervisor and later collaborator), the two great theoretical evolutionary geneticists of the 1950s are Motoo Kimura from Japan and Gustave Malecot from France.

Kimura was a theoretical evolutionary biologist who was best known for proposing the neutral theory of molecular evolution. Some of his work was done in collaboration with others. Kimura's work followed the masterful contributions of the great trio, J.B.S. Haldane, R.A. Fisher, and Sewall Wright, who founded and dominated the field of population genetics for four decades: 1920s to the 1950s. He is remembered in genetics for his innovative use of diffusion equations to calculate the probability of fixation of beneficial, deleterious, or neutral alleles. Combining theoretical population genetics with molecular evolution data, he also developed the neutral theory of molecular evolution in which genetic drift is the main force changing allele frequencies.

NATIONAL INSTITUTE OF GENETICS

In 1949, Kimura joined the National Institute of Genetics in Mishima, Shizuoka-ken. In 1953, he published his first population genetics paper (which would eventually be very influential), describing a "stepping stone" model for population structure that could treat more complex patterns of immigration than Sewall Wright's earlier "island model." After meeting visiting American geneticist Duncan McDonald (who was with the Atomic Bomb Casualty Commission), Kimura entered graduate school at Iowa State College in summer 1953 to study with J.L. Lush. But he found that Iowa State College was too restricting; he moved to the University of Wisconsin to work on stochastic models with James F. Crow and join a strong intellectual community of other geneticists, including Newton Morton but especially Sewall Wright, who was one of the founders of classical population genetics.

COLD SPRING HARBOR SYMPOSIUM

Near the end of his graduate study, Kimura presented a paper at the 1955 Cold Spring Harbor Symposium; although few were able to understand it (both because of mathematical complexity and Kimura's English pronunciation), it received strong praise from Wright and, later, J.B.S. Haldane. His accomplishments at Wisconsin included a general model for genetic drift, which could accommodate multiple alleles, selection, migration, and mutations, as well as some work based on R.A. Fisher's *Fundamental Theorem of Natural Selection*. He also extended the work of Wright with the Fokker–Planck equation by introducing the Kolmogoroff Backward Equation to population genetics, allowing the calculation of the probability of a gene to become established in a population. Kimura received his PhD in 1956, before returning to Japan, where he would remain for the rest of his life, at the National Institute of Genetics in Mishima.

POPULATION GENETICS

Kimura worked on a wide range of theoretical population genetics problems, occasionally in collaboration with Takeo Maruyama. He introduced the "infinite alleles" and "stepwise mutation" models for the study of genetic drift, both of which would be used widely as the field of molecular evolution grew alongside the number of available peptide and genetic sequences. He also created the "ladder model" that could be applied to electrophoresis studies, where homologous proteins differ by whole units of change. An early statement of his approach was published in 1960, in his *An Introduction to Population Genetics.*

Kimura's talent in manipulating the Kolmogorov equations and applying them to significant evolutionary problems was outstanding. Here are a few of Kimura's significant findings. I have already mentioned the "stepping stone" model of population structure, which has been the starting point for investigations by many authors. He discovered the phenomenon of "quasi-linkage equilibrium." He showed that with loose linkage, the population generates just enough linkage disequilibrium to cancel the epistatic variance, so that the additive variance, without epistatic terms, is the best predictor of change under selection. He analyzed a case of meiotic drive in Lilium. He investigated genetic load and wrote a review in 1961. In one study, he showed that the mutation load can be reduced with epistasis, but only when there is sexual reproduction. He was also the first to consider the mutation and segregation loads in finite populations. His first calculations, done when computers were primitive, involved some very inventive, and to some critics dubious, approximations. Later computer work has vindicated them.

Inspired by Haldane's (1927) pioneering work on the probability of fixation of a new mutant, Kimura investigated the average time until fixation, the time until loss, the number of individuals carrying the mutant gene and the number of heterozygotes during the process, the age of a neutral mutant in the population, and, more generally, the moments for the sum of an arbitrary function of allele frequency during the process. A curious result was his finding that the time required for fixation of a selectively favored mutation is the same as for a deleterious one, despite the enormous differences in the probability of occurrence.

With reference to inbreeding, he showed that Wright's maximum avoidance of inbreeding was not the best way to preserve heterozygosity in the long run. He showed how to calculate the selective efficiency of truncation selection and its load-reducing effect. Of particular importance in practical application was the surprising result that a very crude approximation to truncation selection is almost as effective as strict truncation. This made the theory much more realistic for natural populations.

With respect to mutation, Kimura was associated with the origin of three classical models of mutation, widely used in population genetics, the "infinite allele," the "infinite site," and the "ladder" models. He collaborated with Crow in some of these studies (Crow 1970, 1988, 1994, 1996).

Population Genetics, Molecular Evolution, and the Neutral Theory: Selected Papers—This is a collection of important papers of Motoo Kimura. Edited by Naoyuki Takahata. Chicago: University of Chicago Press, 1994. A summary of the book is given in the following.

The section headings, with the number of papers in parentheses, are as follows (summary after Crow):

1. Random gene frequency drift (4)
2. Fluctuation in selection intensity (2)
3. Population structure (3)
4. Linkage and recombination (4)
5. Evolutionary advantages of sexual reproduction (1)
6. Natural selection (2)
7. Meiotic drive (1)
8. Genetic load (3)
9. Inbreeding systems (2)
10. Evolution of quantitative characters (4)
11. Probability and time of fixation or extinction (5)
12. Age of alleles and reversibility (4)
13. Intergroup selection (1)
14. Infinite allele, infinite site, and ladder models (3)
15. Molecular evolution (2)
16. Nucleotide substitutions (3)
17. Molecular clock (3)
18. Neutral theory (10)

The book includes a bibliography of Kimura's major publications, a total of 161 papers.

Altogether, Kimura wrote about 660 papers and six books. Although he had several collaborators, most of these publications are his alone.

NEUTRAL THEORY

Kimura emphasized that the theory applies only for evolution at the molecular level. Phenotypic evolution is controlled by natural selection (as postulated by Charles Darwin). During the 1970s and 1980s, the development of neutral theory was followed by an extensive discussion in terms of the "neutralist-selectionist" controversy, with special reference to molecular divergence and polymorphism (Kimura 1983; King and Jukes 1969).

The neutral theory of molecular evolution holds that at the molecular level, most evolutionary changes and most of the variation within and between species are not caused by natural selection but by genetic drift of mutant alleles that are neutral. A neutral mutation is one that does not affect an organism's ability to survive and reproduce. The neutral theory allows for the possibility that most mutations are deleterious but holds that because these are rapidly purged by natural selection, they do not make significant contributions to variation within and between species at the molecular level. Mutations that are not deleterious are assumed to be mostly neutral rather than beneficial. In addition to assuming the primacy of neutral mutations, the theory also assumes that the fate of neutral mutations is determined by the sampling processes described by specific models of random genetic drift.

The neutral theory can be discussed in two parts:

a. The neutral theory suggests that when one compares the genomes of existing species, the vast majority of molecular differences are selectively "neutral"; i.e. the molecular changes represented by these differences do not influence the fitness of organisms. As a result, the theory regards these genomic features as neither subject to, nor explicable by, natural selection. This view is based in part on the degenerate genetic code, in which sequences of three nucleotides (codons) may differ and yet encode the same amino acid. Consequently, many potential single-nucleotide changes are in effect "silent" or "unexpressed." Such changes are presumed to have little or no biological effect.
b. A second hypothesis of the neutral theory is that most evolutionary change is the result of genetic drift acting on neutral alleles. After appearing by mutation, a neutral allele may become more common within the population via genetic drift. Usually, it will be lost, or in rare cases, it may become established (or fixed) in the population. This stochastic process is assumed to obey equations describing random genetic drift by means of accidents of sampling.

The neutral theory explains that mutations appear at rate μ in each of the $2N$ copies of a gene and fixed with probability $1/(2N)$. This means that if all mutations were neutral, the rate at which fixed differences accumulate between divergent populations is predicted to be equal to the per-individual mutation rate, e.g., during errors in DNA replication; both are equal to μ. When the proportion of mutations that are neutral is constant, so is the divergence rate between populations. This provides a rationale for the molecular clock, although the discovery of a molecular clock predated neutral theory.

Neutral theory does not deny the occurrence of natural selection. There are two main types of natural selection: *stabilizing selection* or *purifying selection*, which acts to eliminate deleterious mutations; and *positive selection*, which favors advantageous mutations.

Positive selection can, in turn, be further subdivided into directional selection, which tends toward fixation of an advantageous allele, and balancing selection, which maintains a polymorphism. The neutral theory of molecular evolution predicts that stabilizing selection is ubiquitous, but that positive selection is rare, and recognizes the importance of positive selection in the origin of adaptations.

"NEUTRAL" VERSUS "NONNEUTRAL"

A heated debate arose when Kimura's theory was published, largely revolving around the relative percentages of alleles that are "neutral" versus "nonneutral" in any given genome. Kimura argued that molecular evolution is dominated by selectively neutral evolution but at the phenotypic level, changes in characters were probably dominated by natural selection rather than genetic drift.

According to the neutral theory of molecular evolution, the amount of genetic variation within a species should be proportional to the effective population size. Levels of genetic diversity vary much less than census population sizes, giving rise to the "paradox of variation." While high levels of genetic diversity were one of the original arguments in favor of neutral theory, the paradox of variation has been one of the strongest arguments against neutral theory.

Tomoko Ohta emphasized the importance of nearly neutral mutations, in particular, slightly deleterious mutations. The population dynamics of nearly neutral mutations is essentially the same as that of neutral mutations unless the absolute magnitude of the selection coefficient is greater than $1/N$, where N is the effective population size with respect to selection. The value of N may therefore affect how many mutations can be treated as neutral and how many as deleterious.

In 1968, when Kimura introduced the neutral theory of molecular evolution—the idea that, at the molecular level, the large majority of genetic change is neutral with respect to natural selection—making genetic drift a primary factor in evolution, he justified his neutral theory partly on Haldane's idea of "cost" of selection where he showed that, in horotelic (standard rate) evolution, the mean time taken for each gene substitution is about 300 generations. This appears to explain the slowness of evolution.

The neutral theory was immediately controversial, receiving support from many molecular biologists and attracting opposition from many evolutionary biologists.

Kimura spent the rest of his life developing and defending the neutral theory. As James Crow put it, "much of Kimura's early work turned out to be pre-adapted for use in the quantitative study of neutral evolution" (Crow 1997).

LAST YEARS

We have the following account from Kimura's close friend and colleague James Crow. "He was complex. I found him a pleasant conversationalist, with a broad knowledge of both Eastern and Western culture. His interests were catholic and he had a wide-ranging curiosity. He enjoyed science fiction, particularly of Arthur C. Clarke. He was impressed by the powerful writing of Sophocles, which he picked up as a paperback. He admired Bertrand Russell. Over the years we had many fruitful discussions, including a number of friendly scientific disagreements. With his close friends he was generous, helpful, and appreciative. But others saw a different side. He could be self-centered, demanding, and dogmatic. As he grew older, his interests narrowed, and he became increasingly concerned for his place in scientific history and more obsessive about his neutral theory. He was becoming recognized throughout the world, and in Japan he was a celebrity. Ironically, these traits increased with his growing scientific recognition. Scientific disagreements became personal and several felt his barbs. But the scars will disappear as people measure them against his sterling accomplishments."

CONCLUSION

Kimura always had a biological problem in mind and pursued one biologically significant question after another. He solved problems, but he also formulated them. Kimura is best known for his neutral theory. Yet, influential as this is and despite the great impact it has had on molecular evolution, many population geneticists probably remain even more impressed by the steady flow of papers in mathematical population genetics, with their inventive solutions to important and difficult problems. He left a nearly completed paper at the time of his death (Kimura 1995). It is not one of his great ones, but it is vintage Kimura. He shows that the number of favorable mutations that can be simultaneously in transit toward fixation, $n < (N/2) \ln (1 - L)$, where N is the effective population number and L is the amount of excess reproduction available for selection of these mutants. If L has the reasonable value of 10%, n is less than 1/20 the population number. This then places a limit on the rate of favorable evolution in a finite population. Like Haldane's (1957) cost of natural selection, this value is independent of selection intensity. But Kimura's formulation is more realistic in taking population number into account and not depending on the initial frequency of the mutation.

In 1968, I planned and organized the JBS Haldane Memorial Symposium during the XIIth International Congress of Genetics in Tokyo, Japan. Motoo Kimura, James Crow, and Sewall Wright took an active part in that symposium. Later, I saw Kimura at the XVIth International Congress of Genetics in 1988 in Toronto, Canada. He appeared to be in good health then, but he was afflicted with ALS, which led to a rapid deterioration of his health in the following years. He died in 1994, when he fell down and injured his head. He was 70 years old. I recall Kimura as a brilliant and intense individual who dedicated his life to science and one who loved nature, plants, and flowers.

REFERENCES

Crow, J.F. (1997) Motoo Kimura. 13 November 1924–13 November 1994. *Biog. Memoirs of the Fellows of the Royal Society, 43*: 255–210.

Crow, J.F. (1988) Sewall Wright (1989–1988). *Genetics, 119*: 1–4.

Crow, J.F. (1994) Hitoshi Kihara, Japan's pioneer geneticist. *Genetics, 137*: 891–894.

Crow, J. (1996) Memories of Motô. *Theoretical Popul. Biol., 49*: 122–127.

Crow, J.F., and M. Kimura (1970) *An Introduction to Population Genetics*. New York: Harper and Row. Reprinted by Burgess International, Minneapolis.

Haldane, J.B.S. (1957) The cost of natural selection. *J. Genet., 55*: 511–524.

Kimura, M. (1950) The theory of the chromosome substitution between two different species. *Cytologia, 15*: 281–294.

Kimura, M. (1953) "Stepping-stone" model of population. *Annu. Rpt. Natl. Inst. Genet., 3*: 62–63.

Kimura, M. (1954) Process leading to quasi-fixation of genes in natural populations due to random fluctuation of evolutionary rate at the molecular level. *Genetics, 39*: 280–295.

Kimura, M. (1968). Evolutionary rate at the molecular level. *Nature, 217*: 624–626.

Kimura, M. (1983) *The Neutral Theory of Molecular Evolution*. Cambridge University Press, Cambridge.

Kimura, M. (1988) Thirty years of population genetics with Dr. Crow. *Jap. J. Genet., 63*: 1–10.

Kimura, M. (1994) *Population Genetics, Molecular Evolution and the Neutral Theory. Selected Papers.* Edited with Introductory Essays by Naoyuki Takahata. Chicago: University of Chicago Press.

Kimura, M. (1995) Limitations of Darwinian selection in a finite population. *Proc. Natl. Acad. Sci. USA, 92*: 2343–2344.

King, J.L., and T.H. Jukes (1969) Non-Darwinian evolution: random fixation of selectively neutral mutations. *Science, 164*: 788–798.

Nagylaki, T. (1989) Gustav Malecot and the transition from classical to modem population genetics. *Genetics, 122*: 253–252.

Nei, M. (1995). Motoo Kimura (1924–1994). *Mol. Biol. Evol., 12*: 719–722.

Steen, T.Y. (1996). Always an eccentric?: a brief biography of Motoo Kimura. *J. Genet., 75*: 19–25.

Section III

Biochemical Genetics

In 1902, a London physician, Archibald Garrod, discovered that patients with alkaptonuria have a defective gene that produces a faulty enzyme that interrupts an important metabolic pathway. This was the first recognition of the possibility that genes direct the assembly of enzymes and, more specifically, that each gene codes for one enzyme. Over the next few years, Garrod discovered three more metabolic diseases that behave like recessive traits, including albinism, and published in 1909 as Inborn Errors of Metabolism. At a time when Mendel was still not well understood, the lectures had little impact, it was generally not appreciated at that time that he had established a direct connection between Mendelian genetics and Darwinian evolution.

In the 1930s, Haldane initiated research on the biochemical genetics of plant petal pigments in England, particularly the chemistry of the main classes of pigments, but the research was interrupted by the Second World War and further research on the synthetic pathways could not be carried out.

Many years later, working with the mold *Neurospora*, George Beadle (and Edward Tatum), reproposed the "one gene, one enzyme" hypothesis in 1941, for which they received the Nobel Prize in 1958. However, their work was essentially a confirmation of Garrod's original discovery. Edward Tatum's pupil, Joshua Lederberg, shared the Nobel Prize with Beadle and Tatum in 1958 for his pioneering research in microbial genetics.

9 Archibald E. Garrod (1857–1936)

It is an interesting fact that the early work of Sir Archibald E. Garrod (1902, 1909) is noted for its farsightedness in pioneering both human genetics and biochemical genetics when the science of genetics itself had barely started. The term "genetics" itself was yet to be coined by William Bateson, which occurred in 1906. Garrod was a contemporary of physiologist John Scott Haldane, biochemist F. Gowland Hopkins, and biologist William Bateson. However, Garrod's contribution was far ahead of its time. His work was not followed by anyone, which is partly due to the gulf that existed between clinicians and biologists. This is surprising because his advisor was the well-known Cambridge University biologist William Bateson (Dronamraju 1992).

INTELLECTUAL BACKGROUND

Garrod's father, Alfred Baring Garrod, was a brilliant physician who obtained his medical degree at the age of 23 years. He discovered the presence of uric acid in the blood of patients with gout and was appointed professor of medicine at University College, London, when he was only 32 years old. Archibald's two older brothers also had brilliant academic careers. Following in their footsteps, Garrod not only met these high standards set by his father and brothers but also easily exceeded their greatness, enjoying a more enduring career.

Archibald Garrod received a broad education, attending a preparatory school in Harrow and entering Marlborough at the age of 15 years. One frequent visitor to the Garrod household was his cousin, Charles Keene, an outstanding illustrator for the magazine *Punch* for over 40 years.

At the age of 10 years, Archibald himself displayed his artistic talent by writing an illustrated booklet, *A Handbook of Classical Architecture*. He was keenly interested in natural history from an early age, displaying interest in what was later termed "genetics." As a tireless collector of butterflies at the age of 12 years, Garrod noted the dearth of females among the specimens collected. This is not usually due to a lopsided sex ratio but to the greater visibility and activity of the males of many species. His early writings contained references and speculations about the possible inheritance of certain characters in mammals.

Garrod's early performance at Marlborough was poor, which was largely due to his lack of interest in classical studies, especially Latin prose and grammar. However, with the encouragement of the master of his college, his performance improved, and he was able to enter Christ Church, Oxford. Garrod graduated in 1878, with a first in Natural sciences. Following in his father's footsteps, in 1880, Garrod entered St. Bartholomew's Hospital in London, for his medical education. He won several scholarships, including the coveted Brackenbury Scholarship. Garrod qualified in 1884 and spent a year in Vienna attending the medical clinics, as was the custom in those days. One consequence of his visit was his very popular book on the laryngoscope (Garrod 1886). After joining his father's medical practice, he wrote a treatise on rheumatoid arthritis in 1890.

BIOCHEMISTRY OF DISEASES

Garrod was appointed assistant physician at the Hospital for Sick Children on Great Ormond Street (London) in 1892, and for the rest of his life he remained deeply interested in diseases of children. He was interested in studies of normal and pathological urine, especially in differences of their coloration. It is of interest that his interest in butterflies and flowers in childhood was also related to color differences and biological variation, and it may well have helped to sharpen his perception in this regard. It should be mentioned further that his work was greatly benefitted from his close friendship with William Bateson, who advised him on the genetic aspects, and with F. Gowland Hopkins, who advised him on the chemistry of pigments. These two individuals were also responsible for profoundly influencing J.B.S. Haldane's work in genetics and biochemistry.

Garrod's interest in joint disease led him to study the chemistry of pigments in urine. While working as a visiting physician at the Great Osmond Street Hospital for Sick Children, he examined a three-month-old boy, Thomas P., whose urine was stained a deep reddish-brown. Garrod's diagnosis was alkaptonuria, which is caused by an abnormal build-up of homogentisic acid, or alkapton. In a normal person, the acid is broken down through a series of chemical reactions into carbon dioxide and water. But in rare cases, the metabolic process is interrupted and the acid is excreted in the urine, where it turns black on contact with the air. According to the germ theory of disease, which had transformed the study of medicine in Garrod's time, alkaptonuria was thought to be a bacterial infection of the intestine. The disorder was almost always diagnosed in infancy, lasted throughout life, and was thought to be contagious. Garrod's training in physical science, however, led him to investigate the disease as a series of chemical reactions. He reviewed 31 cases of alkaptonuria from his own practice and from the medical literature and presented his findings to the Royal Medical and Chirurgical (Surgical) Society of London in 1899. Alkaptonuria, he noted, although rare, tended to appear among children of healthy parents. It was not contagious and seemed to be a harmless error in metabolism.

When a third child with alkaptonuria was born to the parents of Thomas P., Garrod suspected that something more than mere chance was involved. When he learned that Thomas P.'s parents were blood relations—their mothers were sisters—he inquired into the backgrounds of other families with one or more children with alkaptonuria. In every instance, their parents were also first cousins. It was while walking home from the hospital one afternoon that Garrod conceived of the possibility that alkaptonuria might be a disease caused by heredity (genetics). Gregor Mendel's work on the principles of heredity, newly discovered in England, offered a simple explanation. The mating of first cousins apparently created conditions under which a rare, recessive Mendelian factor (or gene) appeared in the offspring. Garrod's classic paper on alkaptonuria was published in *The Lancet* in 1902.

Garrod was greatly interested in both the etiological and the clinical aspects of alkaptonuria. Garrod's chemical studies established the true nature of the disease. He disproved the belief that alkaptonuria was infectious—i.e. attributable to microorganisms responsible for the formation of homogenistic acid in the gut, as had been believed—instead showing, because of its frequent occurrence in siblings, that it was congenital and possibly hereditary. These views were presented before the Medical and Chirurgical Society meeting in 1899. Gowland Hopkins was in the audience and vigorously supported Garrod in his commentary. In the following years, Garrod tirelessly emphasized the chemical individuality of each person, emphasizing its congenital and metabolic nature. Because of advice of William Bateson, Garrod came to appreciate the relationship between consanguinity and the recessive nature of the disease, in the families he studied. His classic paper on the genetics of alkaptonuria appeared in 1902 in *The Lancet* (Garrod 1902). In later years, Garrod regularly attended the meetings of the Genetical Society of Great Britain.

ROLE OF CONSANGUINITY

With reference to the role of parental consanguinity, Garrod (1902) wrote: "The question of the liability of children of consanguineous marriages to exhibit certain abnormalities or to develop certain diseases has been much discussed, but seldom in a strictly scientific spirit. Those who have written on the subject have too often aimed at demonstrating the deleterious results of such unions on the one hand, or their harmlessness on the other, questions which do not here concern us at all. There is no reason to suppose that mere consanguinity of parents can originate such a condition as alkaptonuria in their offspring, and we must rather seek an explanation in some peculiarity of the parents, which may remain latent for generations, but which has the best chance of asserting itself in the offspring of the union of two members of a family in which it is transmitted. This applies equally to other examples of that peculiar form of heredity which has long been a puzzle to investigators of such subjects, which results in the appearance in several collateral members of a family of a peculiarity which has not been manifested at least in recent preceding generations."

Garrod (1902) continued: "It has been recently pointed out by Bateson that the law of heredity discovered by Mendel offers a reasonable account of such phenomena. It asserts that as regards two mutually exclusive characters, one of which tends to be dominant and the other recessive, cross-bred organisms will produce germinal cells (gametes) each of which, as regards the characters in question, conforms to one or other of the pure ancestral types and is therefore incapable of transmitting when a recessive gamete meets one of the dominant type the resulting organism (the zygote) will usually exhibit the dominant character, whereas when two recessive gametes meet the recessive character will necessarily be manifested in the zygote. In the case of a rare recessive characteristic we may easily imagine that many generations may pass before the union of two recessive gametes takes place."

The application of this to the case in question is further pointed out by Bateson, who, commenting upon the above observations on the incidence of alkaptonuria, writes as follows: "Now there may be other accounts possible, but we note that the mating of first cousins gives exactly the conditions most likely to enable a rare, and usually recessive, character to show itself. If the bearers of such a gamete mate with individuals not bearing it the character will hardly ever be seen; but first cousins will frequently be the bearers of similar gametes, which may in such unions meet each other and thus lead to the manifestation of the peculiar recessive characters in the zygote."

Garrod (1902) further wrote: "Whether the Mendelian explanation be the true one or not, there seems to be little room for doubt that the peculiarities of the incidence of alkaptonuria and of conditions which appear in a similar way are best explained by supposing that, leaving aside exceptional cases in which the character, usually recessive, assumes dominance, a peculiarity of the gametes of both parents is necessary for its production. Hitherto nothing has been recorded about the children of alkaptonuric parents, and the information supplied by Professor Osler and Dr. Ogden on this point has therefore a very special interest. Whereas Professor Osler's case shows that the condition may be directly inherited from a parent Dr. Ogden's case demonstrates that none of the children of such a parent need share his peculiarity. As the matter now stands, of five children of two alkaptonuric fathers whose condition is known only one is himself alkaptonuric. It will be interesting to learn whether this low proportion is maintained when larger numbers of cases shall be available. That it will be so is rendered highly probable by the undoubted fact that a very small proportion of alkaptonurics are the offspring of parents either of whom exhibits the anomaly. It would also be extremely interesting to have further examples of second marriages of the parents of alkaptonurics."

INBORN ERRORS OF METABOLISM

According to Graham (1936), the idea that alkaptonuria might be due to a chemical error in metabolism first occurred to Garrod one afternoon while he was walking home from the hospital to 9 Chandos Street in London. He thought of it at once as a congenital defect that persists throughout life. Later, in his lectures of 1903–1904, Garrod extended his observations to a class of metabolic disorders that are congenital and lifelong.

Bearn (1976) noted that Garrod's friendship with F. Gowland Hopkins was critical to the investigation of the chemistry of diseases. In addition to alkaptonuria, Garrod (1909) dealt with several other disorders, such as albinism, cystinuria, porphyria, and pentosuria, calling them the "inborn errors of metabolism." However, Harris (1953) pointed out that not all of the conditions considered by Garrod can be regarded as "true" metabolic errors. For instance, in cystinuria, the excessive excretion of cystine and certain other amino acids may be due to a failure in renal tubular reabsorption, and it would be more correct to call it renal anomaly rather than a metabolic one. However, Garrod's conclusions were essentially correct with respect to both the genetic basis of metabolic disorders as well as the gene–enzyme concept.

GARROD'S INFLUENCE

Garrod's brilliant deductions might have ushered in the field of biochemical genetics. However, his work was totally ignored for many years, by both biochemists and geneticists, despite of the support he enjoyed from both Bateson and Hopkins. Several reasons have been advanced to explain the failure to follow up Garrod's pioneering work. Among the possible reasons are the following: (a) it was regarded as an *isolated* observation and not as the first of a series of complex metabolic disorders that were discovered later (Caspari 1968, pp. 43–50); (b) Garrod's approach combined the ideas and methods from a number of sources, such as genetics, biochemistry, and pathology, an approach that was not readily understood by most scientists of that time; and (c) the principles of genetics had not yet been formed and the term "genetics" itself was yet to be coined by Bateson. In the years following Garrod's work, biologists interested in genetic research were primarily interested in the fruitfly *Drosophila*, corn, and *Oenothera*. Human genetics as a discipline did not become a reality until the 1930s and later. In the order of scientific developments, Garrod's discoveries were too early to be adequately understood and recognized by his colleagues. Bearn and Miller (1979) speculated that physicians were not interested in such rare disorders as alkaptonuria, which they seldom encountered in their daily practice.

In 1920, Haldane discussed the gene–enzyme concept but attributed it to Cuenot (1903). It is hard to obtain Haldane's (1920) paper, but it is included in a book of selected papers of Haldane (Dronamraju 1990, p. 542). During the 1920s and 1930s, research on the genetics of plant pigments was conducted under Haldane's direction, by Scott-Moncrieff (1936, 1981) at the John Innes Institution in England. In 1935, Beadle and Ephrussi adapted the transplantation technique, which was developed by Caspari (1933) for research on the mealmoth, *Ephestia kuhniella*, to *Drosophila*. Through their experiments involving reciprocal translocations of imaginal disks in *Drosophila*, Beadle and Ephrussi (1936) demonstrated the sequential chain of events in a normal metabolic chain. These studies paved the way to an understanding of biochemical mutant phenotypes.

Later, the genetic and biochemical investigations of Beadle and Tatum (1941), using *Neurospora*, led to the firm establishment of the principles of biochemical genetics. These studies were decisive in providing experimental proof for the genetic basis of biochemical phenomena. The convenience of using *Neurospora* mutants, which are haploid and are thus devoid of any complications of recessiveness and dominance and their shorter generation time, made it very easy to pursue research on the genetics of these mutants. Because the mutants are reparable, specific biochemical and genetic investigations could be carried out.

Tatum (1959) has summarized the essential findings of the work of Beadle and Tatum: (a) all biochemical processes in all organisms are under genetic control; (b) these biochemical processes are resolvable into series of individual stepwise reactions; (c) each biochemical reaction is under ultimate control of a different, single gene; and (d) mutation of a single gene results only in alteration in the ability of the cell to carry out a single primary chemical reaction.

In a seminal paper, Haldane (1937) emphasized the genetic basis of human chemical individuality, drawing attention to the pioneering work of Garrod (1902). Beadle (1967) wrote that he was not aware of Garrod's discoveries in biochemical genetics until after his own early work on *Neurospora* was completed (Beadle and Tatum 1941). He recorded that it was Haldane (1941) and Wright (1941) who first brought Garrod's work to his attention. However, Beadle was aware of Scott-Moncrieff's (1936) research on the genetics of anthocyanin pigments, which was carried out under Haldane's direction at the John Innes Institution in England. By the middle and late 1930s, the structure and variation of the main classes of pigments had been established, and genes controlling the relative amounts of the different pigments, the formation and inhibition of co-pigments, the state of oxidation and methylation of the pigments, and the pH of cell sap had been identified in several species. Furthermore, identification of the enzymes and catalytic agents involved in the formation of the plant pigments was planned. However, these plans were interrupted by World War II, and, in the meantime, many of the questions raised by the work on plant pigments were answered by the work of Beadle and Tatum on *Neurospora* (Kay 1989).

WEATHERALL'S EVALUATION

Garrod's contributions in the light of modern genetics have been evaluated by Sir David Weatherall, Regius Professor of Medicine Emeritus at Oxford: "Garrod's contributions seen in the light of current genetics Garrod's concept of biochemical individuality as it relates to single-gene disorders has been amply confirmed in the era of molecular genetics. But what is more remarkable, the ideas that he set out in Inborn factors of disease are almost identical to those that led to current genome searches for genetic variation that underlies susceptibility or resistance to common diseases, ranging from the devastating infectious killers of the developing countries to the major psychoses. Indeed, it is sobering to reflect that there have been relatively few fundamentally new ideas in human genetics since the early part of the 20th century and the work of Garrod, Fisher, Haldane and a handful of others; it is the remarkable technology of the molecular era that has made it possible to bring them to full fruition. The RCP (Royal College of Physicians) should be proud to celebrate the centenary of Garrod's great Croonian lectures, for as well as opening up a completely new field of biological thinking they emphasised the critical importance of the role of physician–scientists, with their unique opportunity to take questions from the bedside into the laboratory. Indeed, he was one of the first genuine physician-scientists and, to encourage young physicians to follow similar careers, was a major force behind William Osler in the establishment of the Association of Physicians of Great Britain and Ireland."

In retrospect, how do we evaluate Garrod's place in the history of biochemical genetics? That he was a great pioneer who first formulated the gene–enzyme concept is not in doubt. However, Garrod's work exercised no influence in founding the gene–enzyme concept. By the time his work was rediscovered, its foundations had already been laid by the experiments of Beadle and Tatum (1941). Haldane (1954) wrote that he himself was partly responsible for the suggestion that a gene makes a particular species of enzyme or antigen (Haldane 1920); but he was led to this hypothesis by Cuenot (1903) and not by Garrod.

In his own time, Garrod's work was ignored by geneticists and physicians. It was Beadle and Tatum's experimental work that was mainly responsible for establishing biochemical genetics as a viable field.

REFERENCES

Beadle, G.W. (1967) Mendelism. In Brink, R.A. (ed.), *Heritage from Mendel.* Madison: University of Wisconsin Press, pp. 335–350.

Beadle, G.W., and B. Ephrussi (1936) The differentiation of eye pigments in *Drosophila* as studied by transplantation. *Genetics, 21*: 225–232.

Beadle, G.W., and E.L. Tatum (1941) Genetic control of biochemical reactions in *Neurospora. Proc. Natl. Acad. Sci. USA, 27*: 499–510.

Bearn, A.G. (1976) *Inborn Errors of Metabolism.* New York: Lettsomian Lectures.

Bearn, A.G., and E.D. Miller (1979) Archibald Garrod and the development of the concept of inborn errors of metabolism. *Bull. Hist. Med., 53*: 315–328.

Caspari, E. (1933) Uber die Wirkung eines pleiotropen Gens bei der mehlmotte *Ephestia kuhniella. Z. Arch. Entwickl. Mech., 130*: 353–391.

Caspari, E. (1968) Haldane's place in the growth of biochemical genetics. In K.R. Dronamraju (ed.), *Haldane and Modern Biology.* Baltimore: Johns Hopkins University Press, pp. 43–50.

Cuenot, L. (1903) Hypothese sur l'heredite des couleurs dans les croisments des souris noirs, gris et blanches. *CR Soc. Biol. (Paris), 55*: 301–302.

Dronamraju, K.R. (1990) *Selected Genetic Papers of JBS Haldane.* New York: Garland.

Dronamraju, K.R. (1992) Biography: profiles in genetics: Archibald E. Garrod (1857–1936). *Am. J. Hum. Genet., 51*: 216–219.

Garrod, A.E. (1886) *An Introduction to the Use of the Laryngoscope.* London: Longmans, Green.

Garrod, A.E. (1902) The incidence of alkaptonuria: a study in chemical individuality. *Lancet, 2*: 1616–1620.

Garrod, A.E. (1909) *Inborn Errors of Metabolism.* 1st ed. London: Frowde, Hodder & Stoughton.

Graham, D.R. (1936) Obituary Notice. London: St. Bart's Hospital Report, 69: 12.

Haldane, J.B.S. (1920) Some recent work on heredity. *Trans. Oxford Univ. Jr. Sci. Club., 1*: 3–11.

Haldane, J.B.S. (1937) Biochemistry of the individual. In J. Needham and D.E. Green (eds.), *Perspectives in Biochemistry.* Cambridge: Cambridge University Press.

Haldane, J.B.S. (1941) *New Paths in Genetics.* London: Allen & Unwin.

Haldane, J.B.S. (1954) *An Introduction to Human Biochemical Genetics.* Cambridge: Cambridge University Press.

Harris, H. (1953) *An Introduction to Human Biochemical Genetics.* Cambridge: Cambridge University Press.

Kay, L.E. (1989) Selling pure science in wartime: the biochemical genetics of G.E. Beadle. *J. Hist. Biol., 22*: 73–101.
Scott-Moncrieff, R. (1936) A biochemical survey of some Mendelian factors for flower colour. *J. Genet., 32*: 117–170.
Scott-Moncrieff, R. (1981) The classical period in chemical genetics: recollections of Muriel Wheldale Onslow, Robert and Gertrude Robinson, and J.B.S. Haldane. *Notes & Rec. Roy. Soc. Lond., 36*: 125–154.
Tatum, E.L. (1959) A case history in biological research. *Science, 129*: 1711–1716.
Weatherall, D. (2008) The centenary of Garrod's Croonian lectures. *Clin. Med., 8(3)*: 309–311.
Wright, S. (1941) The physiology of the gene. *Physiol. Rev., 21*: 487–527.

10 George Wells Beadle (1903–1981)

George Wells Beadle shared the 1958 Nobel Prize for medicine or physiology with Edward L. Tatum (1909–1975) and Joshua Lederberg (1925–2008) for their work in the field of genetic research. Beadle and Tatum were cited for their discovery that genes act by regulating specific chemical processes, and Lederberg was honored for his detailed studies on the genetic crossing of bacteria. Beadle's demonstration that genes affect heredity by determining enzyme structure helped lay the foundation for the field of biochemical genetics. I knew Beadle personally; we were both at the International Congress of Genetics in Tokyo in 1968. He was president of the congress and I organized a plenary session to memorialize my mentor J.B.S. Haldane.

CHILDHOOD AND EDUCATION

George Wells Beadle was born in 1903 of parents who owned and operated a 40-acre farm near the small town of Wahoo, Nebraska. Both had grown up in similarly small communities: father, Chauncey Elmer Beadle, in Kendallville, Indiana, and, mother, Hattie Albro, in Galva, Illinois. The farm was a model for farms of its size and was so designated by the U.S. Department of Agriculture in 1908. This early influence of rural America shaped Beadle's informality and simple friendship that was evident throughout his life. Because of its small size, the farm was highly diversified, growing field crops as well as a number of vegetables and fruits. Numerous farm animals and domestic pets of a great variety were very much a part of Beadle's boyhood in rural Nebraska. His father Chauncey was a stern and demanding taskmaster who expected young George to perform a host of farm chores from dawn to dusk in the summer and before and after school during the remainder of the year. The ability to work long, strenuous hours at demanding tasks remained with Beadle all his life, as did the honesty and modesty nurtured by his rural upbringing. Gardening remained one of his greatest pleasures, and the victory garden he grew around his home later at Stanford during the Second World War produced enough for two families. This garden included beehives, but Beets (as he was known to his close friends) wouldn't eat the honey, saying he had been stung too many times as a boy. He loved corn, on the other hand, and raised several kinds, including a small Mexican variety that gave his garden the distinction of having the earliest sweet corn at Stanford.

When his mother died, he was only four and half years old. Young George and his two siblings—an older brother and a younger sister—were reared by a series of housekeepers. Five years later, his brother Alexander died from a kick by a horse. It was assumed that George would eventually take over the farm, but a young teacher of physics and chemistry at the Wahoo High School, Bess MacDonald, encouraged him to go on to college. She was his close friend and confidante during his years at the local high school and was, as Beadle recalled later, a kind of mother substitute, who often invited him to her house. Against his father's advice, Beadle enrolled at the University of Nebraska College of Agriculture in 1922.

Beadle graduated in 1926 with a BS degree and stayed on for another year to work for a master's degree with Professor Franklin D. Keim of the Agronomy Department. His first scientific publication, with Keim, dealt with the ecology of grasses. At some point along the way, under Keim's beneficent influence, Beets became interested in fundamental genetics and was persuaded to apply to the graduate school at Cornell University instead of going back to the farm. He entered Cornell in 1927 with a graduate assistantship and shortly afterward joined R.A. Emerson's research group on the cytogenetics of maize (Beadle 1930; Horowitz 1995).

RESEARCH AT CORNELL

This was a decisive moment in Beadle's career. As he put it: "Emerson was the perfect employer, graduate advisor, and friend. He turned problems over to me. One of my special assignments was to complete a summary of all genetic linkage studies in maize up to that time.... These were indeed exciting times for all of us working with Emerson" (Beadle 1974).

Corn genetics was new and exciting for Beets, and Emerson and his team—which included Barbara McClintock, Marcus Rhoades, George F. Sprague, H.W. Li, and several others—were truly inspiring. Emerson was the leading plant geneticist of that time and there was much stimulation and excitement in his laboratory. He was also a fellow Nebraskan; Emerson moved from the University of Nebraska College of Agriculture to Cornell University in 1914. Beadle clearly regarded his study under Emerson as the most important aspect of his career in genetics and was obviously deeply influenced by Emerson. Emerson promoted a sense of team work and camaraderie among maize workers worldwide. Among other factors, he freely distributed his stocks of maize mutant seeds, thus establishing a cooperative community. Beadle and another graduate student Marcus Rhoades developed close ties with Barbara McClintock. They recognized early that McClintock's unique skills and methods in cytogenetics would benefit their own research.

Early in his research at Cornell, Beadle pursued two independent lines of investigation. One was concerned with the relationship between corn and a wild, Central American plant, teosinte. Because he performed this part of the work as a paid research assistant to Emerson, it could not be used toward the requirements for his PhD. The other investigation, which became the core of his graduate thesis, concentrated on mutations that were associated with the sterility of corn plants.

Such mutants were collected by earlier workers and were noted for poorly developed anthers, which produced little or no pollen. Beadle's goal was to investigate the formal inheritance patterns of sterility and to use cytogenetics to study the mechanism behind the inability of the mutant plants to produce functional eggs and sperm. In his PhD thesis, Beadle acknowledged the help "so freely given by Dr. McClintock both in matters of technic and of interpretation of material." Beadle performed all the breeding experiments himself, including all the fertilizations, collecting the seeds, and planting them to grow the next generation. He found it a comforting task to grow corn, which reminded him of his childhood, and returned to it for a variety of reasons in his later years, no matter what his job was, including the period when he served as president of the University of Chicago!

The Synapsis Club at Cornell provided an opportunity for scientific and social interactions. Beadle regularly attended their meetings. In the fall of 1927, the beginning of his second year in Ithaca, a new member from California, Marion Hill, began attending the meetings. She was a master's student in botany at Cornell and also enjoyed bird watching and ecological projects. Soon George and Marion fell in love and were married. Their wedding took place in the presence of many Cornell friends and also his sister Ruth Beadle, who was persuaded by her brother George that Cornell was a better college than Nebraska to study nutrition so that she could become a dietician.

To improve his understanding of the chemistry of gene function, Beadle took courses in physical chemistry and biochemistry, the latter given by James B. Sumner. His graduate research further included some cytogenetics, especially the genetic control of meiosis using corn lines, in which chromosome behavior was markedly modified genetically—including such mutants as the asynaptic and polymitotic and, later, the sticky chromosomes as well.

The result was that in the following five years, Beets published no fewer than 14 papers dealing with investigations on maize, all begun while he was a graduate student at Cornell. In 1928, he married Marion Hill, who assisted him with some of his early corn research. Their son, David, was born in 1931.

Marion Beadle put aside her studies after her marriage and obtained a position as a research assistant in pomology. Furthermore, she found time to help her husband in the lab in her spare time. Beadle continued his research on sterility in maize, especially a recessive mutation that he eventually named *polymitotic* because of the many abnormal cell divisions late in meiosis. He discovered that the polymitotic mutation interrupted normal meiosis at a step after synapsis. Beadle showed further that polymitotic was associated with a linkage group and was inherited independently of many genes.

Beets earned his PhD in 1931 and was awarded a National Research Council Fellowship to do postdoctoral work in T.H. Morgan's Division of Biology at the California Institute of Technology. At Caltech, while finishing the work on maize cytogenetics he had started at Cornell—on genes for pollen sterility, sticky chromosomes, failure of cytokinesis, and chromosome behavior in maize-teosinte hybrids (a subject he would return to in his retirement)—Beadle also began doing research on *Drosophila*. Out of it would come one of the most interesting investigations of his career.

RESEARCH AT CALTECH

The highly stimulating environment at Caltech greatly helped Beadle's career. The faculty and associates then included Morgan, Sturtevant, Bridges, Dobzhansky, Schultz, Anderson, Emerson (son of R.A. Emerson), Belar, and the Lindegrens. Among the numerous visitors, Beadle (1974) recalled Haldane, Darlington, and Karpechenko. It was precisely during his visit to Caltech that Haldane deeply influenced Beadle's thinking about biochemical genetics (see Kay 1989). The extensive program of research on the chemical genetics of anthocyanins under Haldane's direction at the John Innes Institution in England was in full swing at that time. In several personal conversations and seminars, Haldane discussed the results with Beadle and others at Caltech (Beadle 1974, Kay 1989, Scott-Moncrieff 1981). Beadle, who had just completed his graduate work, was infected by Haldane's enthusiasm and excitement in biochemical genetics. Shortly afterward, Beadle visited Haldane at the John Innes Institution in England.

Beadle was caught up in the excitement of *Drosophila* genetics at Caltech and started working with Dobzhansky, Emerson, and Sturtevant on genetic recombination. Another significant turning point in his career was the arrival of Boris Ephrussi from Paris on a Rockefeller Fellowship during 1933–1934. Another significant event in Beadle's career was attending the Sixth International Congress of Genetics at Cornell University in Ithaca in 1932. It enlarged his vision of genetics greatly. For the first time, young Beadle came into contact with many famous geneticists from all over the world. The decision to start the Corn Newsletter was made during that Congress.

BEADLE'S CONTROVERSY WITH DARLINGTON

Beadle's stature as a mature geneticist was growing. He was even involved in a controversy with a distinguished scientist from England. The British geneticist C.D. Darlington had, in 1932, reinterpreted Beadle's work on polymitotic, suggesting that the supernumerary divisions after completion of meiosis were in fact like additional meioses and involved not only synapsis but even homologous chromosome exchanges. Beadle responded at once, gathering new data and arguing that Darlington's view was untenable and unsupported by experimental evidence. In the meantime, Darlington had visited Caltech, presenting a diminished image. Beadle's colleague, Anderson, who hosted Darlington, was reported to have said bluntly, that he would "never believe another thing that man says" (Berg and Singer 2003, p. 90). Shortly afterward, Beadle visited Haldane at the John Innes Institution in England, where both Haldane and Darlington worked. It was founded by William Bateson in 1910.

COLLABORATION WITH BORIS EPHRUSSI

The next step in Beadle's career was his collaboration in Paris with Boris Ephrussi on transplantation involving *Drosophila* eye pigments. It arose from Sturtevant's work that the character vermilion eye (absence of the brown component of the two normal eye pigments) was nonautonomous in the sense that if one eye and a small part of the adjacent tissue were vermilion and the remainder wild type, the genetically vermilion eye would produce both pigment eye components. By transplanting genetically vermilion embryonic eye buds in the larval stage to wild-type host larvae, Ephrussi and Beadle showed that an essential part of the brown component of the two normal eye pigments was nonautonomous in the sense that if one eye and a small part of the adjacent tissue were vermilion and the remainder wild type, the genetically vermilion eye would produce both pigment eye components. By transplanting genetically vermilion embryonic eye buds in the larval stage to wild-type host larvae, Ephrussi and Beadle showed that an essential part of the brown pigment system was produced outside the eye and could move to it during development. As Beadle (1974) put it: "It involved two people working cooperatively through paired binocular dissecting microscopes focussed on one recipient larva." By means of reciprocal transplants between the two mutants lacking brown pigment, Ephrussi and Beadle showed that there were two substances involved, one being a precursor of the second. They postulated that one gene was immediately concerned with the final chemical reaction in forming substance 1, and the second, with its conversion to substance 2.

In his recollections, Beadle (1974) explained that the two eye color genes, cinnabar and vermilion, were assumed to directly control the two postulated enzymes. Their work was in fact based on the transplantation method developed earlier by Caspari (1933) for studying the genetics of eye pigments in the mealmoth *Ephestia*. This was also the origin of the "gene–enzyme" concept as Beadle and Ephrussi saw it, although it was not so designated by them at that time. Beadle (1974) recalled that he and Ephrussi were encouraged in formulating their concept by the work of Scott-Moncrieff (1936, 1981) on the genetic control of anthocyanin pigments in higher plants. It is interesting to recall that Jacques Monod, who was a young instructor at the Sorbonne, took great interest in Ephrussi's collaborative work with Beadle and later accompanied Ephrussi as a visiting investigator at Caltech.

COLLABORATION WITH TATUM

Beadle knew Edward L. Tatum's father, then a pharmacologist at the University of Wisconsin, who once expressed his concern that his son would amount to nothing because he was going to be neither a geneticist nor a biochemist. Several individuals expressed doubts concerning the future of the young field of biochemical genetics. The origins of Beadle's interest in *Neurospora* were due to several fortuitous circumstances. The precise moment of realization that a study of mutant organisms, which had lost the ability to carry out specific chemical reactions, would be an easier approach to an understanding of the gene–enzyme relationship came to Beadle while he was listening Tatum's lecture on comparative biochemistry one day at Stanford during 1940–1941. Beadle's (1974) choice of *Neurospora* was based on his knowledge that its cytogenetics had already been worked out by his former Cornell acquaintance B.O. Dodge and the Lindegrens, whom he knew at Caltech. He also knew that many other fungi related to *Neurospora* could grow on chemically defined media that contained a proper balance of inorganic salts, a source of carbon and energy such as a sugar, and one or more vitamins. Beadle first heard of

Neurospora as a graduate student when Dodge had visited Cornell and presented a seminar on the mold's cytogenetics.

The clearest case of a biochemical mutation was discovered when the culture from a single ascospore, whose parent culture had been X-rayed, was found unable to grow in minimal medium but did so with added vitamin B_6 (Beadle 1945). By crossing the mutant strain grown on a supplemented culture medium with the original strain of the appropriate mating type and then testing cultures from the eight single spores derived from a single meiotic event, they showed that four such cultures required vitamin B_6 while four did not—thus indicating change in a single genetic unit (Beadle and Tatum 1941).

More mutants with other requirements were produced for other vitamins and for various essential amino acids. Genes for individual steps in sequences of biosynthetic reactions were identified. In general, the one gene-one biosynthetic step was recognized. Both biochemical and morphological mutants were produced. Beadle (1974) related an account of how Charles Thom, a widely recognized authority on fungi, visited his laboratory and admonished him for having so many "contaminants" in the cultures. He suggested that they ought to hire a good mycologist because what were called "mutants" were, in fact, contaminants!

There is a remarkable parallel between this comment and a similar (Cohn–Koch) dogma that prevailed with respect to bacterial cultures earlier. Any variation was regarded as the result of contamination. Zuckerman and Lederberg (1986) have reviewed this subject.

RESEARCH DURING WARTIME

Beadle's research activities were in full swing when World War II started. By 1941, the United States was at the height of its war "preparedness." The nation's scientific resources were coordinated under an executive order to establish a National Defense Research Committee and later another order to establish the Office of Scientific Research and Development (OSRD).

When Beadle and Tatum (1941) were publishing their early results on *Neurospora*, nonessential research expenditures were being cut back. They stressed the practical significance and utility of their work to such related areas as nutrition and pharmacology. The methodologies established in their research were applicable to other areas of research. Beadle planned a large-scale attack on the gene–enzyme problem by establishing a systematic program. He approached the American Philosophical Society as well as the Rockefeller Foundation for additional support. Fortunately, his work was of some interest to the food and drug industry.

The *Neurospora* bioassays were therefore attractive procedures for commercial houses that dealt with the manufacture of vitamins and amino acids. Beadle was closely involved with Merck and Company and also with Sharp and Dohme in collaborative projects. He was troubled by the restrictions that might be imposed, especially with reference to patent rights and manufacturing procedures. He undoubtedly benefitted from such associations, which provided the badly needed financial support during wartime.

Both Stanford University, where Beadle was then employed, and the Rockefeller Foundation encouraged these contacts. He was forced to turn more and more toward applied research efforts. Some governmental support was also available.

By November 1942, Beadle's program in biochemical genetics was classified as essential to the war effort.

To his great relief, Beadle was informed by Dr. A.N. Richards, the chairman of the Committee on Medical Research, an offshoot of war-created OSRD, that his research work is of such importance that it should not be interrupted in favor of any other war-related work.

According to Kay (1989), during the war years, Beadle's group isolated about 89,000 single spores, of which 500 produced mutant strains that were incapable of carrying out essential syntheses. Over 100 mutant genes that controlled vital syntheses were detected. Beadle was not restricted in publishing his results. Consequently, his work became familiar to his colleagues almost instantly. The majority of the mutants were characterized by loss of the ability to synthesize either a vitamin, an amino acid, or a nucleic acid component. The major contribution of Beadle's group was the elucidation of the relationship between genes and specific metabolic reactions and the enzymes that regulate these reactions.

The Beadle–Tatum collaboration resulted in a significant change in the development of genetics. Under their collaboration, gene action had become a biochemical problem and preceded what later came to be called "molecular biology."

Before their work on *Neurospora*, Haldane's research on the genetics of plant pigments was the only program that made a serious effort to interpret gene action in terms of biochemical pathways. However, Haldane's research was interrupted by political interference caused by Sir Daniel Hall and C.D. Darlington at the John Innes Institution (Dronamraju 2017).

LIFE IN OXFORD

Beadle was invited to be a visiting professor at Oxford University's Balliol College for the 1958–1959 academic year. His request for a sabbatical was approved by the authorities at Caltech. The Eastman Visiting Professorship at Oxford was endowed in 1929 by George Eastman, founder of the Eastman Kodak Company. It was intended to permit senior American scholars of the "highest distinction" to spend a year at Oxford. The list of previous Eastman Professors included Beadle's colleague Linus Pauling and George F. Kennan, the distinguished diplomat. Their experience was not quite what they had expected. Muriel McClure Beadle, George's second wife, wrote a humorous account of their experiences in Oxford (*These Ruins Are Inhabited*, Doubleday, New York, 1961).

Muriel's first impression as their train pulled into Oxford was that "it looks like the outskirts of Peoria." A cottage, which was estimated to be 150 to 300 years old, had been rented for the Beadles by the college. Muriel recorded her first impression: "a no nonsense stone box devoid of frills." Although charming in its own way, it lacked the amenities that Americans had come to expect. A plain old coal stove was intended for both cooking and heating water for the entire house. Learning to take care of the stove was only one of many chores Muriel acquired. Staying warm in the house required tending paraffin stoves, three electric fires, a fireplace, and the stove.

George put his distinctive mark to the garden by planting 10 dozen newly purchased spring bulbs. Their newly purchased car (a German Borgward) proved useless in narrow Oxford Streets and traffic congestion. George preferred to walk a distance of two miles to his lab in the botany department. However, the car helped them to explore the beautiful English countryside.

Their first visit to Balliol was clearly disappointing. According to Muriel, "It was just a chain of dirty, gray stone buildings with a Scottish baronial tower on one end, at the base of which was a door that looked as little as a mousehole." A tour of the college taught Muriel some of the local quirks and idiosyncracies; non-Fellows are not allowed in certain parts of the college grounds, and women are not allowed to enter the dining hall under any circumstances.

Beadle was required to give a regular schedule of lectures and seminars throughout the academic year. He chose the title "Topics in Genetics," which included the emerging concepts of genetics and their impact on society.

There were certain other problems. Beadle never understood the protocol of wearing the gown that was emblematic of an Oxford don, nor was he comfortable with the "silly" regulations about taking an oath before being granted eligibility for reading privileges. His attempts to keep up with the scientific developments were frustrated by the arcane arrangement for locating a book or periodical at the 50 Oxford libraries. Finally, he asked his Caltech office to airmail monthly editions of the *Proceedings of the National Academy of Sciences*.

NOBEL PRIZE AND CONTROVERSY

A Nobel Prize for George was rumored for some time while they were in Oxford. Reporters and photographers hounded Beadle in Oxford, Tatum in New York, and Lederberg in Wisconsin. Finally, Beadle received the official telegram from the Nobel Foundation that he would share the prize in physiology or medicine with Edward L. Tatum and Joshua Lederberg.

Nobel awards have occasionally been followed by counterclaims and controversies regarding those who were not recognized for their achievements. And Beadle's prize was no exception. One claimant was the French geneticist Boris Ephrussi, who was Beadle's collaborator in research on the genetic control of eye color in *Drosophila* during the 1930s. Their collaboration led to the idea that each step in the pathway for producing the mature eye pigments was controlled by a single gene. It was suggested in their paper that the gene might be acting through enzymes. Beadle and Tatum recognized the limited possibilities when using *Drosophila* for proceeding to the next step and decided to use a different organism. Nevertheless, the basic idea of genetic control in such situations originated with Beadle and Ephrussi.

Ephrussi was deeply disappointed at having been excluded from the Nobel selection. In a letter to the Indiana biologist Tracy Sonneborn, Ephrussi wrote: "I must admit I was disturbed these days by the Nobel Prize. I have to admit it: I suddenly felt my life wasted." In his reply, Sonneborn encouraged Ephrussi to see the whole thing in its proper perspective and not evaluate his life as success or failure on the basis of winning or not winning a Nobel Prize. Ephrussi was deeply moved by Sonneborn's letter and assured him that the distress of not sharing the prize was over and he was now looking forward to do more work. Their correspondence was summed up in the book *George Beadle: An Uncommon Farmer* (Berg and Singer 2003, pp. 249–251).

Three years later, as a young student of JBS Haldane, I was accompanying him in France and met Ephrussi at the Biological Station in Roscoff. I saw no signs of a disappointed man. He appeared to be quite cheerful in his conversations with Haldane and myself and was anxious to tell us about his new research in somatic cell genetics. I met him again in 1967 at the American Tissue Culture Association meeting at the Jack Tar Hotel in San Francisco. During our dinner together, I saw no signs of bitterness, only a cheerful man who reminisced about his friendship with Haldane. On neither occasion was Beadle mentioned.

PRESIDENT OF THE UNIVERSITY OF CHICAGO

Returning from Oxford, Beadle was restless and was bored by the predictable issues in the division of biology at Caltech. However, as a Nobel laureate, he was much in demand as a speaker on scientific topics and especially enjoyed speaking on the applications of biology that would benefit the general public. In 1960, he was approached by the University of Chicago's Board of Trustees who were interested in appointing him as the chancellor of the university. Later, at Beadle's request, the title was changed to president. During his initial interviews with the local press, Beadle explained his approach. He considered that the separation between the sciences and the humanities is a fallacy. Science is not opposed to culture any more than culture is opposed to science. Intelligent people seek a balance. The science faculty counted on him to elevate the university's standing to that of Caltech.

Opening a new phase in his career, Beadle had to confront numerous challenges, including extensive fund raising, which the university desperately needed, threats to academic freedom, civil disobedience, student unrest, and especially restoring the university's academic eminence.

A parking area near the president's house was set aside to create a field for him to grow corn. His lifelong love of farming had created a great desire to work the soil, whether it was growing flowers, corn, or some other crops. Visitors to the campus were often surprised when they learned that the man in coveralls tending the president's cornfield was the president himself!

He confronted various other controversial matters that crossed his path with reason, good humor, and diplomacy. In matters of evolution and the origin of life, he rejected the faith-based view that everything in the universe was created by the guiding higher intelligence we call God. He argued that the emergence of life in all its manifestations was inevitable given the properties of the hydrogen atom and the elements formed from it. His son David saw his father as a very moral person, basing his precepts in reason rather than a prescribed faith.

Much of Beadle's time at Chicago, however, was spent on fund raising ($160 million) and settling academic disputes regarding priorities of funding for various faculties and departments. Richard Lewontin, an innovative geneticist, was appointed as associate dean for basic sciences. He was given the authority for reviewing faculty appointments and promotions, as well as access to the divisional "treasury" for administering the subdivision. One of his early decisions was to merge the departments of botany and zoology, to form a new department of biology.

Beadle's impending retirement on his 65th birthday caught his colleagues and the society by surprise. The trustees lauded Beadle's role in restoring the university's faculty to the top rank of American universities. Beadle was important to the university as an intellectual symbol. Personal letters from the faculty praised his service. One in particular came from the director of the university library: "the library probably has made more solid and enduring progress during your administration than at any other time in the history of the university, your astute selection of some distinguished administrators; your recognition of the importance of the physical plant—including grass, your combined patience and firmness in trying to open and maintain better channels of communication with students; your candor and honesty; and by no means least, Mrs. Beadle's hard work on community and related problems; along with other efforts and qualities that have made this a much stronger university than it was when you came."

ONE LAST CONTROVERSY: THE CORN WARS

Beadle was always fascinated by the genetic and evolutionary origin of corn: its historical roots and its relationship to other plant species. The problem first attracted his attention while he was a student at Cornell in 1930. He believed that a wild ancestor of corn actually existed and was even known but was not easily recognizable as the ancestor of corn. This hypothesis was based on the belief that ancient Central American people had derived corn by conscious selection of advantageous mutants of the wild, indigenous plant called teosinte. The name itself suggested the hypothesis because Teosintl means God's corn in the Unto-Azteca language. The idea of this origin had been suggested to Beadle by Professor R.A. Emerson during his student days at Cornell.

After his retirement from Chicago, Beadle returned to this problem. Earlier, Beadle analyzed the corn–teosinte hybrids that Emerson had already produced. He observed under the microscope that during meiosis in the hybrids, the 10 teosinte chromosomes readily paired with the 10 corn chromosomes, a surprising finding if they were really two different species. Not only did the two groups of chromosomes pair perfectly during meiosis in the hybrid, but also they engaged in recombination, in the same way as occurred in pure corn plants (Emerson and Beadle 1932). Not only were corn and teosinte chromosomes and genes very similar, the arrangement of the genes on the chromosomes of the two plants was also the same. Beadle first stated his hypothesis in print in 1939.

An alternative hypothesis proposed by Paul C. Mangelsdorf claimed that an extinct or undiscovered ancestor of corn existed or exists and that teosinte was of more recent origin than corn. They claimed that teosinte was the product of crossbreeding through which some four or five chromosome segments from a plant called tripsacum were inserted into primitive wild corn about 1300 years ago. Tripsacum is native to Central America and Mexico, as are corn and teosinte. Tripsacum, which has 18 chromosome pairs, cannot be easily hybridized with corn. Beadle thought that the Mangelsdorf hypothesis was untenable. Mangelsdorf and his supporters had argued that it would have taken a very long time for ancient people to develop corn by the selection of naturally occurring teosinte. However, Beadle responded that there were so few genetic differences between the two plants that selection could have taken place over a reasonable amount of time (Beadle 1980).

DEATH

In his post-retirement years, Beadle's health deteriorated, which concerned his loving wife Muriel who took care of him to the best of her ability. He suffered from Alzheimer's disease, which steadily got worse as he got older. Muriel decided to move him to a rest home near Pomona in southern California, where George Beadle died on June 9, 1989. His Caltech colleague, James Bonner, eulogized him as the central figure in the dawn of this new golden age of biology. Many years later, James Watson (of *Double Helix* fame) recalled how Beadle unhesitatingly invited him home for dinner on a summer visit to Caltech. This was well before Watson's famous discovery of the DNA structure, whereas Beadle was at the height of his scientific reputation. His unpretentious modesty and simplicity and personal warmth made a deep impression on young Watson, as he recalled in his keynote address at the dedication of the University of Nebraska's George Beadle Center for Genetics and Biomaterials Research in Lincoln.

REFERENCES

Beadle, G.W. (1930) *Genetical and Cytological Studies of Mendelian Synapsis in Zea mays* (PhD thesis). Cornell University.

Beadle, G.W. (1974) Recollections. *Annu. Rev. Biochem., 43(0)*: 1–13.

Beadle, G.W. (1980) The ancestry of corn. *Sci. Am., 242(1)*: 112–119.

Beadle, G.W., and E.L. Tatum (1941) Genetic control of biochemical reactions in *Neurospora. Proc. Natl. Acad. Sci., 27(11)*: 499.

Berg, P., and M. Singer (2003) *George Beadle: An Uncommon Farmer. The Emergence of Genetics in the 20th Century.* Cold Springs Harbor Laboratory Press.

Caspari, E. (1933) Uber die wirkung eines pleiotropen Gens bei der Mehlmotte *Ephestia kuhniella. Z. Arch. Entw. Mech., 130*: 353–381.

Dronamraju, K.R. (2017) *Popularizing Science: The Life and Work of JBS Haldane*. Oxford: Oxford University Press.

Emerson, R., and G.W. Beadle (1932) Studies of *Euchlaena* and its hybrids with *Zea*. II Crossing over between the chromosomes of *Euchlaena* and those of *Zea. ZIAVA, 62*: 305–315.

Horowitz, N.H. (1995) George Wells Beadle. 23 October 1903–9 June 1989. *Biogr. Mem. Fellows R. Soc., 41*: 44–54.

Kay, L.E. (1989) Selling pure science in wartime. The biochemical genetics of George W. Beadle. *J. Hist. Biol. 22*: 73–101.

Scott-Moncrieff, R. (1981) The classical period in chemical genetics: recollections of Muriel Wheldale Onslow, Robert and Gertrude Robinson and J.B.S. Haldane. *Notes Records R. Soc. Lond., 36*: 125–154.

Zuckerman, H., and J. Lederberg (1986) Postmature scientific discovery? *Nature, 324*: 629–634.

11 Joshua Lederberg (1925–2008)

Joshua Lederberg shared a Nobel Prize with George W. Beadle and Edward L. Tatum in 1958 for his pioneering work on genetic recombination in bacteria. His discoveries advanced the field of molecular genetics into the forefront of biological and medical research. They have contributed to genetic engineering, molecular biology, and gene therapy.

Besides molecular genetics, Lederberg's ideas impacted on several other fields. He recognized the potential of computers for the analysis of scientific data and how the logical processes of scientific induction were applicable to the construction of intelligent computer programs. He collaborated with scientists across many disciplines to revolutionize the applications of computers to chemistry, medicine, and information technology. While president of Rockefeller University in his later years, he continued his advocacy for scientific understanding, drew attention to the threat of emerging infectious diseases, and ultimately returned to the laboratory to explore the process of gene mutation.

He advised U.S. presidents and international organizations on a wide variety of issues and devoted a prodigious amount of time and effort to the task of informing policy makers and the larger public on important scientific matters.

Interestingly, Joshua Lederberg cautioned any future biographer who might be interested in studying his life and work. History is an accounting that begins when the witting participants are gone and unable to contradict the inventions of those who had not participated in the events and must rely on extant fragments of evidence. Historian Eugene Garfield (2008) correctly observed that Lederberg treated documentation of the scientific literature in the Talmudic tradition in which he was raised—as a sacred obligation.

Joshua Lederberg was born May 23, 1925, in Montclair, New Jersey, the son of Zvi H. Lederberg, an Orthodox rabbi, and Esther Goldenbaum Lederberg. His parents had emigrated from Palestine to the United States in 1924.

They were penniless when they arrived in the United States. The rabbi and his wife struggled to make ends meet. According to Lederberg, his father was ill during most of his childhood. His mother had to work very hard to support the family. Joshua's younger brother, Seymour, was born on October 30, 1928. According to Josh, his younger brother's presence introduced him to new responsibilities, while their parents were fully occupied in earning a living. Seymour later recalled that Josh treated him as his first student and mentored him accordingly.

Josh was initially expected to become a rabbi like his father; however, while still very young, he developed a keen interest in science. At the age of seven he wrote, "What I would like to be. I would like to be a scientist of mathematics like Einstein. I would discover a few theories in science."

Josh requested the book *Introduction to Physiological Chemistry* by Meyer Bodansky as his Bar Mitzvah present. He was a precocious child who found little stimulus for his interests in the New York City public grammar and junior high schools he attended. Much of his scientific knowledge was self-taught by selective reading in the Washington Heights public library and later in the Cooper Union Library stacks, where he read voraciously. It was almost entirely self-directed and eclectic. Like many of his contemporaries, Josh's interest in biology was stimulated by Paul de Kruif's book *The Microbe Hunters*.

One of his teachers, Mrs. Fanny S. Rippere, remembered him as a 12-year-old boy. She wrote: "Early in 1937 I had a most unusual pupil whom I still remember vividly. I can still remember how he prepared a paper on the classification of Protozoa using a graduate text for a reference."

In 1938, he entered Stuyvesant High School, which emphasizes science, mathematics, and technology. Stuyvesant had a highly selective admissions process based on competitive examinations. There, he finally found a sympathetic intellectual environment. With the facilities available at Stuyvesant, he was able to begin experiments in cytochemistry and receive a fuller education in the sciences.

Graduating from Stuyvesant in 1941, Josh wanted to continue his education at Columbia University. However, he was too young to be admitted to Columbia until the autumn. In the meantime, he was successful in obtaining laboratory privileges at the American Institute Science Laboratory, which enabled him to continue his experiments there until he could matriculate. The American Institute Science Laboratory had been established after the 1939 World's Fair in New York with support from Thomas J. Watson, the first president of IBM, and it occupied an IBM showroom on Fifth Avenue. Its purpose was to allow high school students to engage in after-school scientific research. Here, Lederberg was able to focus his studies on the chemistry of the nucleolus, continuing his investigations until September 1941, when he was admitted to Columbia University at the age of 16 years. Although he had applied to Cornell also, he had not been able to receive financial support there.

COLUMBIA UNIVERSITY

Two factors made it possible for him to attend Columbia, located at the Morningside Heights campus in New York City: he could live at home and he received a $400 tuition scholarship from the Hayden Trust. His previous autodidactic reading allowed him to take several graduate-level science courses.

At Columbia, Josh met the most important man of his life who opened the doors to further his career. He was Francis J. Ryan. Ryan had received his PhD from the Zoology Department at Columbia in 1941 and then spent a year as a postdoctoral fellow at Stanford University in California. While at Stanford, he worked in the laboratory of Edward L. Tatum and George W. Beadle studying growth properties of *Neurospora crassa.*

Josh worked part-time in Ryan's laboratory at Columbia, preparing culture medium and propagating strains of *N. crassa*. Josh recounted, "Professor Ryan took a callow underclassman from Washington Heights, brash and argumentative as precocious students often are, and turned me into a scientist" (Ligon 1998).

When Ryan returned to Columbia in 1942 as a faculty member in the Department of Zoology, he renewed his research with *Neurospora.*

Ryan was the first working scientist whom Lederberg had met, and he introduced Lederberg to experimentation in biochemical genetics, taking him on as a protégé and part-time laboratory helper. From Ryan, Lederberg learned how to conduct experiments and record the results in a disciplined and professional way. While at Columbia, Lederberg joined the Navy V-12 program. This program, which ran from July 1, 1943, to June 30, 1946, was designed to provide officer candidates for the U.S. Navy. The navy had expanded enormously as a result of World War II, and the Naval Academy at Annapolis, Maryland, could not meet the resulting increase in demand for officers. At the same time, colleges and universities were suffering a severe attrition in enrolment because of the large number of men who had gone into the military services. In response to this situation, the navy created a program to support students at various colleges, paying tuition and stipends for those who passed the competitive examinations and enlisted in the V-12 program. For those like Lederberg who were training to be medical officers, the program compressed premedical training to about 18 months and medical training to three years. Lederberg was able to continue his studies at Columbia and his research activities with Ryan, alternating them with periods of duty at the U.S. Naval Hospital, St. Albans, Long Island. There, he was primarily assigned to the clinical pathology laboratory, screening samples from U.S. Marines returning from Guadalcanal: stool samples for parasite ova and blood samples for malaria. He later remarked that this experience contributed to his thinking about microbial life cycles.

DISCOVERY OF GENETIC EXCHANGE IN BACTERIA

Having obtained a BA in Zoology from Columbia in October 1944, Lederberg was reassigned to the Columbia College of Physicians and Surgeons at the Washington Heights campus to begin his medical courses, but he maintained his contacts and activities at the Morningside Heights campus.

An important paper by Avery, MacLeod, and McCarty (1944), documented experiments showing that DNA was the active agent that could cause the conversion of one form (rough) of the bacterium *Pneumococcus* into another form (smooth). These results provoked vigorous discussion because they were the first to show that DNA is the genetic material, and they inspired Lederberg to propose experiments to look for DNA-mediated transformation of *Neurospora*. As it happened, the mutant chosen to be the recipient in his experiments proved to be too revertible for the purpose;

however, a study of the reverse mutation phenomenon resulted in Lederberg's first publication with Ryan.

Josh had become aware of research on the transfer of bacterial genes by nucleic acid preparations and persuaded Ryan to allow him to pursue this in *Neurospora*. The nutritional mutants of *Neurospora* were not stable, but he and Ryan were able to complete a study on reverse mutations, which was published in the *Proceedings of the National Academy of Sciences* (Ryan and Lederberg 1947).

After a series of disappointing experiments to demonstrate transfer of genetic information by nucleic acid preparations, Ryan encouraged Josh to apply to the Department of Microbiology and Botany at Yale University to pursue a genetics project with bacteria under the direction of Edward L. Tatum, who had recently relocated from Stanford University. The original plan was for Josh to spend three to six months at Yale University and then return to Columbia's Physicians and Surgeons Medical School. Within six weeks, using nutritional mutants of *Escherichia coli* K-12 that Tatum had prepared, Josh conducted experiments establishing that gene transfer occurred in bacteria (Zuckerman and Lederberg 1986). The initial reports in 1946 and 1947 were not universally accepted, it being argued by senior international authorities that the results could be attributed to exchange of released nutrients by the mutants, coupled with back mutation during the resulting limited growth. Max Zelle from Cornell University in Ithaca, New York, proposed that the issue be resolved by studies on single cells isolated by micromanipulation, a task that was not completed until 1950.

YALE UNIVERSITY

The GI Bill was in full operation after the war, and the military-sponsored physician-training program had passed. After devoting two years with Tatum at Yale University in research that was building a new foundation for bacterial genetics, Ryan and Tatum convinced their academic administrators to award graduate credit for Josh's prior studies and work at Columbia's Physicians and Surgeons Medical School and at Yale University. Josh was obliged to assume the financial obligation for retroactive tuition to establish his academic residence at Yale University. Josh was awarded his PhD degree with Tatum as his research adviser at Yale University in 1947. Josh was faced with many challenging decisions: recently married, in debt, the possibility of returning to medical school, and the potential loss of one of his key mentors, Tatum, who was negotiating a return to Stanford University. Tatum learned of and recommended Josh for a position in genetics at the University of Wisconsin in Madison. Until that time, Josh had always envisioned himself as a medical scientist, and the idea of an appointment in the College of Agriculture was not particularly appealing. Moreover, Josh, reared in New York as the son of an Orthodox rabbi, was apprehensive about relocating to the Midwest, where he would be separated from his cultural support system and mentors.

UNIVERSITY OF WISCONSIN

With confidence and anticipation, Josh was drawn to the faculty of the University of Wisconsin, which included scientific giants such as enzymologists David E. Green and Henry A. Lardy and geneticists such as Royal Alexander Brink and M.R. Irwin, and the College of Agriculture offered financial support through its resources and those of the Wisconsin Alumni Research Foundation. Other luminaries joined the faculty later. They include the future Nobel laureates Har Gobind Khorana, Howard Temin, and Oliver Smithies, as well as Sewall Wright, who was one of the founders of population genetics, and the brilliant geneticist James F. Crow (Anonymous 2005).

There was some concern on the University of Wisconsin side as well: could a young New Yorker with a passion for medical research adjust to an agricultural college in America's heartland, and should the faculty offer a position to a 22-year-old with dubious academic credentials based upon research that was still questioned by many authorities in fall of 1947? These skepticisms continued to be expressed during the dissertation research by Josh's wife, Esther Miriam Zimmer Lederberg. Esther had received her master of arts degree at Stanford University in 1946, under the direction of George W. Beadle, and then returned to New York City to work with Norman Giles at Yale University on *Neurospora*. There she met and married Josh and they moved to Wisconsin in the span of one year. An interesting coincidence noted by Tatum in his Nobel lecture is that Esther Zimmer assisted in the isolation of the *E. coli* K-12 nutritional mutants at Stanford University that were used by Josh and Tatum at Yale University. In Madison, Esther entered the graduate program in bacteriology at the University of Wisconsin, pursuing research on the genetic control of mutability in the bacterium *E. coli*. She completed her PhD degree in 1950.

YOUNG PROFESSOR

Joshua Lederberg joined the faculty of genetics at the University of Wisconsin in Madison as an assistant professor in the fall of 1947. His research career began before the era of big science. Indeed, his early work was funded piecemeal and at a very modest level. The Wisconsin Alumni Research Foundation initially funded his research and the University of Wisconsin Graduate School provided support for his graduate students. By 1954, he had received a contract from the Atomic Energy Commission and an extramural grant from the National Cancer Institute.

A number of his trainees were supported by fellowships from the National Science Foundation and the National Research Council. He advanced rapidly through the academic ranks, becoming an associate professor in 1950 and a full professor in 1954.

ZINDER AND TRANSDUCTION

Josh was sympathetic to the application of Norton Zinder for graduate study in genetics, both because Zinder had unsuccessfully aspired to enter medical school and because he had been recommended by Francis Ryan. Zinder joined the laboratory in the fall of 1948 and was immediately steered to a project on *Salmonella typhimurium*, with the expectation that bacterial conjugation in *E. coli* could be extended to include this closely related pathogen. Zinder, in a relatively short period, was able to isolate single mutants and complete the standard conjugation experiment. The promising results led to the demand for double nutritional mutants, with somewhat confusing results. There were many twists and turns as Zinder and Josh ruled out back mutation, complementary cross-feeding of nutrients, and even the need for actual cell contact. Ultimately, the critical bacteriophage vector was discovered, and the mechanism of generalized genetic transduction, that is, the unilateral transfer of a limited number of genes by a viral vector, began to unfold (Zinder and Lederberg 1952).

Josh received the Eli Lilly Award as the outstanding young bacteriologist, presented at the Society of American Bacteriologists annual meeting in 1953. Concurrently, there was a swirl of controversy about the process of bacterial conjugation. The Lederbergs, in cooperation with their longtime friend, Italian geneticist Luigi (Luca) Cavalli-Sforza from Milan, found that the asymmetry in bacterial recombination was controlled by a fertility factor on the cell surface that was readily transmissible.

Another addition to Josh's lab, M. Laurance Morse, a mature student with a master's degree and work experience at Oak Ridge National Laboratory, came to Josh's lab as a graduate student to work with lambda phage. Esther Lederberg, by this time, had described the K-12 variants of *E. coli* that were lysogenic, sensitive, and resistant to lambda phage. It was also known that the locus for maintaining the lambda genome was linked to the galactose locus. Morse suggested that it would be possible to use induced lambda phage to transfer the galactose locus to lambda-sensitive *E. coli*. Indeed, Morse, Esther Lederberg, and Josh developed the gene transfer system that continues to underpin molecular genetics and genetic engineering. Josh clearly understood the implications of this discovery and anticipated that parallel systems would be developed for gene therapy in humans.

INDIA AND AUSTRALIA

By the 1960s, Josh was an established senior scientist who was serving on study sections of the National Science Foundation and the National Institutes of Health. Josh recognized the need to improve communication among scientists, the general public, and the private sector. Josh was elected to the National Academy of Sciences in 1957.

Josh received a Fulbright Visiting Professorship to go to Frank Macfarlane Burnet's department at Melbourne University in Australia, expecting to spend a few months exploring influenza virus genetics.

On the way, Josh stopped in Calcutta to see the famous scientist J.B.S. Haldane, who had moved from University College London to the Indian Statistical Institute in Calcutta in July 1957 (Dronamraju 1995, 2017). Their discussion was mainly focused on the political and defense implications of the Sputnik, which the Soviets had recently placed in orbit. Josh's concerns included the possible contamination of space, especially the moon and other heavenly bodies!

By the time of his arrival in Australia, Josh found that Burnet had redirected all research in the Walter and Eliza Hall Institute toward immunology. Gustav Joseph Victor Nossal, a new medical graduate, had planned to start a virology project for his PhD degree. Josh and Nossal were thrown together to develop a project on immunology. This provided Josh with the opportunity to address a truly biomedical question, the onset of the immune response, by collaborating with Nossal to test Burnet's theory of clonal selection of antibody producing cells. Josh used his knowledge of antigenic phases of bacterial flagella and his experience with single cells under the light microscope to develop with Nossal a direct experiment to test if individual antibody-producing cells produced one or more antibody specificities. They labored through 62 antibody-producing cells and found no cell that produced more than one antibody specificity (Lederberg and Nossal 1960). The experience in Australia had a second profound effect on Josh. The Soviet Union launched the satellite Sputnik, which was visible in Australia for several days before it was seen over the United States. Josh immediately recognized that this provided a new tool to understand the universe and at the same time posed the threat of cross-contamination of life forms.

Upon his return to the United States, Josh persuaded Detlev Bronk and Frederick Seitz of the National Academy of Sciences to express officially a concern about interplanetary cross-contamination. Josh was intrigued about the possibility of extraterrestrial life ever since he had listened to Orson Wells's radio broadcast of "War of the Worlds" in 1938 (Lederberg 1987). In particular, he wanted to test the hypothesis by Svante Arrhenius that life arrived on Earth from bacterial spores traveling through space. Harlyn Halvorson was working on the germination of bacterial spores at the University of Wisconsin, so Josh contacted Halvorson about collaborating on an experiment to determine whether spores could survive exposure to outer space. Josh and Halvorson designed a miniaturized apparatus that would unwind a tape to which spores could attach and, after a period of time, rewind the strip and spray it with germinating agents. The refractive index of the exposed tape was to be measured to detect changes in the characteristic of germinating spores. They prepared and submitted a grant request to the U.S. Air Force, but it was not approved. At the same time, Josh was advocating that this type of research should be conducted by a civilian agency, not the military.

Georges Cohen from the Pasteur Institute came to the University of Wisconsin to spend 1958 in Halvorson's laboratory. Josh learned from Cohen that Dean B. Cowie of the Carnegie Institute of Terrestrial Magnetism was designing an experiment to capture particles in space, place these on Petri dishes, and look for spores. Josh suggested that he collect Moon dust, which had been collecting particles for eons. Someone mentioned that the Soviet Union was sending dogs into space, and if one of these landed on the Moon, it would contaminate the Moon. Josh and Cowie immediately drafted a letter, which they sent to Science and the New York Times, urging that all spacecraft be first sterilized.

Eminent population geneticist James F. Crow said that during his first few years in Madison, Josh was the most important intellectual influence in his life. His course in microbial genetics was widely respected and attended by postdoctoral trainees and faculty from the entire Madison campus as well as by graduate students. Josh required the auditors to take his examinations along with the students, a requirement not entered into lightly by his faculty peers. Josh's microbial genetics course addressed segregation in fungi and ill-defined variation in a range of microorganisms as well as the current state of gene transfer in bacteria.

His laboratory was a magnet that attracted visiting scientists and guest workers worldwide, such as Sydney D. Rubbo from the University of Melbourne, Luigi Luca

Cavalli-Sforza from the Istituto Sieroterapico in Milan, and the immunogeneticist Ruggero Ceppellini from the University of Turin.

MEDICAL GENETICS

Josh increasingly felt the need to be associated with the medical school and encouraged John H. Bowers, dean of the University of Wisconsin Medical School, to develop a program in medical genetics, which it did in 1957, with Josh as its founding chair (Lederberg 1997). The Rockefeller Foundation provided funding for three years to launch the program, and Josh enthusiastically set out to expand the department and to offer research opportunities for medical students. The university was not immediately forthcoming with funding for new faculty positions, and Josh began to entertain seriously an offer from Stanford University, which was poised to expand research in its medical school. In his resignation letter, Josh wrote to University of Wisconsin president and biochemist Conrad Elvehjem, "Genetics and biochemistry are rapidly converging on the fine structure and biosynthesis of nucleic acid" (Halvorson 2007).

Josh had done almost all the work leading to his share of the Nobel Prize at Wisconsin, and he was leaving for Stanford University in 1958 during the interval between notification and presentation of the prize. To acknowledge the support that the University of Wisconsin had provided for the research honored by the Nobel Prize, Josh donated the gold medal to the regents of the University of Wisconsin. With a sense of a renewed mission to pioneer new biomedical fields, Josh moved on, rarely looking back at the fields he established.

STANFORD UNIVERSITY

Joshua Lederberg and Esther moved to Palo Alto, California (Ligon 1998), and he rarely returned except to accept an honorary degree from the University of Wisconsin in 1967. Josh had evolved into a confident, poised 33-year-old who had learned the value of social graces. Josh was driven by a strong social conscience, nurtured by a father whose faith coexisted with a broad humanist consciousness and a tolerance for divergent views.

OTHER ACTIVITIES

Josh continued to feel the need to address larger problems that will improve the future of humankind. The opportunity to return to a research-intensive medical school environment was attractive and was reinforced by the strong tradition of Stanford University. In addition, Josh did not enjoy refining the details of recombination in bacteria and the clash of personalities of newcomers to the field, who were trying to establish their reputations as leaders in the field. Bernard D. Davis (1987) noted the collegial spirit of Josh and the pioneers of microbial genetics and observed that the workers in the field became intensively competitive over time. Josh told Barbara Hyde in a 1992 interview, "We urgently need to dispel the idea that the primary motivation of researchers is to beat their competitors. I firmly believe that idealism and the excitement of discovery are necessary parts of science" (Lederberg 1992).

Stanford University's medical school was moving from San Francisco to Palo Alto in 1958 and was on a campaign to recruit outstanding scientists. Among the first new hires were Arthur Kornberg and Joshua Lederberg. At Stanford, Josh took on more administrative duties and his extracurricular activities expanded. He was commissioned to create a department of genetics and was named professor and executive head in 1959. One of his first hires was Leonard Herzenberg. He acknowledged that his experiments with mammalian cells were largely inspired by Josh's work on bacterial genetics. He commented, "So, I was doubly delighted when Dr. Lederberg took an interest in my work and invited me to join the new genetics department he was founding at the Stanford University School of Medicine." (Herzenberg 2006, Kyoto Prize Commemorative Lecture).

The move to Stanford University provided the right opportunities and challenges in Josh's scientific career. He was a remarkable man with a remarkable intellect. He was especially drawn to policy matters and the "big" issues of science and public policy.

He was not by nature an experimental scientist. He was surely right that simple experiments are the key but occasionally one needs a bit of technology. Technology is not necessarily good science, but good science without good technology is quite difficult. At Stanford, much of Josh's creative effort was cross-disciplinary and directed toward collaborative work with intellectual peers in chemistry, engineering, and computer science. Carl Djerassi recalled, "When I came to Stanford in 1960, Josh was probably my closest friend, professionally and personally, and within about two years we started to collaborate, I as a chemist, he as a person who works in every field conceivable, including even chemistry" (Shortliffe and Rindfleisch, 2000).

For his work in biomedical computing, the American College of Medical Information awarded Josh its highest award for lifetime achievement, the Morris R. Collen Award, in 1999. He also served as a consultant on health care policy to John F. Kennedy's transition team and on the President's Panel on Mental Retardation. In this latter role, he met and earned the respect and confidence of President Kennedy's sister and brother-in-law, Eunice and Sargent Shriver. The Shrivers, deeply interested in the neurological and genetic causes of mental illness, established the Joseph R. Kennedy Jr. Laboratory for Molecular Medicine at Stanford University. Josh became director of the Kennedy Laboratories for Molecular Medicine in 1962. He was also increasingly concerned about the public's understanding of science and committed to writing a weekly column for *The Washington Post* from 1966 to 1971. In these articles, he addressed a range of topics, from population control to cloning to scientific ethics to space biology.

EXOBIOLOGY

In spring 1958, the National Academy of Sciences established the Space Science Board at the request of the U.S. National Committee for the International Geophysical Year. Lederberg was one of its founding members and served on the board until it was dissolved in 1974. The board's mission was to assess the scientific aspects of space exploration, of interplanetary probes, manned spaceflight, and space stations and to propose directions for science experiments and the search for life in space. It urged that "great care" be taken in sterilizing spacecraft before launch and recommended a "stringent" quarantine for samples returned from other planets until it could be determined that they were harmless. The National Aeronautics and Space Administration (NASA) adopted both recommendations prior to its first flight to the moon in 1969.

Lederberg's urgent warnings about interstellar contamination and his call for the scientific study of life beyond earth's atmosphere—for which he coined the term *exobiology*—tapped into popular fascination with the dawning of the space age and brought him international media attention. "I was the only biologist at the time who seemed to take the idea of extraterrestrial exploration seriously," Lederberg remembered. "People were saying it would be a hundred years before we even got to the moon." He, however, "was convinced that once the first satellite was up the timetable would be very short, and [his] fear was that the space program would be pushed ahead for military and political reasons without regard for the scientific implications." He collaborated with the well-known astronomer Carl Sagan in establishing exobiology as a scientific discipline and educated the public on the biological implications of space exploration in his weekly "Science and Man" columns in *The Washington Post* during the 1960s.

By publicly promoting exobiology, Lederberg almost single handedly gained a place for biologists in the burgeoning U.S. space program, as well as a share of its ample research funds. He pressed upon NASA the need to include biological science in its mission and research designs and represented the interests of biologists on the agency's Lunar and Planetary Missions Board between 1960 and 1977. In this role, he helped define the scientific objectives for the Mariner Mars missions, launched between 1964 and 1971 to map the planet's surface and study its atmosphere from close-in orbits.

Lederberg participated directly in engineering experimental devices that were to decide mystifying and intriguing questions about the possibility of life on Mars. As principal investigator of the NASA-funded Instrumentation Research Laboratory at Stanford University Medical School from 1961 onward, he helped develop an automated biomedical laboratory that would allow scientists to examine the soil of Mars for traces of life as part of the Viking mission to the red planet in 1975. Like Sagan and the geneticist Herman J. Muller, Lederberg assumed that if life existed on Mars—which they considered quite possible—it must be in the form of microorganisms, because only they could withstand the planet's hostile environment and destructive radiation.

The device Lederberg designed with the help of the Instrumentation Research Laboratory's director, engineer Elliott Levinthal, consisted of a conveyer belt that scooped up samples of Martian soil and deposited them within a computer-controlled mass spectrometer. Inside the spectrometer, the soil substance was bombarded with electrons, producing a fragmentation pattern, which sorted the electrified particles (ions) according to their mass. This pattern was transmitted to earth, where scientists could analyze it for evidence of organic compounds and microbial life.

The data on microbial life on Mars produced by the Viking Mars Lander were inconclusive and have since been discounted by further exploration of the planet's surface. Whether liquid water and, with it, life existed on Mars in the distant past remains an open question. Even so, the Mariner and Viking missions proved to

Lederberg the effectiveness of automated space probes, robotics, and computers in acquiring valuable scientific data about space. By contrast, he regarded NASA's growing emphasis on returning man to space and preoccupation with the Space Shuttle during the second half of the 1970s as a threat to instrumented space flight and to the agency's scientific endeavors. After the launch of the twin Voyager probes to the outer solar system in 1976, NASA's last major instrumented mission for over a decade, Lederberg in 1977 ended his role as a consultant to the U.S. space program.

Josh continued his interest in space, serving on the lunar and planetary mission boards of NASA from 1960 to 1977 and as an intellectual contributor and adviser to the Viking Mission to Mars. Josh argued passionately that it would be more cost-effective to use technology for space exploration rather than manned missions. In particular, he expressed concern about the reciprocal infection of planets by agents carried on spacecraft. He was an advocate for worldwide reduction of nuclear weapons and warned of the potential use of biological and chemical agents by extremists.

Josh was also an adviser to the arms control administration that was negotiating the Biological Weapons Convention and made several visits to Geneva during 1970–1972. Josh spoke openly about the need for arms control and measures to prevent the proliferation of nuclear and biological weapons. Esther Lederberg remained dedicated to lab research while Josh was engrossed in global issues. This divergence in interests, among other reasons, contributed to their divorce in 1966.

Esther continued at Stanford, becoming director of the Plasmid Reference Center, and Josh focused his energies on exobiology, information science, and public policy. In 1968, Josh married Marguerite Stein Kirsch, who was born in Paris and had lived as a refugee in southern France during World War II. After the War, she immigrated to the United States, was educated at the Lycéc Français de New York, Bryn Mawr College, and Yale University Medical School and later became a psychiatrist. Josh and Marguerite had one daughter, Anne, born in 1974.

NEW YORK AGAIN

In 1978, Josh was appointed president of Rockefeller University and served in that capacity until he retired in 1990. He became an active member of the New York Academy of Sciences and was, in quick succession, made an honorary life member in 1980 and honorary life governor in 1982. As president, he directed his energies into resources and scientists to conduct outstanding research, using the tools of molecular biology and computing to address problems with medical applications. His achievements earned him further international recognition by being named a foreign member of the Royal Society of London in 1979. Throughout his tenure at Rockefeller University, Josh served on many governmental advisory committees, which for his entire career spanned nine U.S. presidential administrations.

RETIREMENT

Upon retirement in 1990, he assumed the role of Raymond and Beverly Sackler Foundation Scholar and head of the Laboratory of Molecular Genetics and Informatics, where he once again challenged his nemesis, mutation, with new experimental tools and insight. He focused on how activation of genes affects their susceptibility to mutation. Josh's final contribution to the scientific literature addressed the proposition "that the growth rate of bacteria may not be only determined by factors of nutrition and the environment, but Josh thought that it might be 'self controlled' in bacterial populations by cellular communication or quorum sensing" (Nackerdien et al. 2008). Stephen S. Morse stirred Josh's interest in emerging infectious diseases at a Rockefeller University reception in 1987. After several exchanges with Morse, Josh concluded, "The problem of emerging viruses is one that must be addressed at the highest levels" (Morse 2008).

In 2001, Josh testified before the U.S. Senate Committee on Foreign Relations on biological warfare, asserting that "per kilogram of weapons, the potential lives lost approach those of nuclear weapons, but less costly and sophisticated technology is required." He concluded, "Studies of hypothetical scenarios document the complexity of managing bioterrorist incidents and the stress that control of such incidents would impose on civil order." He argued that first responders needed more training to cope with attacks by extremists. In 2004, Josh, in the Rhoda Goldman Health Lecture to the University of California, Berkeley, Goldman School of Public Health, observed, "We are at a greater risk today in a globalized world than we were in 1918" and cautioned that "the U.S. should be concerned with such epidemics not only because it would be a 'betrayal of humanity' if we did not, but we should also be concerned for our own safety and security, if such 'diseases are so rampant.'" His other, great passions were the threat of bioterrorism and disarmament of all weapons of mass destruction—nuclear, chemical, biologic.

Josh was apprehensive that the goals of political extremists are to interrupt regular normal life and influence policy. Josh broadened his interest in computer modeling of scientific reasoning and bioinformatics. This brought him into close collaboration with colleagues at the National Library of Medicine, which was cataloging his extensive archives. Josh contributed to the terminology of science, inventing exobiology, euphenics, prototroph, and microbiome, the last to signify the ecological community of commensals, symbiotic, and pathogenic microorganisms that literally share our body space and have been all but ignored as determinants of health and disease (Lederberg and McCray 2001).

Josh's early fascination with information science led him to propose by analogy to binary units (bits) the terms "mits" for mutational units or genes, "rits" for recombinational units or genes, and "phits" for physiological units or genes. He also proposed the term "eugram" for electron mail. Josh anticipated that there would be controversy about the applications of genetic technology in medicine and introduced the term "euphenics" to contrast with eugenics, which had taken on a negative connotation. He was proposing interventions that restored the health phenotype, as distinct from interventions that altered the genotype. Joshua Lederberg not only changed the course of life science but also changed the lives of his associates. Josh did not fear to go where no one had gone before. He had the ability to see the answer to a problem, buried in a vast literature. His personal experimentation was direct and simple, but he readily entered into collaborations calling for sophisticated technology. Josh understood the theoretical and applied implications of his discoveries, and his vision extended far into the future.

Above all, Josh was a man of passion for science and was driven to improve the lot of humankind.

PUBLIC POLICY

Lederberg began his long career as a government advisor in 1957, when he joined the President's Science Advisory Committee, a panel made up of several of the nation's leading scientists who provide advice and analysis to the president and the federal government on a wide range of scientific and technological matters, in particular those related to nuclear weapons and national security. He served on President John F. Kennedy's White House transition team as a consultant on health care policy and on the President's Panel on Mental Retardation from 1961 to 1962. In this role, Lederberg forged a close professional relationship with President Kennedy's sister, Eunice Shriver, and her husband, Sargent Shriver, who shared his interest in the neurological and genetic causes of mental illness. They decided to establish the Joseph P. Kennedy Jr. Laboratories for Molecular Medicine, a facility that focused on the study of mental retardation, at Stanford University Medical School, Lederberg's home institution.

Throughout his life, Lederberg retained his interest in medicine and human welfare, and it is in this area in which his scientific and public policy interests combine most closely. He continued to build connections between basic science, medicine, and health care, both nationally and internationally, as a member of the National Mental Health Advisory Council from 1967 to 1971 and later as chairman of the President's Cancer Panel from 1979 to 1981, under President Jimmy Carter. He joined the Advisory Committee for Medical Research of the World Health Organization during the 1970s and again during the 1990s.

He continued to advise the government on scientific issues during the 1980s and 1990s. He chaired Congress's Technology Assessment Advisory Council between 1988 and 1995. Between 1988 and 1993, he was cochairman of the Carnegie Corporation's Commission on Science, Technology, and Government (unofficially entitled "The Commission on Everything" by its members because of its broad mandate), which examined the ways in which science informs governmental decision-making in the United States.

Among his engagements in nonprofit organizations, Lederberg served on the Board of Trustees of the Natural Resources Defense Council from 1972 to 1984. He resigned his position in disagreement over the Council's growing emphasis on litigation, rather than lobbying and public education, as its primary strategy for protecting the environment.

Committed more fully than most notable scientists to bringing developments in science and their social implications to the attention of the public and of Capitol Hill, Lederberg wrote a weekly editorial column entitled "Science and Man" for *The Washington Post* between 1966 and 1971. In it, he commented on the state of science education and reporting, on scientific freedom and the role of scientists in society, and on the ethically challenging scientific issues of the day, from population control and intelligence testing to the regulation of recombinant DNA technology.

One of the issues Lederberg repeatedly addressed in his column was the role modern genetics should play in enhancing the health and biological fitness not only of individuals but also of human populations. Advances in molecular genetics during the 1960s, in particular the deciphering of the genetic code, promised a better understanding of how genes regulate human development and impart desirable mental and physical traits. At the same time, modern genetics raised the specter of new forms of discrimination against people considered to be at heightened genetic risk for disease or mental abnormality.

Lederberg distinguished himself in this discussion by contrasting *eugenics*—the controversial and often coercive practice of fostering desired genetic traits within a population by encouraging selective mating and controlling reproduction by those deemed genetically inferior—with "euphenics," a term he coined to denote a full range of modern medical technologies and health policies designed to promote individual human development to its fullest genetic potential. (In proposing this term, Lederberg drew an analogy to the distinction between *genotype*, an organism's genetic constitution, and *phenotype*, its physical and physiological make-up when fully developed.) Among the measures he espoused were fine-detail mapping of the human genome and location of genes that predispose an individual towards disease, genetic counseling of prospective parents, prenatal diagnosis, genetic modification and therapy, and the assessment of the effects of environmental hazards such as chemical pollutants and radioactive radiation on the human genome.

Lederberg has been most deeply involved in public policy debates regarding national security and national defense, disarmament, and preparedness. To a geneticist, bacteriologist, and virologist, the dangers to human health and welfare posed by biological warfare, in particular, and later by bioterrorism were obvious and had to be prevented. Although biological warfare agents were considered by some to be similar to nuclear weapons, Lederberg pointed out in his editorials and other places that biological weapons were in fact uncontrollable, unpredictable, and indiscriminate in their effect on soldiers and civilians alike. Unlike nuclear weapons, he noted, microbial agents could not be tested without directly injuring humans or animals. His work on bacterial genetics led him to see that once introduced into the environment, self-replicating biological weapons agents threatened the entire human population, especially if they consisted of bacteria or viruses engineered to be more virulent or drug resistant. Biological weapons research itself was irresponsible, Lederberg argued, as disease organisms could escape from the laboratory and imperil the public health even in times of peace. Moreover, such research perverted the aims of medical science by using scientific knowledge to infect and to kill, rather than to promote human well-being.

To help contain the threat of biological warfare, Lederberg became a consultant to the U.S. Arms Control and Disarmament Agency during negotiations in Geneva over the international Biological Weapons Convention, signed by the Soviet Union, the United States, and Great Britain in April 1972. As a member of the National Academy of Sciences' Committee on International Security and Arms Control since its founding in 1980, he met regularly with representatives of the Academy of Sciences of the U.S.S.R. between 1985, when President Mikhail Gorbachev introduced reforms of the Soviet government, and the mid-1990s. The two groups conferred in an attempt to uncover the extent of the Soviet biological weapons program, to help secure the agents it had produced, and to facilitate the transition of biological weapons scientists to civilian research. Also during the mid-1990s, Lederberg took up the growing threat of terrorist attacks with biological weapons, urging policy-makers to act against the proliferation of such weapons and to enhance preparedness by strengthening the nation's public health capabilities.

In the broader area of defense policy, Lederberg has served on the Defense Science Board for over two decades, beginning in 1979. The Board advises the Secretary of Defense on developments in science and technology that affect military weaponry and strategy, military manpower policy, and arms control. In this capacity, he chaired a task force that assessed the medical evidence and policy dimensions of Gulf War Syndrome. It concluded that current medical science was unable to identify a pattern of symptoms that would indicate a specific new disease among veterans of the Gulf War in 1991.

Through his work in bacterial genetics, exobiology, and biological warfare, Lederberg has gathered insights into infectious diseases and other bacterial and viral health threats for many decades. His experience has led him most recently to decry complacency in the never-ending contest between man and microbe. In speeches and publications, he has warned of the likelihood—in fact, the certainty—that old infectious diseases, scourges of mankind like epidemic influenza and tuberculosis, will reemerge as humans lose their immunity and strains become drug resistant and that new ones will emerge as pathogens continue to evolve. Firm in his lifelong belief that scientific knowledge and enlightened public policies can overcome such challenges and ameliorate the human condition, he has called upon scientists and policymakers alike to further research on the molecular biology of infectious agents as well as on vaccines and drugs and to build a public health system able to contain a future outbreak.

DEATH

At the time of his death, Joshua Lederberg was university professor and president emeritus of Rockefeller University. He died of pneumonia on February 2, 2008, at New York-Presbyterian Hospital. Surviving him are his wife, Marguerite Stein Kirsch Lederberg; his children Anne Lederberg and David Kirsch; two grandchildren; and his younger brothers Seymour and Dov. His first wife Esther Zimmer Lederberg predeceased him, dying in December 2006.

BRIEF CHRONOLOGY

- 1925—Joshua Lederberg born May 23 in Montclair, New Jersey, to Zvi Hirsch Lederberg, a rabbi, and Esther Goldenbaum Lederberg, a homemaker
- 1938–1941—Attended Stuyvesant High School, a selective science and technology school in Manhattan
- 1941–1944—Undergraduate studies at Columbia University, leading to a BA in zoology. Examined genetics of *Neurospora* (a common bread mold) with Professor Francis J. Ryan
- 1943–1945—Military service in the U.S. Naval Reserve's V-12 program, a compressed premedical and medical curriculum, at St. Albans Naval Hospital, Long Island

- 1944–1946—Medical Student at Columbia University College of Physicians and Surgeons and research assistant in Professor Ryan's zoology laboratory
- 1946–1947—Research Fellow at Yale University with Professor Edward L. Tatum. Discovered mating and genetic recombination in the bacterium *E. coli*, making *E. coli* available as an experimental organism for genetic research. Received his PhD from Yale with a thesis on his discovery
- 1947–1959—Professor of genetics at the University of Wisconsin. Conducted research in the genetics of *E. coli* and *Salmonella* as well as on antibody formation. Discovered and named plasmids, particles of DNA in bacterial cells that replicate separately from chromosomal DNA
- 1950–1998—Member of various panels of the President's Science Advisory Committee
- 1951—Discovered, with Norton Zinder, the exchange of genetic material in bacteria through viral vectors, a process he called transduction. Their discovery has important applications in bacterial genetics and biotechnology
- 1957—Elected to the National Academy of Sciences
- 1957–1959—Founder and chairman of the Department of Medical Genetics at the University of Wisconsin
- 1958—Shared Nobel Prize in Physiology or Medicine with Tatum and George W. Beadle "for his discoveries concerning genetic recombination and the organization of the genetic material of bacteria"
- 1958–1977—Investigated the possibility of life on other planets and of interplanetary contamination as a member of several National Academy of Sciences and NASA committees on space biology, and as organizer of the Instrumentation Research Laboratory at Stanford
- 1959–1978—Founder and chairman of the Department of Genetics, Stanford University School of Medicine. Began research in the genetics of *Bacillus subtilis* (1959) and in splicing and recombining DNA (1969)
- 1961–1962—Member of President John F. Kennedy's Panel on Mental Retardation
- 1964—Together with computer scientist Edward A. Feigenbaum Lederberg launched DENDRAL, a computer program designed to emulate inductive reasoning in chemistry and medicine through Artificial Intelligence
- 1966–1971—Published "Science and Man," a weekly column on science, society, and public policy in *The Washington Post*
- 1969–1972—Consultant to the U.S. Arms Control and Disarmament Agency during negotiations for the Biological Weapons Convention in Geneva
- 1973–1978—Helped establish SUMEX-AIM, a nationwide time-share computer network hosting biomedical research projects
- 1976—U.S. Viking I and Viking II spacecraft explored Mars with the help of instruments for soil analysis designed by Lederberg and his associates at the Instrumentation Research Laboratory. The spacecraft find no clear signs of life
- 1978–1990—President of Rockefeller University in New York City, a graduate university specializing in biomedical research
- 1979–1981—Advisor to President Jimmy Carter on cancer research as chairman of the President's Cancer Panel

- 1979–2008—Trustee of the Sackler Medical School, Tel-Aviv University, Israel, the Carnegie Corporation, New York, and other academic, research, and environmental institutions. Member of the U.S. Defense Science Board, which advises the Secretary of Defense on scientific developments affecting the military and national security
- 1989—Awarded the National Medal of Science by President George H.W. Bush
- 1990–2008—Professor emeritus and Raymond and Beverly Sackler Foundation Scholar at Rockefeller University
- 1994—Headed the Defense Department Task Force on Persian Gulf War Health Effects, which concluded that there is insufficient epidemiological evidence for a coherent Gulf War "syndrome"
- 2005—Continued to conduct laboratory research on bacterial and human genetics, and to advise government and industry on global health policy, biological warfare, and the threat of bioterrorism
- 2006—Awarded the Presidential Medal of Freedom by President George W. Bush
- 2008—Died at New York-Presbyterian Hospital in New York, February 2

REFERENCES

Anonymous (2005) Interview with Professor James Crow. *BioEssays, 28(6)*: 660–678.

Avery O.T., C.M. MacLeod, and M. McCarty (1944) Studies on the chemical transformation of Pneumococcal types. *J. Exp. Med., 79*: 137–158.

Davis, B.D. (1987) This week's citation classic. *Curr. Contents, 30(33)*: 17.

Dronamraju, K.R. (ed.) (1995) *Haldane's Daedalus Revisited*. Oxford: Oxford University Press.

Dronamraju, K.R. (2017) *Popularizing Science: The Life and Work of JBS Haldane*. Oxford: Oxford University Press.

Garfield, E. (2008) Saying goodbye. *Scientist, 22(4)*: 31–32.

Halvorson, H.O. (2007) Development of molecular biology at the University of Wisconsin, Madison. *Biol. Cell, 99*: 717–724.

Herzenberg, L.A. (2006) The more we learn. The 2006 Kyoto Prize Commemorative Lectures: Advanced Technology. Inamori foun-namori Foundation. http://www.inamori-f.or.jp/laureates/k22_a_leonard/img/lct_e.pdf. Accessed April 2009.

Lederberg, J. (1987) Sputnik+ 30. *J. Genet., 66*: 217–220.

Lederberg, J. (1992) Bacterial variation since Pasteur. Rummaging in the attic: Antiquarian ideas of transmissible heredity, 1880–1940. *ASM News, 58*: 261–265.

Lederberg, J. (1997) Some early stirrings (1950 ff.) of concern about environmental mutagens. *Environ. Mol. Mutagen., 30*: 3–10.

Lederberg, J., and R.E. Wright (1958) Extranuclear transmission in yeast heterokaryons. *Proc. Natl. Acad. Sci. USA, 43*: 919–923.

Lederberg, J., and G.J.V. Nossal (1960) Antibody production by single cells. *Nature, 181*: 1419–1420.

Lederberg, J., and A.T. McCray (2001) 'Ome sweet 'omics—a genealogical treasury of words. *Scientist, 15*: 8.

Ligon, B.L. (1998) Joshua Lederberg, PhD: Nobel Laureate, geneticist, and president emeritus of the Rockefeller University. *Semin. Pediatr. Infect. Dis., 9*: 34–355.

Morse, S.S. (2008) Joshua Lederberg (1925–2008). *Science, 319*: 1351.

Nackerdien, Z.E., A. Keynan, B.L. Bassler, J. Lederberg, and D.S. Thaler (2008) Quorum sensing influences *Vibrio harveyi* growth rates in a manner not fully accounted for by the marker effect of bioluminescence. *PLoS ONE, 3(2)*: e1671.

Ryan, F.J., and J. Lederberg (1947) Reverse-mutation and adaptation in leucineless *Neurospora. Proc. Natl. Acad. Sci., USA, 32*: 163–173.

Shortliffe, E.H., and T.C. Rindfleisch (2000) Presentation of the Morris F. Collen Award to Joshua Lederberg, PhD. *J. Am. Med. Informatics Assoc., 7*: 326–332.

Zinder, N.D., and J. Lederberg (1952) Genetic exchange in *Salmonella. J. Bacteriol., 64*: 679–699.

Zuckerman, H., and J. Lederberg (1986) Postmature scientific discovery? *Nature, 324*: 329–3311.

OTHER PUBLICATIONS

(With F.J. Ryan.) Reverse-mutation and adaptation in leucineless *Neurospora. Proc. Natl. Acad. Sci. USA, 32*: 163–173. 1946.

(With E.L. Tatum and E. Lederberg.) Gene recombination in the bacterium *Escherichia coli. J. Bacteriol., 53*: 673–684. 1947.

(With M.R. Zelle.) Single-cell isolations of diploid heterozygous *Escherichia coli. J. Bacteriol., 61*: 351–355. 1951.

(With E. Lederberg.) Replica plating and indirect selection of bacterial mutants. *J. Bacteriol., 63*: 399–406. 1952.

(With N.D. Zinder.) Genetic exchange in *Salmonella. J. Bacteriol., 64*: 679–699. 1952.

(With L.L. Cavalli and E.M. Lederberg.) Sexual compatibility in *Escherichia coli. Genetics, 37*: 720–730. 1953.

(With E.M. Lederberg.) Genetic studies in lysogenicity in *Escherichia coli. Genetics, 38*: 51–64. 1953.

Bacterial protoplasts induced by penicillin. *Proc. Natl. Acad. Sci. USA, 42*: 574–577. 1956.

(With M.L. Morse and M. Lederberg.) Transduction in ransduction in *Escherichia coli* K-12. *Genetics, 41*: 142–156. 1957.

(With R.E. Wright.) Extranuclear transmission in yeast heterokaryons. *Proc. Natl. Acad. Sci. USA, 43*: 919–923. 1958.

(With G.J.V. Nossal.) Antibody production by single cells. *Nature, 181*: 1419–1420. 1960.

Exobiology: approaches to life beyond the earth. *Science, 132*: 393–400. 1960.

(With E.W. Nester.) Linkage of genetic units of Bacillus subtilis in DNA transformation. *Proc. Natl. Acad. Sci. USA, 47*: 52–57. 1961.

(With C. Sagan and E.C. Levinthal.) Contamination of Mars. *Science, 159*: 1191–1196. 1967.

(With G.L. Sutherland, B.G. Buchanan, E.A. Feigenbaum, A.V. Robertson, A.M. Duffield, and C. Djerassi.) Application of artificial intelligence for chemical inference. I. The number of possible organic compounds. Acyclic structures containing C, H, O, and N. *J. Am. Chem. Soc., 91*: 2973–2976. 1969.

(With C. Sagan, J. Veverka, P. Fox, R. Dubisch, R. French, P. Gierasch, L. Quam, E. Levinthal, R. Tucker, and B. Eross.) Variable features on variable features on Mars. 2. Mariner 9 global results. *J. Geophys. Res., 78*: 4163–4196. 1973.

(With R.G. Dromey, B.G. Buchanan, D.H. Smith, and C. Djerassi.) Application of artificial intelligence for chemic inference. XIV. A general method for predicting molecular ions in mass spectra. *J. Org. Chem., 40*: 770–774. 1975.

Digital communications and the conduct of science: the new literacy. *Proc. IEEE, 66*: 1314–1319. 1978.

(With R.K. Lindsay, B.G. Buchanan, and E.A. Feigenbaum.) *Applications of Artificial Intelligence for Organic Chemistry: The Dendral Project*. New York: McGraw-Hill. 1980.

(With R.E. Shope and S.C. Oaks Jr., eds.) *Emerging Infections: Microbial Threats to Health in the United States*. Washington, DC. 1992.

Section IV

Molecular Biology

In 1944, Avery et al. published their paper "Studies on the Chemical Nature of the Substance Inducing Transformation of Pneumococcal Types: Induction of Transformation by a Desoxyribonucleic Acid Fraction Isolated from Pneumococcus Type III," suggesting that DNA, rather than protein as widely believed at the time, may be the hereditary material of bacteria and could be analogous to genes and/or viruses in higher organisms. In 1952, Hershey and Chase used radioactive isotopes to show that it was primarily DNA, rather than protein, that entered bacteria upon infection with bacteriophage; it was soon widely accepted that DNA was the material. This was closely followed by Watson and Crick's discovery of the molecular structure of DNA, in 1953, which proved to be a very important landmark in the history of biology. Simultaneously, the phage school, founded by Max Delbruck (in collaboration with Salvadore Luria and Alfred Hershey), launched the molecular biology research program in the late 1930s. They shared the 1969 Nobel Prize in Physiology or Medicine for their discoveries concerning the replication mechanism and the genetic structure of viruses.

Quite independently, physicist Erwin Schrödinger argued in his book *What Is Life?* (1944) that the genetic material had to have a nonrepetitive molecular structure. He claimed that this structure flowed from the fact that the hereditary molecule must contain a "code-script" that determined the entire pattern of the individual's future development and of its functioning in the mature state. Schrödinger's book had a profound impact on those interested in the molecular structure of DNA, especially James Watson and Francis Crick.

12 Oswald Theodore Avery (1877–1955)

Oswald Thodore Avery (1877–1955) belonged to that small group of scientists who were primarily neither geneticists nor biologists of any kind but nevertheless contributed significantly to the advance of molecular genetics at a crucial moment in its progress. Indirectly, he inspired several early leaders of molecular biology. A small group of physicists and chemists, which includes Niels Bohr, Erwin Schrödinger, Oswald Avery, Linus Pauling, Max Delbruck, Max Perutz, and Francis Crick, among others, have contributed important ideas and methods, which have laid the foundation of certain branches of genetics. Avery's discovery of the chemical nature of the transforming process in *Pneumococcus* opened the gateway to the elucidation of the nature of the gene, which led to the important discoveries of Watson and Crick and others in the following years. In this sense, Avery's contribution to genetics differed from those of Muller, Haldane, Fisher, Wright, Beadle, and others who spent all their lives in genetic research. Many of these were identified as biologists during their lifetime. In retrospect, it can be clearly seen that the remarkable success of the science of genetics has been due to the contributions of both kinds of scientists.

Avery was born on October 21, 1877, in Halifax, Nova Scotia. However, when he was 10 years old, his father—a clergyman—moved to New York City, where he was engaged in the missionary work in the Bowery. Almost all of Avery's life was spent in New York. He became a U.S. citizen in 1917. At first glance, it appears that there was nothing remarkable in Avery's education and early life. He attended Colgate University, receiving a BA degree in 1900. Pursuing medical education at the Columbia University's College of Physicians and Surgeons, Avery received the MD in 1904. His early work was in bacteriology, which he pursued at the Hoagland Laboratory in Brooklyn until 1913, when he joined the Rockefeller Institute for Medical Research (now The Rockefeller University). He became Member Emeritus in 1943 but continued his research at the Institute for another five years, before retiring finally in 1948 to Nashville, Tennessee, to join his brother, Roy Avery. He died of cancer of the liver on February 20, 1955, in Nashville, Tennessee.

EARLY LIFE

Avery seldom spoke of his childhood, but he was often eloquent about his years at the Colgate Academy and Colgate University, which were located in Hamilton, New York. Both were founded in 1819 by the Baptist Education Society of the State of New York. The intellectual climate of the Academy appears to have been extremely liberal at the time of Avery's attendance, and it may have nurtured his intellectual independence and free spirit of inquiry. It is certainly of interest that a group of six students, including Avery, approached a young professor to organize for them a special course in metaphysics to examine the credibility of the Christian faith.

At Colgate, Avery scored an average of 8.5 (out of 10) for the freshman year and exceeded 9 for the remaining three years. He majored in the humanities, taking courses in modern philosophy, ethics, history, English literature, history of art, economics, political economy, and public speaking (interestingly, with debate and oration added). Strangely enough, he displayed no keen interest in science, taking only those courses that were compulsory while ignoring many elective courses in science that were offered at Colgate.

Avery graduated in 1900 and immediately entered the College of Physicians and Surgeons of Columbia University in New York. Avery's choice of medical education seems to have been rather unexpected. His mother encouraged him to enter the ministry. As mentioned earlier, he preferred to study humanities at Colgate with some emphasis on public speaking. Science was virtually ignored. Rene Dubos suggested that Oswald may have rebelled against "the kind of bibliolatry and theology" he had been taught. Furthermore, medicine fitted well into the mood of the Baptist community at that time and was also a fitting subject for Mr. Rockefeller's philanthropic interests. However, it is surprising that he found his brief foray into practice of medicine quite frustrating; otherwise, he could have served the public, especially the poor. He chose instead a research career that provided him a rather closed private life, involving no contact with the public.

HOAGLAND LABORATORY

Interestingly, Avery's course grades were good, except in bacteriology and pathology! There is no evidence to indicate that Avery derived much satisfaction from his medical education. He seldom spoke of his medical school years. His real interests were in research and laboratory work. In 1907, when he came into contact with Benjamin White, Director of the Hoagland Laboratory in Brooklyn, Avery expressed his interest in research and was hired immediately as the associate director of the laboratory. The six years spent at the Hoagland Laboratory were of crucial importance in the maturity and growth of Avery as a medical scientist.

Benjamin White trained him well in laboratory techniques and chemical methods. Avery gained valuable experience both in research and teaching in a variety of related disciplines. All bacterial cultures were subjected to a careful scrutiny, which stood him well throughout his scientific career. During his years at the Hoagland Laboratory, Avery received a very broad practical training in various fields of bacteriology and immunology. Rene Dubos characterized those years of Avery's life as a "self-training period." Much of his work during that period was scholarly but not of a strikingly original nature. For some time, he conducted systematic clinical research on tuberculous patients, especially a study of secondary infections in pulmonary tuberculosis, while "vacationing" at the Trudeau Sanatorium in Saranac Lake in the Adirondacks. It was a fortunate development in Avery's life because it caught the attention of Dr. Rufus Cole, Director of the Rockefeller Institute Hospital, eventually leading Avery to the scientific climate that matched his talents in a most fitting manner. Among other activities, Avery worked on the bacteriology of postsurgical infections, acted as a consultant to Brooklyn area clinicians on their bacteriological problems ("teaching by conversation"), and ran a course for student nurses.

THE ROCKEFELLER INSTITUTE

At the invitation of Dr. Cole, Avery joined the Rockefeller Institute in 1913. He was, in part, attracted by the opportunity to participate in the Institute's research program on the lobar pneumonia, a disease that killed his mother in 1910. He collaborated with A. Dochez, who was investigating the types of pneumococci present in pneumonic patients and in carriers. A significant outcome of this program was the discovery of specific soluble substances of pneumococcus origin in the blood and urine during lobar pneumonia.

Before discussing the impact of Avery's experiments on genetics, it is desirable to understand the range and quality of Avery's scientific activities that led up to his most important scientific work. A number of discoveries were related to the role played by capsular polysaccharides in determining the immunological specificity and virulence of pneumococci.

a. Encapsulated bacteria become avirulent when they lose the capacity to produce an ectoplasmic layer.
b. The pneumococci are protected against the defense mechanisms of the host body (e.g., phagocytosis by the cells of the blood and tissues), by the capsule.
c. Encapsulated pneumococci differ chemically in the composition of their capsules; the capsule is made up of a polysaccharide but each has immunologic specificity for each pneumococcal type.
d. The antiphagocytic property of the capsules is neutralized by the antibodies produced in the blood serum against the capsular polysaccharides. The protective mechanism is specific for each pneumococcal type.
e. Specificity can be precisely characterized in terms of molecular chemistry, for polysaccharides as well as other types of chemical substances.

GENETICS

Fred Griffith, a medical officer of the British Ministry of Health, discovered the S and R forms of pneumococci in 1923. Furthermore, he showed that when large numbers of avirulent R cells are injected into mice, it is possible to find, in the heart blood of these animals, S cells that are fully virulent. These S cells possess a capsular polysaccharide with the same immunological type as the S cells from which the R cells were initially derived. Griffith developed techniques that made it possible to transform one colonial form of pneumococci into another, both *in vivo* as well as *in vitro*. He concluded that the transformation process involved a reversible mutational change. Later, in 1932, Griffith clearly demonstrated that pneumococci can be made to change types, a discovery which Dubos described as follows: "It…provides an illustration of the fact that, while progress in science often depends on interesting accidents, these can be recognized and exploited only by investigators endowed with theoretical knowledge and experimental skills." In Pasteur's words, "chance favors only the prepared mind."

However, Griffith misinterpreted his findings. He believed that all pneumococci represented different breeds of a single species, providing a Lamarckian explanation of the phenomenon. At first, Avery found it difficult to believe that pneumococci could be made to change their immunological specificity. He assigned M.H. Dawson to investigate the conditions most favorable for the occurrence of transformation. Dawson prepared "pure line strains," that is to say, colonies derived from single R cells, and subcultured these clones in several different media. He concluded that the great majority of avirulent R cells have the ability to revert to virulence by the presence in the culture medium of growth-promoting factors and of serum-containing antibodies to pneumococcus proteins. Dawson repeated Griffith's experiments, confirming the occurrence of transformation, both *in vivo* and *in vitro*.

In 1930, when Dawson moved to the College of Physicians and Surgeons, Avery assigned another young investigator, J.L. Alloway, to continue the study of the phenomenon *in vitro*. Alloway found that transformation can be brought on by whole, killed S cells, as well as with a soluble fraction prepared from S pneumococci and that the active material could be obtained as "a thick syrupy precipitate" by precipitating with alcohol. According to Dubos, "Alloway was thus the first person to handle the active, fibrous substance that was to be identified as DNA 10 years later." Alloway's work is important historically because it produced the first soluble preparations of the transforming substance. Then in 1932, Alloway, like Dawson, left Avery's group for "greener" pastures. Now, Avery himself started devoting some of his attention to the transformation phenomenon. His goal was to study, as much as possible, the properties of the transforming substance, preparing it for the quantitative assay of its activity. His first task was to isolate and purify it and identify its chemical nature. He was often frustrated, which led him to remark: "Disappointment is my daily bread; but I thrive on it." Later, he recalled: "Many are the times we were ready to throw the whole thing out of the window."

But Avery persisted, and his stubbornness paid off.

The next significant development was the addition of Colin MacLeod to the staff, who joined the department in 1935. MacLeod first killed the organisms by heat and then treated them with sodium deoxycholate (instead of dissolving the living pneumococci in deoxycholate, which Alloway had followed). MacLeod's method yielded a more active and stable substance because it avoided being inactivated by the pneumococcal enzymes. By selecting a strain that gave consistent results and by modifying the culture medium, MacLeod successfully developed a dependable assay dilution technique. In 1936, in a conversation with Hotchkiss, Avery speculated that the transforming agent could neither be a carbohydrate nor a protein, but it might possibly be a nucleic acid (but there was no mention of deoxyribonucleic acid [DNA]). Some work on ribonucleic acid was in progress in the laboratory at that time, and Avery's reference to nucleic acid seemed to have no special meaning at that time.

As late as 1940–1941, in the Annual Report to the Board of scientific Directors of the institute, there was no mention of DNA.

MacLeod, too, departed in 1941, and Avery appointed Maclyn McCarty, a young pediatrician and biochemist, to replace him. But MacLeod, who joined the New York University School of Medicine, frequently returned to Avery's department and continued to collaborate with the team. The immediate result of their collaboration was the discovery that the transforming activity was centered in a highly viscous fraction which consisted almost entirely of polymerized DNA.

AVERY, MACLEOD, AND MCCARTY

Avery's originality did not lay in the development of new laboratory methods but in the intellectual style which he adopted for solving specific problems. Dubos stated that the successful collaboration of Avery, MacLeod, and McCarty was due to their skill in using a great diversity of methods ranging from the most physical to the most biological.

Avery et al. (1944) stated that their paper was concerned with a more detailed analysis of the phenomenon of transformation of specific types of pneumococcus. They wrote: "The major interest has centered in attempts to isolate the active principle from crude bacterial extracts and to identify if possible its chemical nature or at least to characterize it sufficiently to place it in a general group of known chemical substances. For purposes of study, the typical example of transformation chosen… represents the transformation of a non-encapsulated R variant of Pneumococcus Type II to Pneumococcus Type III."

The successive sections of the paper dealt with the preparation of the nutrient bath, selection and testing of suitable sera or serous fluids, selection of a suitable R strain (R36A), and extraction, purification, and identification of the chemical nature of the transforming principle.

The pneumococci (type III) were heated at 65°C to destroy the pneumococcal enzyme that was known to inactivate the transforming material. Later, McCarty found that this enzyme requires magnesium, and it is ineffective in the presence of citrate, which binds this metal. When citrate is added to the medium, it resulted in much larger yields of the transforming substance. The heated cells were repeatedly washed with salt solution to remove proteins, polysaccharides, and ribonucleic acid. To extract the active material, the washed cells were then shaken with 0.5 deoxycholate. It was then separated from the deoxycholate solution by precipitation with alcohol. Solutions of the active material were then treated with a series of enzymes, which can hydrolyze proteins, ribonucleic acid, and the type III capsular polysaccharide, and finally shaken with chloroform to remove the last traces of proteins.

Finally, the extract was repeatedly precipitated by the dropwise addition of one volume of absolute ethyl alcohol, while stirring constantly. The active material then separated into long, white, extremely fine fibrous strands. The preparation consisted of highly polymerized DNA, and it was able to bring about the transformation of R pneumococci into encapsulated type III pneumococci, even in a dilution of less than 1 in 100,000,000. The R pneumococci that underwent transformation retained the newly acquired immunological specificity without the need to add any further transforming material.

AVERY'S PERSONALITY AND RESEARCH METHODS

Avery was a man of few words, both in writing and in speech. His scientific papers were characterized by brevity, precision, and intellectual self-discipline. In their 1944 paper, Avery et al. expressed the significance of their discovery in a restrained manner, but Avery was fully aware of its importance. He found the endless speculations of various biologists and others, who were freely expressing their opinion of the nature of the transforming principle, to be irritating and tedious.

In his annual report to the Board of Scientific Directors for 1946–1947, Avery wrote: "Various interpretations have been advanced as the nature of this phenomenon. However, those of us actively engaged in the work have for the most part left matters of interpretation to others and have chosen rather to devote our time and thought to experimental analysis of the factors involved in the reaction. This is not to say that we are indifferent and have not among ourselves indulged in speculation and discussion of the relation of the problem to other similar phenomena in related fields of biology." According to his pupil, Rene Dubos, "His intellectual puritanism won him the admiration of those who were in direct contact with him, but it prevented him from gaining full recognition of his achievements by the outside world."

PRIVATE THOUGHTS OF AVERY

Avery's private thoughts and speculations regarding the nature of the transforming principle were revealed in a letter to his brother Roy, dated May 26, 1943: "Sounds like a virus—may be a gene. But with mechanisms I am not now concerned—one step at a time—and the first is, what is the chemical nature of the transforming principle? Someone else can work out the rest."

Avery went on to express his inner, more speculative, ideas, which he would not utter in a more public setting. He continued: "Of course, the problem bristles with implications. It touches the biochemistry of the thymus type of nucleic acids which are known to constitute the major part of the chromosomes but have been thought to be alike regardless of origin and species. It touches genetics, enzyme chemistry, cell metabolism, and carbohydrate synthesis, etc. Today it takes a lot of well-documented evidence to convince anyone that the sodium salt of desoxyribose nucleic acid, protein-free, could possibly be endowed with such biologically active and specific properties and this evidence we are now trying to get. It's lots of fun to blow bubbles—but it's wiser to prick them yourself before someone else tries to. So there's the story Roy—right or wrong it's been good fun and lots of work—Talk it over with Goodpasture but don't shout it around—until we're quite sure or at least as sure as present method permits. It's hazardous to go off half cocked—and embarrassing to have to retreat later."

According to his associate, Rene Dubos, Avery's public reticence in discussing the broader implications of his discovery tended to diminish his impact as a scientist and limited his influence on his fellow scientists. It also hindered a much broader understanding and appreciation of the importance of the discovery of Avery and his colleagues.

Avery himself keenly followed the developments in genetics and also the implications of his own discovery within the broader context of genetics. His associate, Rollin Hotchkiss, later wrote: "Although Avery's transformation papers include only the barest mention of genes and viruses, I can testify that he was well aware of the implications of DNA transforming agents for genetics and infection."

OTHER DEVELOPMENTS IN GENETICS

It is an unfortunate coincidence that Avery's greatest work was overshadowed by the more glamorous and exciting developments in phage genetics. There was a fashionable trend at that time for physicists to enter biology and bring new methods and insights to attack problems in biology.

In the 1940s, two brilliant physicists, Erwin Schrödinger and Max Delbruck, introduced ideas and methods that have had a lasting impact on the development of genetics. In 1946, the Nobel Prize for H.J. Muller (a biologist) for his discovery of the radiation-induced genetic mutations ignited much debate about the nature of the gene and gene action. This magnified the dangers of atmospheric radiation, which was already in the public mind because of the explosion of atomic bombs over Japan during the preceding year. Consequently, Avery's approach to the study of biological problems was considered by some as old fashioned. There was a gulf between the geneticists, especially the theoreticians of genetics, and Avery. He made no attempt to convert anyone to his point of view. Indeed, no one outside his small group even knew what his point of view was with respect to the work in progress in his laboratory. He was not given to philosophizing in any fashion. Dubos wrote: "He did not express irritation at criticisms of his work, even when these were unjustified."

REACTION TO THE DISCOVERY

Shortly before the publication of the 1944 paper, geneticist Theodosius Dobzhansky visited Avery's laboratory and told Avery that the so-called transformation phenomenon, in fact, represented the process of mutation that is well known to geneticists, similar to the mutants of *Drosophila*. In his 1941 classic, *Genetics and the Origin of Species*, Dobzhansky wrote: "We are dealing with authentic cases of induction of specific mutations by specific treatments—a feat which geneticists have vainly tried to accomplish in higher organisms."

Biochemist Alfred Mirsky, a colleague of Avery at The Rockefeller Institute, argued that certain proteins are resistant to the proteolytic enzymes used by Avery and his colleagues and that other proteins resist enzymatic action until they have been denatured. Furthermore, Mirsky claimed that the so-called pure nucleic acid could contain significant amounts of protein that cannot be detected by histochemical tests. As late as 1947, Mirsky asserted that experimental evidence was not supportive of the claim that the specific agent in transforming bacterial types is a desoxyribonucleic acid.

Nobel laureate Herman J. Muller stated that Avery's discovery involved a nucleoprotein, not a nucleic acid. In his contribution to the symposium, "Genetics in the 20th Century," which was held in 1950 to celebrate the golden jubilee of genetics, Muller made no reference to Avery's discovery. In a letter to cytogeneticist Darlington in 1946, Muller wrote: "Mirsky gave reasons for believing that Avery's so-called nucleic acid is probably nucleoprotein after all, with the protein too tightly bound to be detected by ordinary methods, and that what he had was free chromosomes, or pieces of chromosomes." In response to the results of phage experiments, reported by Delbruck in 1946, Muller suggested that "to my mind this suggests strongly that in both Delbruck's and Avery's cases what really happens is a kind of crossing over between chromosomes or protochromosomes of the inducer strain and those of the viable strain."

On the other hand, there were many others including some geneticists who recognized the true importance of Avery's discovery immediately. In his Silliman lecture (1948), George Beadle wrote: "…Pneumococcus type transformations, which appear to be guided in specific ways by highly polymerized nucleic acids, may well represent the first success in transmuting genes in predetermined ways."

Additional support came from Paris, where Andre Boivin (with his collaborators, Roger Vendrelyand Yvonne Lehoult) obtained similar results with the colon bacillus *Escherichia coli*. They found evidence for the extensive multiplicity of antigenic types among the colon bacilli, each type possessing its own polysaccharide, characterized by its own unique chemistry and serologic specificity. Each type remains stable through successive cultures and can undergo antigenic degradation changing from smooth form (S) to rough form (R) by losing its polysaccharide. In 1947, Boivin participated in the Cold Spring Harbor Symposium, reading a paper on bacterial transformation and its relationship to Beadle and Tatum's work on *Neurospora*.

In 1943, MacFarlane Burnet visited Avery and regarded the discovery of transformation in *Pneumococcus* as nothing less than the isolation of a pure gene in the form of desoxyribonucleic acid. Many years later, in 1968, he wrote: "the discovery that DNA could transfer genetic information from one pneumococcus to another… heralded the opening of the field of molecular biology…" In 1948, in a Paris symposium, Andre Lwoff expressed similar sentiments.

Lederberg wrote that his own immediate response to reading the Avery et al.'s paper was "unlimited in its implications…. Direct demonstration of the multiplication of transforming factor…Viruses are gene-type compounds… What could be done to incorporate this dramatic finding into the mainstream of biological research; how could one further advance these new hints about the chemistry of the gene? These questions suggested to me the merits of attempting a similar transformation by DNA in *Neurospora*… In mid-Spring 1945, I brought this suggestion to Francis Ryan, who welcomed it… However, we soon discovered that the *Neurospora* mutant *leucineless*…would spontaneously revert to prototrophy. We did not therefore have a reliable assay for the effect of DNA in *Neurospora*."

Nevertheless, the leading figures of mainstream genetics at that time were too slow to recognize the significance of Avery's discovery. They include H.J. Muller and T. Dobzhansky. One reason for this failure was the gulf that existed between geneticists and bacteriologists. If Avery had been a geneticist, there is no doubt that geneticists would have been less reluctant to accept his discovery. Another was the prevailing dogma of the central role of proteins that was advanced by Mirsky and was conveyed to a wider audience by Muller through his Pilgrim Trust Lecture (delivered in 1946; published in 1947). Muller wrote: "…to what extent the given specificity depends on the nucleic acid polymer itself, rather than upon the protein…, must as yet be regarded as an open question."

As early as 1920, J.B.S. Haldane wrote that the chemist may regard genes as large "nucleoprotein" molecules, which possess the capacity for self-reproduction. In 1945, Sewall Wright commented that DNA might be a chromosome fraction with a genetic function.

RECOLLECTIONS OF AVERY

Rene Dubos worked in the laboratory next to Avery's at The Rocklefeller Institute from 1927 to 1942. Dubos recorded for posterity his impressions, which give us some understanding of how Avery functioned and some idea of the single-minded devotion which brought him success in scientific research. Most of the time, Avery worked in his laboratory with its door open or occasionally spent some time in a small office attached to it. It appears that Avery often behaved as if he were alone. During their brief conversations, his close associates addressed him as "Fess" (short for Professor). Avery was a man of slender build, who dressed in a neat and conservative style. He possessed sparkling and prominent eyes, which dominated the bulky dome of his head and frail body. Dubos described the personal encounters of visitors as follows: "He transformed even the most casual conversation into an artistic performance by a panoply of words and gestures that managed to be simultaneously spirited and restrained… The extroverted playfulness of his nature, and his phenomenal empathy for every person or situation that engaged his interest, made any contact with him an intellectually rewarding experience, always entertaining and often enchanting."

Dubos (1976) described that Avery combined an exquisitely gentle-mannered exterior with a tough-minded interior; "he was indeed ruthless with regard to what he elected to do or not to so… He had been endowed by nature with many intellectual gifts, great sensitivity, and an immense skill in dealing with people, and could thus have been successful in many different activities and environments."

Avery's reticence and extreme caution in later life, in reaching conclusions on scientific problems, may have been due to his early experiences. According to Dubos, Avery had been guilty of unwarranted statements during his youth and early adulthood. Dubos wrote: "In college, he had engaged in brash talk on almost any subject. At the Rockefeller Institute, he had published in 1916 and 1917 hasty conclusions that were soon proved to be erroneous… He therefore knew from experience the human propensity to use facts solely for the sake of rhetorical effects and to ignore facts when they stand in the way of one's prejudices."

Another of Avery's associate, Rollin Hotchkiss, worked closely with him from 1936 to 1942. Hotchkiss narrated how Avery used to enact discourses for the benefit of visitors in his quiet office. These were well-rehearsed presentations of pneumococcus research or some other aspect of their work in progress, which came to be known as "Red Seal Records" within their close circle. During those sessions, Avery successively played the roles of a narrator, expositor, loyal opposition, and finally attorney-in-summation. Hotchkiss (1965) wrote: "These gems of perfection were continually revised and repolished. The highly organized presentation was a kind of debate with himself, punctuated with rhetorical questions like, 'now, why should that be?' or 'what does that all mean?' The auditor who was moved to try to respond, however, quickly found himself overwhelmed—and indeed suppressed—by the ongoing flow of well-rehearsed logic, that even in the voice of the man who seemed merely its spokesman, would brook no interference."

One reason for Avery's success as a scientist is his ability to make sharply distinct decisions and to follow the selected course with determination. He seems to have followed this method quite ruthlessly both in his scientific work as well as private life. He remained a life-long bachelor, pursuing his career with a single-minded determination. Early in his life, he considered joining the ministry and studied humanities but switched to the study of medicine, perhaps hoping that medical practice would be a proper substitute for his Baptist background. Disappointed in the quality of medicine, as it was practiced at the turn of the century, Avery once again changed course, choosing a research career that satisfied his intellectual curiosity and a desire to follow experimental medicine. In his later life, he deliberately decided to devote his entire time to laboratory work, avoiding all forms of distractions. In 1944, he was awarded an honorary degree by Cambridge University, and in 1945, the Copley Medal by the Royal Society of London. Any other scientist would have gladly made the effort to travel to England; however, Avery refused to go to England, on both occasions, citing expense and poor health as excuses. However, the real reason was his reluctance to take time away from his laboratory work.

SCIENTIFIC PHILOSOPHY

Avery seldom spoke of his scientific philosophy or method. However, it was evident in his presidential address to the Society of American Bacteriologists, which was never published. Much of the address was traditional and conventional. He followed Pasteur and said: "science, in obeying the law of humanity, will always labor to enlarge the frontiers of knowledge." Avery believed that microbiological studies have illuminated a wide range of biological problems ("Go to the microbe, thou scientist, consider its ways and be wise"). It expressed his scientific philosophy as a biologist, the same philosophy that eventually led Avery to the discovery of the chemical nature of the transforming principle in *Pneumococcus.* He believed that the chemical nature of all life can be best understood through a study of microorganisms. But he himself was reluctant to make sweeping generalizations about the nature of "life," which he knew that his discovery would lead to. He left it to others to indulge in speculations. About certain broader aspects of life, Avery believed that they are outside the domain of science because they cannot be formulated in such a manner as to be put to the test of verification or falsification. He also avoided any discussion about the scientific method, believing that effective scientists intuitively go about their business without bothering to formulate it in philosophical terms.

REFERENCES

Avery, O.T., C.M. MacLeod, and M. McCarty (1944) Studies on the chemical nature of the substance inducing transformation of pneumococcal types—induction of transformation by a desoxyribonucleic acid fraction isolated from *Pneumococcus* type III. *J. Exp. Med. 79(2)*: 137–158.

Dubos, R. (1976) *The Professor, the Institute, and DNA: Oswald T. Avery, His Life and Scientific Achievements*. New York: Rockefeller University Press.

Hotchkiss, R.D. (1965). Oswald T. Avery: 1877–1955. *Genetics 51*: 1–10.

13 Erwin Schrödinger (1887–1961)

Hence this life of yours which you are living is not merely a piece of the entire existence,
but is in a certain sense the whole;
only this whole is not so constituted that it can be surveyed in one single glance.
This, as we know, is what the Brahmins express in that sacred, mystic formula which is yet really so simple and so clear: *Tat tvam asi*, this is you. I am this whole world. (From Erwin Schrödinger's *My View of the World*)

INTRODUCTION

Erwin Schrödinger was a Nobel Prize winning physicist whose little book, *What Is Life?*, was greatly influential in inspiring several pioneers of molecular genetics. Among those who acknowledged their debt to Schrödinger's book are Max Delbruck, James D. Watson, Francis Crick, Maurice F. Wilkins, Seymour Benzer, and many others. On the other hand, there were some other eminent scientists, such as Linus Pauling and Max Perutz, who were critical of the contribution of *What Is Life?* to biology.

Pauling (1987) wrote: "When I first read this book, over 40 years ago, I was disappointed. It was, and still is, my opinion that Schrödinger's book made no contribution to our understanding of life.... Several physicists and biologists with whom I have discussed this question have disagreed with me...Schrödinger's discussion of thermodynamics is vague and superficial to an extent that should not be tolerated even in a popular lecture."

Thermodynamics, to which Pauling referred, was defined by Boltzmann (1886) as nature decaying toward a certain death of random disorder (second law of thermodynamics). This is in contrast with the theory of Darwin, which emphasized gradual evolution through increasing complexity, molded by natural selection. Schrödinger (1944) recognized two fundamental processes in life: (a) order from order and (b) order from disorder. He wrote: "Life seems to be orderly and lawful...based partly on existing order that is kept up" (p. 3).

Max Perutz (1987) from the Cavendish laboratory wrote: "Sadly, however, a close study of his book and of the related literature has shown me that what was true in his book was not original, and most of what was original was known not to be true even when the book was written. Moreover, the book ignores some crucial discoveries that were published before it went into print.... The apparent contradictions between life and the statistical laws of physics can be resolved by invoking a science largely ignored by Schrödinger. That science is chemistry."

However, there is a general agreement among most biologists today that Schrödinger, like Oswald Avery, belonged to that small circle of scientists, whose primary research was not in genetics, nevertheless had a decided impact in initiating the development of molecular genetics. Schrödinger's popular book was written from a physicist's point of view. It was a summary of three public lectures, which were delivered to a general audience at Trinity College, Dublin, in February 1943, to fulfill a statutory obligation to the Institute for Advanced Studies (Dronamraju 1999).

EARLY LIFE AND EDUCATION

Erwin Schrödinger was born on August 12, 1887, in Vienna. His maternal grandmother was English. His close relationship with his father, Rudolf, played a most important role in his growth in childhood and teenage years. Rudolf was in oilcloth business, but his major hobby was in botany, especially plant breeding and evolution. There were frequent scientific discussions between the father and son.

About his father's influence, Schrödinger wrote: "To my father I am thankful for...he gave us a very comfortable life, and assured for me an excellent upbringing and a carefree university education,... an unusually broad culture...and a series of publications on plant phylogenetics. To his growing son, he was a friend, teacher and inexhaustible conversation partner.... My mother was very good, cheerful by nature,... I have her to thank, I believe, for my regard for women" (Moore 1989, p. 17).

"Young Erwin was a rather delicate child who experienced all the childhood diseases and accidents. He was raised almost exclusively in the company of adults. He had never experienced the rough and tumble socialization of the schoolroom and playground. He received the customary education, attending the Gymnasium, whose sole purpose was to graduate a human being whose faculties enable him to form a clear and definite conception of the actual world,... This is a man in the full sense of the word; this is true humanistic culture." The education designed to achieve this noble aim was based on an intensive study of the Greek and Roman classics. Erwin was a good student in all subjects, especially mathematics and physics.

Erwin had to wear increasingly strong glasses from about the age of 12 to correct his nearsightedness. During his free time in the afternoons, he continued to practice English and French, subjects not taught at school. Another subject not included in the curriculum was the Darwinian theory of evolution. He was able to discuss this in depth with his father during their long walks and botanical excursions. On the basis of his botanical studies, his father informed him that the melding of natural selection and the survival of the fittest with the de Vries mutation theory had not yet been made.

One book of great interest was Darwin's *The Origin of Species*, which remained Schrödinger's favorite book all his life. His mother, who came from England, and her sister taught him the English language from an early age. This may have influenced his decision in later years to go to Oxford and Dublin.

As the only child of a well-to-do family, Schrödinger received the best education available. His scholastic performance at the Wiener Akademisches Gymnasium and the University of Vienna was above average in all subjects, but he showed a special aptitude for mathematics. His mathematical professors, Wirtinger and Kohn, the experimental physicist Franz Exner, and especially the theoretical physicist Fritz Hasenohrl all had a deep impact on young Schrödinger's intellectual development. Hasenohrl, who succeeded Ludwig Boltzmann, had such a deep impact on Schrödinger that he regarded both Hasenohrl and his father having had an equally great influence on his intellect. Throughout his life, Schrödinger maintained a deep interest in the relationship between physics and philosophy.

According to Erwin Schrödinger's daughter, Ruth Braunizer, Erwin's father possessed a large library, which his son used at random "practically from the day he was able to read." She added: "One of the few true regrets I heard my father voice later on was the loss of that library, which, in a moment of carelessness, he had decided to sell after his father's death" (Braunizer 1995, p. 177). Because of the broad education that he had received, Erwin Schrödinger dreaded specialization. According to his daughter, Schrödinger strived to be a generalist in every way (Braunizer 1995, p. 176). One of the problems that interested him was to breed hornless cattle. Not being a geneticist himself, he sought the advice of my mentor J.B.S. Haldane. And I approached the population geneticist James F. Crow to comment on their correspondence. The result was an elegant paper by Crow (1992).

Schrödinger introduced an approximation that amounts to assuming Hardy–Weinberg proportions and treated the process as continuous in time. The rate of change of x with respect to generation number n is then dx dn 2[1 + (1 – x)] " (1 – x)2, which integrates to n = 2[(1 – xJ–' – In(1 – x,) – (1 – x,,)–' + In(1 – xo)l. For x0 = 0.5 and x, = 0.95, the time n is 40.6 generations. The exact answer is that, after 40 generations, x40 = 0.9516. Actually, the simple asymptotic approximation 1 – x, 2/n does very well, giving x40 = 0.9500. Schrödinger concludes his letter to Haldane by saying, "Well, I wonder whether I have told you anything new! At any rate it has given me great pleasure to work the problem out-though I know you wanted something better." Now what is the question to which this is the answer? Here is one situation that Schrödinger's equations describe. Suppose a breeder wants to get rid of his dangerous horned cattle. He can't afford not to breed each cow, but he can easily afford to discard some bulls, so each generation he mates only hornless bulls. This leads directly to Equations 1 and 2, in which x is the frequency of the dominant allele for hornlessness and y is the frequency of the hornless phenotype. Haldane had an interest in various selection problems, and selection on only one sex was one of them. He noted that selection at half the intensity in both sexes is very nearly equivalent to full intensity in one. In this case, eliminating half the horned individuals of each sex for 40 generations gives x40 = 0.9510.

While Schrödinger was serving in the First World War in Italy, he learned of the theory of relativity, which was being developed by Einstein. However, the postwar years brought difficult times for the Schrödinger family. His father's business was closed. Money and food were difficult to find. At first, he was offered the post of Associate Professor of Theoretical Physics in Czernowitz, but that city was not a part of Austria after the war. His father died in 1919 and any remaining family money became useless because of a steep increase in inflation. However, a subsequent offer of assistant professorship at Jena (under Max Wien) in 1920 and his marriage turned life around for Schrödinger, launching his life and career on a long distinguished path.

In the following years, Schrödinger held brief teaching posts at Stuttgart and Breslau, but he found his niche at Zurich, when he was offered the chair that was previously held by Einstein during the years 1909–1911. His inaugural lecture was titled "What Is a Law of Nature?" However, his most important work, the theory of wave mechanics, was formulated later in 1925 and was at once acclaimed by the scientific community. He was offered the chair at Berlin as a successor to Max Planck and was invited on a lecture tour of United States during the winter of 1926/1927. There, he found several distinguished colleagues to his liking, including Einstein, Planck, Hertz, and Otto Hahn. In 1935, Schrödinger was granted leave to go to Oxford University as a guest professor and resigned his position at Berlin while at Oxford.

Schrödinger was invited to Oxford by Frederick Lindemann, later Lord Lindemann, who advised Winston Churchill during Second World War, causing the Bengal famine, among other disasters. At first, Schrödinger was a Fellow at Magdalen College (October 1933–September 1936). It was during that period that Schrödinger was awarded the Nobel Prize, which he received on December 10, 1933. During 1934–1935, he visited Princeton University, and later Spain, lecturing in English and Spanish, respectively. In 1936, he was offered professorships in Edinburgh and Graz, choosing the latter in spite of the unstable political situation then prevailing in Austria. Later, in his autobiographical sketch, Schrödinger called it "an extremely foolish thing to do." While at Graz, he simultaneously held the position of "Honorary Professor" at the University of Vienna. In 1938, Schrödinger moved once again (because of the worsening political situation in Austria), first going to Rome and finally to Dublin to head the School of Theoretical Physics, which was a part of the newly established Dublin Institute for Advanced Studies. He described the 16 years that he spent in Ireland as very good years. He enjoyed full freedom to pursue his research and teaching activities, producing a prolific output of books and papers on several important topics, including a unified field theory. It was during that period that Schrödinger wrote *Statistical Thermodynamics*, *What Is Life?*, and *Nature and the Greeks*, among others. He found time to indulge in various literary and artistic activities, for instance, poetry writing and translating Homer from ancient Greek into English and old provencal poetry into German.

Schrödinger's final move in his long career came in 1956, when he accepted a professorship in Vienna. For his inaugural lecture, he chose the topic "The Crisis of the Nuclear Concept." He taught for two and half years, but soon, illness forced him to retire from his chair at the end of September 1958. During his last years, he wrote an autobiographical account, *My Life and My View of the World*. He died on January 4, 1961, after a brief period of illness.

WHAT IS LIFE? THE PHYSICAL ASPECT OF THE LIVING CELL

In 1939, Schrödinger arrived in Dublin, Ireland, to serve as the first director of the school of theoretical physics at the newly established Dublin Institute of Advanced Studies. He was personally invited by the Prime Minister of Ireland Eamon de Valera, who took part in designing the institute. The invitation followed his abrupt dismissal from the Chair of Theoretical Physics at the University of Graz after the *Anschluss*. Dublin suited Schrödinger well. He quickly became a leading intellectual in the life of the city.

Schrödinger had broad intellectual interests. In addition to his broad scientific interests, he also pursued studies of Eastern philosophies and aspects of Hindu religion.

In February 1943, Schrödinger delivered the statutory public lectures at Trinity College, Dublin. He decided to prepare semipopular lectures on biology, especially the fundamental question of the nature of life and the underlying mechanism of heredity.

Schrödinger, characterizing himself as a "naive physicist," wondered whether living systems could be thought of in terms of the physical sciences. Nothing was known about the composition of genes at that time. Schrödinger proposed that a gene could be thought of as an "aperiodic crystal." He also discussed the possibility of a genetic code. The very fact that a distinguished physicist was viewing biology in physical terms had an enormous impact upon multiple disciplines. He inspired numerous scientists, especially the younger ones such as Watson and Crick, consider biological questions in physico-chemical terms.

The chapters of Schrödinger's book *What Is Life?* are as follows:

1. The Classical Physicist's Approach to the Subject
2. The Hereditary Mechanism
3. Mutations
4. The Quantum-Mechanical Evidence
5. Delbruck's Model Discussed and Tested
6. Order, Disorder and Entropy
7. Is Life Based on the Laws of Physics?

At the outset, Schrödinger posed the question: "How can the events *in space and time* which take place within the spatial boundary of a living organism be accounted for by physics and chemistry?" Schrödinger continued: "The preliminary answer which this little book will endeavour to expound and establish can be summarized as follows: The obvious inability of present-day physics and chemistry to account for such events is no reason at all for doubting that they can be accounted for by those sciences."

In *What Is Life?*, Schrödinger focused attention on two topics in biology: (a) the nature of the hereditary material and (b) the thermodynamics of living systems.

In a review of the state of knowledge of genetics at that time, Schrödinger considered a number of topics related to genetic phenomena. These are briefly considered later.

Schrödinger considered two themes based on his view of heredity and thermodynamics. First, he considered what is usually termed the "order from order" theme and discussed how organisms pass on information from one generation to the next. He used the information from the well-known paper by Timofeef-Ressovsky et al. (1935) on mutation damage to fruit flies from which the size of the gene was calculated to be about 1000 atoms. The problem faced by the cell was how a gene of this size could survive thermal disruption and still able to pass on information to succeeding generations. To avoid this problem, Schrödinger proposed that the gene was most likely to be an aperiodic crystal that stored information in the form of a code in its structure, or "codescript" as Schrödinger called it. This prophetic statement has led to the great discovery of the molecular structure of DNA, which has revolutionized biology. The second theme mentioned by Schrödinger was "order from disorder." This problem refers to the ability of the organisms to retain their ordered structure in the face of the second law of thermodynamics. Schrödinger pointed out that organisms retain their internal order by creating disorder in their environment. Both Jim Watson and Francis Crick acknowledged that Schrödinger's book had a decisive influence persuading them to study the molecular structure of the gene (Watson and Crick 1953). But it was puzzling. It is a book written by a physicist who did not know any chemistry. But it suggested that biological problems could be understood in terms of physical sciences and that exciting things in this arena were just around the corner.

Distinguished population geneticist James F. Crow commented that he was not sure what impressed him at that time. He wrote: "Perhaps it was Schrödinger's characterization of the gene as an 'aperiodic crystal'. Perhaps it was his view of the chromosome as a message written in code. Perhaps it was his statement of life 'feeds on negative entropy'. Perhaps it was his notion that quantum indeterminacy at the gene level is converted by cell multiplication into molar determinacy. Perhaps it was his emphasis on the stability of the gene and its ability to perpetuate order. Perhaps it was his faith that the all too obvious difficulties of interpreting life by physical principles need not imply that some super-physical law is required, although some new physical laws might be."

During the decades that followed the publication of *What Is Life?*, the successful impact of Schrödinger's theme "order from order" can be seen in the rapid success of molecular biology. On the other hand, his other theme, "order from disorder," has generally been considered to be of less significance; however, future work on the thermodynamics of living systems may change this point of view. Indeed, future historians may see *What Is Life?* as prophetic for its treatment of the thermodynamics of living systems rather than for the prediction of the structure of the gene and the genetic code.

When Schrödinger prepared his lectures for publication, he added an epilogue: "On Determinism and Freewill." He wrote: "Immediate experiences in themselves, however various and disparate they be, are logically incapable of contradicting each other.… Every conscious mind that has ever said or felt 'I'—am the person, if any, who controls the 'motion of the atoms' according to the Laws of Nature,…

in Christian terminology to say: 'Hence I am God Almighty' sounds both blasphemous and lunatic. But please disregard these connotations for the moment and consider whether the above inference is not the closest a biologist can get to proving God and immortality in one stroke."

Schrödinger then related this idea to the expression in Hindu scriptures, the earliest Upanishads: Atman = Brahman, the personal self is identical with the all-comprehending universal self. Far from being blasphemous, this idea, the essence of Vedanta, is to him the grandest of all thoughts. Much of what he wrote in the epilogue is a repeat of what he wrote 18 years earlier in *Meine Weltansicht*. As his biographer Walter Moore wrote in *Schrödinger: Life and Thought*: "He was a lifelong believer in Vedanta,[1] and he never wavered in this belief. Here, however, he lashed out quite savagely at 'official western creeds', accusing them of 'gross superstition" in their belief in individual souls. Plurality of selves is merely an illusion produced by Maya, like a deception seen in a gallery of mirrors" (Moore 1989, p. 401).

Schrödinger's statement on Western Christian beliefs in the last chapter of the book was seen by the local Irish establishment as derogatory, and publisher Cahill & Co. in Dublin refused to publish his book. However, with some help from a friend in London, it was published by the Cambridge University Press in 1944. This was indeed fortunate for the future of biology and science at large because it was widely read by many scientists, especially younger ones such as Watson and Crick, who were stimulated to pursue their important work on the molecular structure of DNA.

REVIEWS OF *WHAT IS LIFE?*

The Place of *What Is Life?* in Biology

Schrödinger's book received mixed reviews.

Delbruck concluded that mutations are quantum transitions resulting from either random thermal fluctuations or the absorption of radiant energy, spontaneous mutations arising predominantly from thermal fluctuations rather than from natural radiation. Schrödinger, although relying heavily on Delbruck's work, failed to mention the discoveries of H.J. Muller on radiation-induced mutagenesis or the important role of complementariness in the specific attraction between molecules and their enzymatic synthesis, which was already suggested by Haldane (1937) and Pauling and Delbruck (1940).

Negative Entropy

Perutz (1987) pointed out that what was true in Schrödinger's account was already known. Indeed, he was paraphrasing the works of others, especially the article by Timofeef-Ressovsky et al. (1935) on the mutagenic effects of X-rays and γ-rays on the fruit fly *Drosophila melanogaster*. Even Schrödinger's most famous hypothesis, that the gene is like an aperiodic one-dimensional crystal, is a reformulation of Delbruck's suggestion that "the gene is a polymer that arises by the repetition of identical atomic structures." In a chapter entitled "Order, Disorder and Entropy," Schrödinger stated that living organisms do not approach thermodynamic equilibrium (defined as a state of "maximum entropy"), which they achieve by feeding on negative entropy. Schrödinger wrote, "How does the living organism avoid decay? The obvious answer is: By eating, drinking, breathing and (in the case of plants) assimilating. The technical term is metabolism...a living organism continually increases its entropy—or, as you may say, produces positive entropy—and thus tends to approach the dangerous state of maximum entropy, which is death. It can only keep aloof from it, i.e., alive, by continually drawing from its environment negative entropy—which is something very positive.... What an organism feeds upon is negative entropy. Or, to put it less paradoxically, the essential thing in metabolism is that the organism succeeds in freeing itself from all the entropy it cannot help producing while alive." Schrödinger further stated that entropy is not a hazy concept but a measurable physical quantity: "At the absolute zero point of temperature (roughly –273°C) the entropy of any substance is zero."

Perutz and others have argued that we live on free energy and that there was no necessity to postulate negative entropy. Pauling (1987) commented that when Schrödinger was discussing a change in the entropy of the system, he never defined the system. Pauling wrote, "Sometimes he seems to consider that the system is a living organism with no interaction whatever with the environment; sometimes it is a living organism in thermal equilibrium with the environment; and sometimes it is the living organism plus the environment, that is, the universe as a whole." Pauling wrote that Schrödinger failed to recognize the most important question: "How biological specificity is achieved; that is, how the amino-acid residues are ordered into the well-defined sequence characteristic of the specific organism."

Lederberg (personal communication, 1999) suggested that Schrödinger's intended meaning may have been "elements of crystallinity" or "near-crystal" rather than "aperiodic crystal." With hindsight, one can say: "DNA has elements of periodicity (the backbone), and of aperiodicity (the message in the base sequence)." This is Lederberg's surmise based on what Schrödinger may have meant at that time.

Code and Entropy

In Schrödinger's time, the genetic material was generally considered to be a protein, rather than a nucleic acid. It was already known that a protein, especially a particular protein, human hemoglobin, has a well-defined sequence of amino-acid residues in its polypeptide chains. Pauling (1987) traced Schrödinger's argument as follows: "Schrödinger seems to have asked himself the question: 'what is the process that leads to the production of these well-defined polypeptide chains, with their low entropy?' He seems to have answered the question, in a rather vague way, by saying that the organism 'feeds upon negative entropy', attracting, as it were, a stream of negative entropy upon itself. The real question about the nature of life, which Schrödinger failed to recognize, is the question as to how biological specificity is achieved; that is, how the amino-acid residues are ordered into the well-defined sequence characteristic of the specific organism."

However, Pauling concluded that the development of molecular biology has resulted almost entirely from the introduction of the new ideas into chemistry that were stimulated by quantum mechanics. He wrote, "It is accordingly justified...

to say that Schrödinger, by formulating his wave equation, is basically responsible for modern biology."

Is life based on the laws of physics? The final chapter in *What Is Life?* is entitled "Is Life Based on the Laws of Physics?" It is assumed that the gene is a molecule, but the bond energies in molecules are of the same order as the energy between atoms in solids, for example, in crystals, where the same pattern is repeated periodically in three dimensions, and a continuity of chemical bonds extending over large distances exists. Following this argument, Schrödinger suggested that the gene is a linear one-dimensional crystal, which lacks the periodic repeat, i.e., an aperiodic crystal. Under the influence of the theoretical physicist Ludwig Boltzmann (1886), Schrödinger concluded, "We are faced with a mechanism entirely different from the probabilistic one of physics, one that cannot be reduced to the ordinary laws of physics.... Living matter, while not eluding the laws of physics...is likely to involve other laws of physics hitherto unknown...." Unfortunately, Schrödinger was advised by the cytogeneticist C.D. Darlington of Oxford that genes are likely to be proteins, a belief that was prevalent among biologists at that time. However, Schrödinger stopped short of mentioning that proteins are long-chain polymers (with 20 different links with aperiodic patterns). Quite understandably, Schrödinger was not aware of the important discovery of Avery et al. (1944) that genes are made of DNA, a finding that had just been published while his book was in press. The lectures were delivered in February 1943, and the book was published in 1944. It was successful beyond the author's expectations. It was translated into seven languages, and the total sales exceeded well over 100,000.

Haldane (1945b) reviewed it favorably: "I wonder if posterity will find crossing-over as interesting as exchange of energy or mutation as atomic transition. However,... every geneticist will be interested in Schrödinger's approach to his or her science."

Muller's (1946) review was highly critical of the epilogue. He predicted that this little volume should prove valuable in furthering the much needed liaison between the fundamentals of the physical and the biological sciences. With a great deal of foresight, Muller wrote that an elucidation of the structure and function of the gene will "entail not only physics but also a good deal of chemistry." Muller pointed out that Schrödinger's limited knowledge of genetics (evidently learned quickly from reading a few publications) led him to make certain incorrect statements, e.g., in explaining why mutation occurs in only one allele at a time, the estimation of the size and minimum number of genes, and so on. However, Muller also stated that Schrödinger's book should render a valuable service in focusing attention on some important problems in biology. Muller's review was further helpful in explaining Schrödinger's terminology in terms that are more familiar to biologists. For instance, "negative entropy" is "potential energy," "aperiodicity" is "complexity," etc.

However, Muller's most critical comment was with reference to the Epilogue. Muller wrote, "It is...legitimate to hold the author to account very rigorously when he sails off, in his epilogue...to use his foregoing conceptions as the means of projecting his boat on the sea of straight old-fashioned mysticism.... If the collaboration of the physicist in the attack on biological questions finally leads to his concluding that 'I am God Almighty', and that the ancient Hindus were on the right track after all, his help should become suspect. It is hoped, however, that the unfortunate revelation of this physicist's inner urge will not keep the relatively sound expositions in the body of the present book from being taken seriously, and that an increasingly useful rapprochement between physics, chemistry and the genetic basis of biology is at last on the way."

In their introduction to the volume dedicated to Max Delbruck on his sixtieth birthday, Stent and Watson (1966) evaluated Schrödinger's contribution to biology. He stated that it probably had little influence on professional biologists. But it had a great impact on physical scientists, who were only too happy to focus their intellect on a new and refreshing problem in the post-war years. Stent pointed out that there were significant gaps in Schrödinger's genetic knowledge. He made no attempt to include chemistry in his discussion. Schrödinger's knowledge was derived from a few published papers and earlier conversations with Delbruck. He conducted no experiments in genetics. His grasp of genetics was outdated. In *What Is Life?*, Schrödinger made no mention of the latest advance in genetics, namely, the "one gene–one enzyme" hypothesis that was supported by the discoveries of Beadle and Tatum (1941), which ultimately contributed to molecular biology. He emphasized research on *Drosophila* even though by the 1940s, genetics was taken over by those working with microorganisms, *e.g.*, *Neurospora*. Even though it was Delbruck's model that inspired Schrödinger's interest in genetics, ironically, Schrödinger did not seem to know that Delbruck had been working with bacterial viruses for the previous five years—leading to the phage school in genetics, which later provided the answers to many of the questions that Schrödinger raised in *What Is Life?*. The significant date was 1940, when Alfred D. Hershey, Salvador E. Luria, and Max Delbruck founded the phage group at Cold Spring Harbor—long before Schrödinger was preparing his lectures in Dublin that eventually became *What Is Life?*.

Jim Watson read the book in the spring of 1946, while he was an undergraduate at Chicago, and was determined to find out the nature of the gene. He applied for graduate work at several leading universities. However, Indiana offered him financial support. Indiana also had a great biology department with such luminaries as Sonneborn, Luria, and the new Nobel laureate Herman J. Muller. The eastern universities such as Harvard were antisemitic and refused to employ such distinguished Jewish professors. Watson chose to do his graduate work at Indiana, at first with Muller and later with Luria (Watson 1968; Baldwin 1994).

Francis Crick (1988), whose training was in physics, also thought that Schrödinger's book was quite influential. Both Crick and Maurice Wilkins thought that the book got them interested in biological problems. Seymour Benzer was captivated by Schrödinger's description of the gene as an "aperiodic crystal." Molecular biology may have evolved without *What Is Life?*; however, its enormous impact on both biologists and physicists is very obvious. There are numerous scientists who know nothing about Schrödinger's Nobel Prize work on wave mechanics, but they are readily familiar with his book *What Is Life?*

a. The hereditary code-script (chromosomes):

Schrödinger used the term "pattern" to include not only the structure and function of that organism but also its ontogenetic development from the fertilized egg to the adult. He then considered the role of the nucleus, especially the role of chromosomes in heredity. He wrote: "It is these chromosomes, probably an axial skeleton fibre of what we actually see under the microscope as the chromosome, that contain in some kind of code-script the entire pattern of the individual's future development and of its functioning in the mature state. Every complete set of chromosomes contains the full code; so there are, as a rule, two copies of the latter in the fertilized egg cell, which forms the earliest stage of the future individual." Schrödinger suggested that one can predict the precise phenotype from the code script. But he acknowledged that the term "code-script" is too narrow. He described chromosome structures as "law-code and executive power" or they are "architect's plan and builder's craft" in one.

Schrödinger wrote: "In calling the structure of the chromosome fibres a code-script we mean that the all-penetrating mind, once conceived by Laplace, to which every causal connection lay immediately open, could tell from their structure whether the egg would develop, under suitable conditions, into a black cock or into a speckled hen, into a fly or a maize plant, a rhododendron, a beetle, a mouse or a woman." He explained further: "However little we understand the device we cannot but think that it must be in some way very relevant to the functioning of the organism, that every single cell, even a less important one, should be in possession of a complete (double) copy of the code-script" (Schrödinger 1944, pp. 22–23).

After considering mitosis, meiosis, and crossing-over, Schrödinger dealt with the maximum possible size of a gene and the permanence of a gene. He considered the gene to be a large protein molecule, in which every atom, every radical, and every heterocyclic ring play an individual role. Understandably, Schrödinger's discussion was based on the work of Haldane, Muller, Darlington, and other eminent geneticists of that period. As a physicist, he had to depend on the findings of others, eminent biologists and geneticists, who were the leading authorities at that time.

SCHRÖDINGER AND INDIAN PHILOSOPHY

Erwin Schrödinger was familiar with Indian philosophies as early as 1905, when he was 18 years old.

His early readings on the subject included Richard Garbe (1894) on the Sankhya doctrines, Paul Deussen (1906) on Vedanta, and Max Muller (1880) about the religions of India.

Quite early in his life, he was impressed with the basic conviction in Vedanta: "This unity of knowledge, feeling and choice are essentially eternal and unchangeable and numerically one in all men nay in all sensitive beings.... This life of yours that you are living is not merely a piece of the entire existence, but is in a certain sense the whole.

Plurality of minds in the living bodies, and plurality of things in the material world, 'what seems to be a plurality is merely a series of aspects of one thing, produced by deception (the Indian Maya),' Myriads of suns, surrounded by possibly inhabited planets, multiplicity of galaxies, each one with its myriads of suns....

According to me (Schrödinger), all these things are Maya, although a very interesting Maya with regularities and laws."

"An Indian metaphor refers to the plurality of almost identical aspects which the many facets of a diamond gives of a single object, say the sun."

"The personal self equals the omnipresent, all-comprehending eternal self."

"Atman equals Brahman" ("Deus factus sum")

Schrödinger on Ethical Lacunae of Science

"Science gives a lot of factual information, puts all our experience in a magnificently consistent order, but it is ghastly silent about all and sundry that is really near to our heart, that really matters to us. It cannot tell us a word about red and blue, bitter and sweet, physical pain and physical delight; it knows nothing of beautiful and ugly, good or bad, God and eternity. Science sometimes pretends to answer questions in these domains, but the answers are very often so silly that we are not inclined to take them seriously."

In ethics, materialism leads to a sort of utilitarian book of recipes. By contrast, the Vedantic doctrine of identity, and the Buddhist critique of the substantial self, incorporates what Francisco Varela called an "embodied ethics."

When you know by direct intuitive evidence either (a) that you are one with every sentient being or (b) that nothing substantial makes you distinct from the other sentient beings, being good with others is a *matter of course*.

Schrödinger's Holistic Style in Theoretical Physics

"Particles are but wave-crests, a sort of froth on the deep ocean of the Universe" (1925). The same holistic philosophy led him to suggest that a universal code-script resides in each cell of a living being, that every single cell, even a less important one, "should be in possession of a complete (double) copy of the code-script" (Schrödinger 1944, pp. 22–23).

MY VIEW OF THE WORLD

In 1964, Schrödinger published a small book entitled *My View of the World*, which contained two philosophical essays separated by 35 years. The first and longer one was written shortly before his appointment as Max Planck's successor in Berlin and shortly before he commenced his important work on wave mechanics. The second and shorter one dates from two years after his appointment as Professor Emeritus at the University of Vienna, shortly before his death. In his Foreword, Schrödinger wrote: "I do not know whether it is presumptuous of me to suppose that readers will be interested in my 'view' of the world. The critics, not myself, will decide on this. But a gesture of decorous modesty is usually in fact a disguise for arrogance. I should prefer not to be guilty of this. Anyway, the total (I have counted) is about twenty-eight to twenty-nine thousand words, not an excessive size for a view of the world."

Concerning the role of metaphysics, Schrödinger wrote: "Speaking as a scientist, it seems to me that it is our uncommonly difficult task, as post-Kantians, on the one hand step by step to erect barriers which will restrain the influence of metaphysics on the presentation of facts seen as true within our individual fields—while on the other hand preserving it as the indispensable basis of our knowledge, both general and particular. It is the apparent contradiction in this which is our problem. We might say, to use an image, that as we go forward on the road of knowledge we have *got* to let ourselves be guided by the invisible hand of metaphysics reaching out to us from the mist, but that we must always be on our guard lest its soft seductive pull should draw us from the road into an abyss…metaphysics does not form part of the house of knowledge but is the scaffolding, without which further construction is impossible" (Schrödinger 1964, pp. 4–5).

Schrödinger pointed out that, in the last hundred years, the western world has achieved an enormous development in the direction of technology. However, he did not consider it to be the most significant event during that period. He wrote that this age of technology will be described in the future as the age of the evolutionary idea and of the decay of the arts. Other lines of development in culture and knowledge, in the western mind, have been neglected and "indeed allowed to decay to a greater degree than ever before." Referring to the great success of science, Schrödinger wrote: "Rising to their feet after centuries of shameful servitude imposed by the Church, conscious of their sacred rights and their divine mission, the natural sciences turned against their ancient tormentress with blows of rage and hatred; heedless that, with all her inadequacies and derelictions of duty, she was still the one and only appointed guardian of our most sacred ancestral heritage…. Most of them (people) have nothing to hold on to and no one to follow. They believe neither in God nor gods; to them the Church is now only a political party, and morality nothing but a burdensome restriction which,… is now without any basis whatever" (Schrödinger 1964, pp. 5–6).

Schrödinger presented a long discourse, discussing his views on the merits of the Hindu caste system, the concept of reincarnation, transmigration of souls, and numerous other aspects of Hinduism and Indian philosophies. He wrote: "Anyone who wishes today to adopt the Vedantic view of the world will above all be well advised to leave out the theme of the transmigration of souls. Not because Christianity denies it…. The idea of the transmigration of souls was by no means alien to the early Greeks, as we know from the Pythagorean tradition. But it is logically meaningless, if combined with that of a complete obliteration of memory. And in fact that school of thought did attribute to an exceptional person such as Pythagoras the fantastic power of remembering his earlier births; he is even supposed to have proved it by recognising things and places which he had never seen!… But in any case it seems bizarre that a human being who is living now and undergoing great distress,… should be atoning for the misdeeds of an evil-doer, now dead, of whom he he…has no recollection whatever. We cannot but eliminate this doctrine of *special* identity, and with it go both the aristocratic attitude and the idea of salvation by enlightenment from the cycle of births, because there is no such cycle. But we still have the lovely thought of unity, of belonging unqualifiedly together,… at the same time… this idea does not naturally lead to any ethical consequences" (Schrödinger 1964, p. 103–104).

ENDNOTE

1. Vedanta, one of the six systems (*darshans*) of Hindu philosophy. The term "Vedanta" means in Sanskrit the "conclusion" (anta) of the *Vedas*, the earliest sacred literature of India. It applies to the Upanishads, which were elaborations of the Vedas, and to the school that arose out of the study (mimamsa) of the Upanishads. Thus, Vedanta is also referred to as Vedanta Mimamsa ("Reflection on Vedanta"), Uttara Mimamsa ("Reflection on the Latter Part of the Vedas"), and Brahma Mimamsa ("Reflection on Brahman"). The three fundamental Vedanta texts are the Upanishads (the most favored being the longer and older ones such as the Brihadaranyaka, the Chandogya, the Taittiriya, and the Katha); the Brahmasutras (also called Vedanta-sutras), which are very brief, even one-word interpretations of the doctrine of the Upanishads; and the Bhagavadgita ("Song of the Lord"), which, because of its immense popularity, was drawn upon for support of the doctrines found in the Upanishads.
   No single interpretation of the texts emerged, and several schools of Vedanta developed, differentiated by their conceptions of the nature of the relationship, and the degree of identity, between the eternal core of the individual self (atman) and the absolute (brahman). Those conceptions range from the nondualism (Advaita) of the eighth-century philosopher Shankara to the theism (*Vishishtadvaita*; literally, "Qualified Non-dualism") of the eleventh–twelfth-century thinker Ramanuja and the dualism (Dvaita) of the thirteenth-century thinker Madhva.

REFERENCES

Avery, O.T., C.M. Macleod, and C.M. McCarty (1944) Studies on the chemical nature of the substance inducing transformation of pneumococcal types. Induction of transformation by a deoxyribonucleic acid fraction isolated from pneumococcus type III. *J. Exp. Med., 79*: 137–158.

Baldwin, J. (1994) *DNA Pioneer, James Watson and the Double Helix.* New York: Walker and Co.

Beadle, G.W., and E.L. Tatum (1941) Genetic control of biochemical reactions in Neurospora. *Proc. Natl. Acad. Sci. USA, 27*: 499–506.

Boltzmann, L. (1886) Der zweite Hauptsatz der mechanischen Warmtheorie. Sitzungsber. Kaiserl. Akad. Wiss., Wien. (Not seen in original; quoted from Perutz 1987).

Braunizer, R. (1995) Reminiscences. In: *What is Life? The Next Fifty Years, Speculations on the Future of Biology.* (eds.) M.P. Murphy and L.A.J. O'Neill, Cambridge: Cambridge University Press, pp. 175–179.

Crick, F. (1988) *What Mad Pursuit: A Personal View of Scientific Discovery.* New York: Basic Books.

Crow, J.F. (1992) Erwin Schrödinger and the hornless cattle problem. *Genetics, 130*: 237–239.

Dronamraju, K.R. (1999) Erwin Schrödinger and the origins of molecular biology. *Genetics, 153*: 1071–1076.

Haldane, J.B.S. (1937) Biochemistry of the individual. In *Perspectives in Biochemistry* (eds.) J. Needham and D.E. Green. Cambridge: Cambridge University Press, pp. 1–10.

Haldane, J.B.S. (1945a) Personal communication to E. Schrödinger, from the J.B.S. Haldane Archives at University College, London.

Haldane, J.B.S. (1945b) A physicist looks at genetics. *Nature, 156*: 375–376.

Moore, W. (1989) *Schrödinger: Life and Thought.* Cambridge: Cambridge University Press.

Muller, H.J. (1946) A physicist stands amazed at genetics. *J. Hered., 37*: 90–92.

Pauling, L. (1987) Schrödinger's contribution to chemistry and biology. In: *Schrödinger: Centenary Celebration of a Polymath* (ed.) C.W. Kilmister. Cambridge: Cambridge University Press, pp. 225–233.

Pauling, L., and M. Delbruck (1940) The nature of the intermolecular forces operative in biological sciences. *Science, 92*: 77–79.

Perutz, M. (1987) Erwin Schrödinger's What Is Life? and molecular biology. In: *Schrödinger: Centenary Celebration of a Polymath* (ed.) C.W. Kilmister. Cambridge: Cambridge University Press, pp. 234–251.

Schrödinger, E. (1944) *What Is Life? The Physical Aspect of the Living Cell.* Cambridge: Cambridge University Press.

Schrödinger, E. (1964) *My View of the World.* Cambridge: Cambridge University Press.

Stent, G.S., and J.D. Watson (eds.) (1966) *Phage and the Origin of Molecular Biology.* Cold Spring Harbor Laboratory Press, Cold Spring Harbor, NY.

Timofeef-Ressovsky, N.W., K.G. Zimmer, M. Delbruck (1935) Über die Natur der Genmutation und der Genstruktur. *Nachr. Ges. Wiss. Göttingen, Fachgr. 6 N.F. 1., 13*: 190–245.

Watson, J.D. (1968) *The Double Helix: A Personal Account of the Discovery of the Structure of DNA.* New York: Atheneum.

Watson, J.D., and F.H.C. Crick (1953) Molecular structure of nucleic acids. *Nature, 171*: 737–738.

14 Max Delbruck (1906–1981)

Max Delbruck (1906–1981) was a founder of molecular biology. For their novel and important contributions to phage genetics, Max Delbruck, Salvador Luria, and Alfred Hershey were awarded the Nobel Prize in Physiology or Medicine in 1969.

Delbruck encouraged and inspired young researchers of many nationalities and from many disciplines who came to work with him on bacteriophage at the California Institute of Technology or to attend his famous "Phage Course" at the Cold Spring Harbor Laboratory, Long Island, New York, and to whom his intellectual approach to biological problems became an inspiration for their own.

FAMILY BACKGROUND AND EARLY LIFE

Max Delbruck came from a family of German intellectuals who excelled in academic achievements. His father, Hans Delbruck, was Professor of History at Berlin University, specializing in the history of the art of war, as well as sole editor for at least 30 years of a monthly journal, *Preussische Jahrbilcher*, for which he wrote a column commenting on German politics.

Three of his father's first cousins were Professor of German Literature at Jena, Chief Justice of the Imperial Supreme Court, and Minister of State.

His maternal great grandfather was the famous Justus von Liebig, Professor of Chemistry at Giessen and Munich, Foreign Member of the Royal Society, and Copley Medalist. His mother's brother-in-law, Adolf von Harnack, was Professor of Theology at Berlin University, a church historian, and director of the Prussian State Library and, in 1910, became cofounder and president of the Kaiser–Wilhelm–Gesellschaft. Their friends included the Bonhoeffer family and the Max Planck family, who lived nearby. The Delbruck family enjoyed a modest degree of affluence, and apparently, their life until 1914 was pretty free and very hospitable.*

CHILDHOOD

Delbruck's main boyhood interests were astronomy and mathematics. In retrospect, some 40 years later, he considered that he chose astronomy as a means of finding and establishing his own identity in an intimate society of so many able and strong personalities, all of them older than himself; but only he was an astronomer, and proclaimed himself one during his last two to three years at the Grunewald Gymnasium. He read popular books on the subject, was the enthusiastic possessor of a 2-inch telescope, and sometimes woke the whole household with the loudest of alarm clocks in the small hours of the morning when he had an appointment with the stars!* His knowledge of astronomy grew under the guidance of Karl Friedrich Bonhoeffer. Although Delbruck intended to study astronomy at the university, when Max Born, who was Professor of Theoretical Physics in Berlin, offered a teaching assistantship, he accepted it and followed Heitler's suggestion that he extend to lithium the quantum mechanical theory of the homopolar bond that had just been developed for hydrogen. He later recalled that this topic turned out to be a nightmare for him because of the complexity of the mathematics involved, but nevertheless, it won him his PhD degree in 1930.

EARLY CAREER IN PHYSICS

A visitor to Gottingen in 1929 was John E. Lennard-Jones, Professor of Theoretical Physics at the University of Bristol, England. He offered Delbruck a Postdoctoral Fellowship at Bristol, which Delbruck eagerly accepted. During the 18 months Max spent at Bristol, he established friendships with some brilliant young men who later became distinguished scientists. They included Patrick M.S. Blackett, later to become president of the Royal Society, and Paul A.M. Dirac, both of whom were later to win Nobel Prizes in physics.

One of them described young Max as a cheerful, outgoing person and one who rapidly established a reputation as a theoretician who was always ready to discuss problems of any kind with experimentalists who needed help and advice.

COPENHAGEN

Delbruck was awarded a Rockefeller Foundation Fellowship to study with the famous physicist Niels Bohr in Copenhagen, where he spent the spring and summer of 1931 and then spent another six months with the quantum physicist Wolfgang Pauli, in Zurich. While in Copenhagen, he collaborated with George Gamow on a nuclear physics project, and they became lifelong friends. In his contribution to a George Gamow Memorial Volume (1972), Delbruck described a lighthearted and facetious account of the gaiety and practical jokes of those days. Also working with Bohr at that time was Victor Weisskopf. It was customary in Copenhagen, at each of the early conferences organized by Niels Bohr, to have what was called a session of "comic physics." It was always Max Delbruck who was the most spirited leader in these activities with his humor and intellectual fantasy. His visit to Copenhagen proved far more important than he could have imagined,

* Reprinted with permission from *Max Ludwig Henning Delbruck 1906–1981: A Biographical Memoir by William Hayes* Copyright 1993 by the National Academy of Sciences. Courtesy of the National Academies Press, Washington, D.C.

because of Bohr's formulation of the complementarity concept as a generalized extension of Heisenberg's uncertainty principle. It marked the turning point in Delbruck's life because of its profound impact on his career and his philosophical outlook.

Delbruck was particularly impressed by the question posed by Bohr whether this new dialectic wouldn't be important also in relation to biology, in discussing the relation between life on the one hand, and physics and chemistry on the other—whether there wasn't an experimental mutual exclusion, so that you could look at a living organism either as a living organism or as a group of molecules. There might very well exist a mutually exclusive feature, analogous to the one found in atomic physics. Delbruck then decided to look more deeply, specifically into the relation between atomic physics and biology. This may have ultimately led him to the study of phage and the genetics of phage.

Delbruck discussed his admiration for Bohr and his idea of complementarity in biology in an oral interview that is part of the Delbruck archives at Caltech. The following is an excerpt from that interview:

> I came to Copenhagen, I think, February 1931, and stayed there five months. During that time, as I mentioned last time, I associated mostly with George Gamow doing some work on nuclear physics. I had come with notions of working on relativistic theory of the electron, on spinors, but that evaporated very quickly. During that time, and during all those years, Bohr incessantly worked and reworked his ideas on the deeper meaning of quantum mechanics. Quantum mechanics had been discovered as a technique in 1925 by Heisenberg, matrix mechanics, and in 1926 the other technical form of quantum mechanics had been discovered by Schrödinger, wave mechanics; the interconvertibility of these two forms of quantum mechanics had been shown very quickly. In 1927 Heisenberg had formulated the uncertainty principle as the real root of meaning of the quantum of action, and Bohr in a lecture at Como had given his version of what the deeper meaning was, and had formulated what was called the "complementarity argument." The essence of this argument was that for any situation in atomic physics, it is impossible to describe all aspects of reality in one consistent space-time-causal picture. The various experimental approaches that you use will reveal one or another aspect of reality, but these various experimental approaches are mutually exclusive; that means they are such that you cannot get the information that you get out of one arrangement, and simultaneously use the other arrangement to get other information. So these various experimental arrangements stand in a mutually exclusive relationship. The nature of the formalism of quantum mechanics is to permit you to derive the predictions for the outcome of the experiment of one kind from the results of experiments made with the mutually exclusive arrangement (if they are done successively); these predictions are of a statistical, probabilistic nature. This feature of atomic physics, expressed in the way Bohr expressed it, or in the more popular way that Heisenberg expressed it as an uncertainty relation, was, of course, a total shock to everybody concerned; in fact, so much a shock that Einstein never got over it. During the rest of his life Einstein tried somehow to get back to the classical picture where reality is just one reality, and if you can't get at the full reality with present methods, then presumably there must be other methods to get at reality; whereas Bohr was insistent on saying that this limitation to the classical picture of reality was not a preliminary stage to be replaced by a return to classical notions, but was an advance over classical notions—that we now had arrived at a new dialectical method to cope with the feature of reality that was totally unexpected. That was the formulation of Heisenberg in 1927, and Bohr in maybe the same year, maybe the next year. But Bohr continued to elaborate and restate his position year in and year out until he died thirty years later, innumerable lectures. Harding: Were you

interested in the idea of complementarity when he first…? Delbruck: Enormously. I was interested—well, anybody who was at all interested in the result of the questions couldn't help but be fascinated. It also motivated me to look at the writings of Kant on causality to see how Kant, who was so clever and thoughtful, could have overlooked this possibility. So for the first time, and with a real motivation, I looked at Kant, and it was very clear that this situation was just utterly removed from anything that Kant had thought of—so there was no doubt that the physicists had been pushed into an epistemological situation that nobody had dreamed of before. Bohr then very vigorously asked the question whether this new dialectic wouldn't be important also in other aspects of science. He talked about that a lot, especially in relation to biology, in discussing the relation between life on the one hand, and physics and chemistry on the other—whether there wasn't an experimental mutual exclusion, so that you could look at a living organism either as a living organism or as a jumble of molecules; you could do either, you could make observations that tell you where the molecules are, or you could make observations that tell you how the animal behaves, but there might well exist a mutually exclusive feature, analogous to the one found in atomic physics. He talked about that in biology and in psychology, in moral philosophy, in anthropology, in political science, and so on, in various degrees of vagueness, which I found both fascinating and very disturbing, because, it was always so vague. It was vague largely because the basic situation wasn't clear enough, and also in many respects Bohr wasn't sufficiently familiar with the status of the science. So it was intriguing and annoying at the same time. It was sufficiently intriguing for me, though, to decide to look more deeply specifically into the relation of atomic physics and biology—and that means learn some biology. So when the question came up of what job I would take after this year with Bohr and Pauli (and another half year in Bristol), and I had the choice of either going to Berlin to become an assistant of Lise Meitner or to Zurich to be an assistant of Pauli, I chose to go to Berlin because of the vicinity of the Kaiser Wilhelm Institutes for biology to the Institute I was going to work in.

BIOLOGY

After spending another six months with Lennard-Jones at Bristol, Delbruck accepted an appointment as assistant to Lise Meitner at the Kaiser Wilhelm Institute for Chemistry in Berlin in the autumn of 1932 because of its proximity to the Kaiser Wilhelm Institute for Biology. But before returning to Berlin, he paid a short visit to Copenhagen to hear Bohr deliver his famous address, "Light and Life," to the opening meeting of the International Congress on Light Therapy in August, in which he explicitly stated his views on complementarity in biology. This lecture confirmed his decision to turn to biology.

BERLIN YEARS (1932–1937)

Not long after the beginning of Max's Berlin period, Max organized a private group of five or six theoretical physicists to join in fairly regular discussions among themselves, often at his mother's house. At his suggestion, some biochemists and biologists also joined the group, including K.G. Zimmer, whose interest was the dose effect of ionizing radiation on biological systems, and N.W. Timofeeff-Ressovsky, a Russian geneticist from the Kaiser Wilhelm Institute for Brain Research, who had been collaborating with Zimmer on the genetic effects of radiation. Timofeeff-Ressovsky's experimental organism was the fruit fly, *Drosophila*.

The result of all these discussions was a paper by Timofeeff-Ressovsky, Zimmer, and Delbruck (1935) on the nature of gene mutation and gene structure, in which Max was mainly responsible for the theoretical interpretation.

He supposed that the molecules from which genes are made must have a very unusual atomic constitution, since they show such remarkable stability in a cellular environment otherwise subject to constant chemical change. This stability suggested that each atom of the gene molecule is fixed in its mean position and electronic state by being sunk in "energy wells," so that discontinuous changes in their state, expressed as mutations, could arise only by the acquisition of very high energies such as ionizing radiations would impose.

Ten years later this paper became famous through the publication in 1945 of Erwin Schrödinger's little book, *What Is Life?*, in which he maintained that Delbruck's model of the gene was the only possible one and went on to put forward the romantic and paradoxical idea that from Delbruck's picture of the hereditary substance it emerges that living matter, while not eluding the "laws of physics" as established up to date, is likely to involve hitherto unknown "other laws of physics."

Schrödinger's book was influential in attracting into biology many physicists, curious to solve the paradox.

In 1937, the Rockefeller Foundation offered Delbruck an unsolicited Fellowship (Biology) to travel abroad, so he took this opportunity to visit the California Institute of Technology, especially the *Drosophila* group led by Thomas Hunt Morgan.

EARLY DAYS AT CALTECH

Delbruck's initial introduction to biology at Caltech was disappointing, despite the help of A.H. Sturtevant and Calvin Bridges, with whom he was especially friendly; he found the highly specialized *Drosophila* jargon too difficult and exacting to grasp in a reasonable time. However, when Emory Ellis showed him the very rudimentary materials and the simple techniques required for his experiments with bacteriophages, Max decided to join Ellis immediately in phage research. After one year, Max continued alone giving continuity to phage research by founding and guiding an expanding number of phage workers. In their brief collaboration, Ellis and Delbruck (1939) invented and greatly refined the one-step growth curve and devised the single-burst experiment, which permitted a comparison of phage multiplication in individual cells.

VANDERBILT UNIVERSITY

Delbruck was on the faculty of Vanderbilt University from 1940 to 1947 but had no students of biology. At the end of 1940, he met Salvador Luria, an Italian microbiologist, who was working on phage at the College of Physicians and Surgeons in New York. Their collaboration led to the founding of molecular biology. At first, they collaborated in experiments with mixed infections by phages Tl and T2 in the summer of 1941 at Cold Spring Harbor. Luria first conceived the idea of comparing the numbers of resistant bacteria arising in otherwise identical *independent* cultures, initially seeded with only a few sensitive cells, with the numbers from equivalent samples from a single culture.

If resistance was induced by contact with the phage, then variation in the numbers of resistant cells would, in either case, be within the limits expected by random sampling. In contrast, the occurrence of resistant *mutants*, which might arise spontaneously and begin to multiply at any time during the growth of each independent culture, would lead to a much wider variation. A fluctuation greater than the sampling error, in the numbers of resistant bacteria from independent cultures, indicates that these variants arose as clones in the cultures before they were exposed to the phage and, therefore, were mutants.

The paper by Luria and Delbruck (1943) reporting their findings and conclusions is a landmark in the history of molecular biology, for it provided the first real evidence that bacterial inheritance, like that of the cells of higher organisms, is mediated by genes and not by some Lamarckian mechanism of adaptation as was widely held at the time. It laid the foundation of bacterial genetics, which became a basic tool for exploring the molecular basis of life.

PHAGE GROUP

Delbruck and Luria had become interested in some papers on phage by Alfred H. Hershey, a microbiologist at the Medical School of Washington University, in St. Louis, and these three, Delbruck, Luria, and Hershey, formed the nucleus of the Phage Group.

PHAGE TREATY

Delbruck negotiated a "phage treaty," which is based on the agreement that research should be limited to a set of seven phages (T1–T7), all of which infected the same host, *Escherichia coli* strain B.

COLD SPRING HARBOR

In 1950, Hershey joined the Department of Genetics of the Carnegie Institution of Washington, which was also located at Cold Spring Harbor.

PHAGE COURSES

In 1945, Max organized the first of 26 successive annual Phage Courses at Cold Spring Harbor, and was the principal instructor in the first three of them. This was made possible through the vision and enterprise of Milislav Demerec, director of the Laboratory from 1941 to 1960. Demerec was a classical geneticist who foresaw the potential of bacteria and their phages as genetic tools. The course was devised not only for biologists but also for biochemists and physicists, and the students ranged from young postdoctorals to eminent physicists such as Leo Szilard, who took the course in 1947.

The importance of a quantitative and statistical approach to the new biology was stressed.

The popularity of these courses may be inferred from the fact that the total number of students over the years was well over 400, including many from abroad. Moreover, of some 130 students who attended the first 10 courses, over 30 became recognized phage workers or bacterial geneticists.

PHAGE MEETINGS

In addition to these courses, Delbruck also organized a series of Phage Meetings at Cold Spring Harbor from 1950 onward and the meetings continued there annually, without interruption, through 1981, attended by hundreds of participants. In the early 1950s, he became interested in sensory perception and transduction and chose, particularly, to study the phototropic response of the large aerial sporangiophores of the fungus *Phycomyces*. As in the case of phage, he became the leader of a *Phycomyces* Group, interested in various aspects of tropic behavior in this organism. From 1965 onward, he organized the first of a series of eight *Phycomyces* workshops, held at Cold Spring Harbor over the next 12 years. Each lasted about two months, and they attracted, all told, more than 100 people, and Max led or participated in all of them.

A recently completed major extension of the Davenport Laboratory, the site of so much of Delbruck's research as well as of the Phage and *Phycomyces* courses at Cold Spring Harbor, was dedicated as the Max Delbruck Laboratory in August 1981.

CALTECH AGAIN

When the war was over, Max received several important job offers from prestigious institutions, such as the Cold Spring Harbor Laboratory, the California Institute of Technology, and the Universities of Illinois and Manchester, England. Vanderbilt University responded by promising him everything he wanted. However, when the offer of a Chair of Biology at Caltech arrived on December 11, 1946, it proved irresistible and was accepted on December 27. This was the first faculty appointment in biology made by George W. Beadle, who had recently succeeded T.H. Morgan as Chairman of the Biology Division.

WHAT IS LIFE?

It is of interest that, during what Stent has called the "Romantic Period" of molecular biology (up to 1953), about the same proportion of recruits to the phage field came from the physical sciences as from the biological sciences. It is likely that an appreciable proportion of the former was motivated by Schrödinger's imaginative prediction about the nature of the gene in his book *What Is Life?* Indeed, a colleague at Caltech at this time was Neville Symonds, who came from postdoctoral studies on wave mechanics with Schrödinger, then working in Dublin. James Watson, on the other hand, whose interests and undergraduate background were in biology, was drawn by the "legendary figure" of Delbruck evoked by Schrödinger's book.

Elie Wollman of the Pasteur Institute, Paris, was one of the early converts. Andre Lwoff, who was head of the Service de Physiologie Microbienne at the Pasteur Institute, had attended the 1946 Cold Spring Harbor Symposium and had met Delbruck. Wollman was his first ambassador to Caltech, and thereafter, many members of the American Phage Group worked for a time at the Pasteur Institute.

Ole Maaloe from Copenhagen University and Jean Weigle, who was head of the Physics Institute in the University of Geneva; constituted a very small "Class of 49" that graduated under Delbruck's supervision.

Weigle's account of his Caltech experiences inspired the electron microscopist Edward Kellenberger to make a study of phage. Weigle himself subsequently resigned his Geneva professorship for a Caltech research appointment.

Maaloe also embarked on phage research in Copenhagen, and Watson worked with him there during the first year of his Fellowship in Europe in 1950, as well as with the Danish biochemist Herman Kalckar, who had attended the first Phage Course in 1945.

DNA AS GENETIC MATERIAL

Delbruck's early work on the one-step growth curve had shown that, following phage infection of bacterial cells, a latent period of about 20 minutes elapses before the cells begin to burst and liberate a hundred or more progeny particles. As mentioned earlier, mutation had also been revealed by Salvador Luria as the cause of variation in phage, as well as in bacteria.

Delbruck and W.T. Bailey (1946) and A.D. Hershey independently demonstrated genetic recombination when bacteria were doubly infected with phages that differed in two characters. This was the finding that led, about 10 years later, to the ultimate genetic analysis of gene structure by Seymour Benzer. However, nothing whatsoever was known about the number or nature of the presumptive precursors inside the infected bacteria during the latent period.

In 1946, Delbruck remarked in a Harvey Lecture that "it should be our first aim to develop a method of determining the number of virus particles which are present in a bacterial cell at any one moment. Here I, and those who have been associated with me in this work, have to make the first admission of failure." Such a method was first developed between 1949 and 1952 by A.H. Doermann, who disrupted cells at intervals after infection but failed to find any plaque-forming entities during about the first 12 minutes; thereafter, infective intracellular particles began to appear and increased linearly. This "eclipse period" clearly showed that the phage changes its state immediately after infection, while the subsequent linear rather than exponential increase in phage numbers implied that this increase is not due to successive replications of its complete organism but is more compatible with an assembly of its component parts.

Niels Bohr's speech, "Light and Life", and his concept of complementarity in biology continued to influence Delbruck's outlook for many years. In an address entitled "A Physicist Looks at Biology," delivered at the thousandth meeting of the Connecticut Academy of Arts and Sciences in 1949, Delbruck commented: "It may turn out that certain features of the living cell, including perhaps even replication, stand in a mutually exclusive relationship to the strict application of quantum mechanics, and that a new conceptual language has to be developed to embrace this situation. The limitation in the applicability of present day physics may then prove to be, not the dead end of our search, but the open door to the admission of fresh views of the matter. Just as we find features of the atom, its stability, for instance, which are not reducible to mechanics, we may find features of the living cell which are not reducible to atomic physics but whose appearance stands in a complementary relationship to those of atomic physics."

In 1952, Hershey and Chase published their famous experiment, in which they infected cells with phage in which the DNA and protein were differentially labeled with radioactive phosphorus and sulfur, respectively; they found that the DNA entered the cells but that most of the protein, in the form of empty heads, remained outside. The eclipse was therefore the period during which the phage DNA was replicating and directing the synthesis of nascent phage protein. Thus, it turned out that the genetic material was DNA and that the genetic material alone entered the cell to initiate a new viral generation.

OSWALD AVERY AND DNA

As early as 1944, Oswald Avery and his colleagues at the Rockefeller Institute, in New York, had published their important discovery that the "transforming principle" of pneumococci, which transfers the hereditary ability to synthesize a polysaccharide characteristic of one type to bacteria of other types, is highly polymerized DNA. Why, then, did the Phage Group seemingly ignore this obvious clue to the chemical nature of the gene until a member of the group itself came to the same conclusion by a less rigorous experiment? In fact, both Delbruck and Luria were very interested in Avery's work a considerable time before its publication, visited him at the Rockefeller Institute, and admired him as a person. In mid-1943, Avery wrote a long letter to his brother Roy, who was a microbiologist at Vanderbilt University and knew Max Delbruck and showed him the letter that explained the results of Oswald's research and suggested, very cautiously, that DNA might be the genetic material.

Although pneumococcal transformation was certainly seen as a very interesting phenomenon by Delbruck and Luria, there were understandable reasons for failing to recognize its *genetic* importance. The phenomenon appeared to be uniquely restricted to polysaccharide production by a single bacterial species and seemed remote from the problems that beset phage workers. At that time, bacterial genetics did not exist, while DNA was generally regarded as an unimportant molecule consisting simply of repeating tetrads of the same nucleotides, which could hardly carry complex information. It was only later that the possibility of contamination of transforming preparations with small amounts of protein, then favored as the most likely genetic material, could be excluded.

FAILURE TO APPRECIATE THE IMPORTANCE OF AVERY'S EXPERIMENT

However, the foremost reason for not appreciating the importance of DNA in transformation was probably that it appeared as a biochemical problem, revealed by biochemical techniques. Luria pointed out that, "People like Delbruck and myself, not only were we not thinking biochemically, I don't think we attached great importance to whether the gene was protein or nucleic acid. The important thing for us was that the gene had the characteristics that it *had* to have" (Stent 1968). But others, including Linus Pauling and Jim Watson, had realized the importance of DNA, confirmed by the Hershey–Chase experiment. Watson and Crick soon succeeded in elucidating its structure, which was narrated in a light-hearted but able style by Watson in his bestseller, *The Double Helix*.

As soon as the model had been built, Jim Watson revealed it first in a letter to Delbruck, who was fascinated and thought it obviously right. Max then wrote to Bohr about the model, saying that he thought it equaled Rutherford's discovery of the nucleus of the atom.

The really important achievement of the group during this romantic phase of the growth of molecular biology was the introduction into microbial genetics of previously unknown standards of experimental design and deductive logic.

PHYCOMYCES PERIOD (1953–1981)

Delbruck was basically a theoretician who lived to search for neat models and hypotheses to explain complex phenomena. In about 1950, after the discovery of the phage eclipse phase but before the Hershey–Chase experiment, he became interested in sensory perception and its transduction into physiological activity—in a simple fungus, *Phycornyces*.

The Cold Spring Harbor workshops, each lasting about two months and beginning in 1964, attracted many participants from abroad.

Max was involved in *Phycomyces* from 1953 onward, but he did not lose touch with phage research. Working with N. Visconti, he developed a mathematical model of phage recombination based on multiple rounds of mating during the eclipse period (1953), and later, he got interested in theoretical problems of DNA replication (1945, 1957) and the genetic code (1958).

COLOGNE (1961–1963)

In the post-war years, Delbruck returned to Germany on several occasions, first in 1947 and then in 1954, when he visited Gottingen for three months.

In 1956, he spent three months at Cologne, introducing molecular genetics to a new institute, and helped its growth, embracing several independent, integrated groups headed by professors but having many facilities in common and an emphasis on research. This was a very novel concept for Germany, and Max agreed to become the first director so that his reputation could be used in negotiations with the Government. The Institute was formally dedicated in June 1962, with Niels Bohr as the principal speaker. His lecture, entitled "Light and Life—Revisited," commented on the original one of 1933, which had been the starting point of Max's interest in biology.

Max's Cologne period was beneficial to German biology as a whole, not only on account of the courses he instituted but also because of his extensive travelling and lecturing. Later, he was persuaded to serve as an adviser in natural science on the Founding Committee of the new University of Constance. This led to a natural sciences faculty that was essentially all molecular biology.

PERSONALITY

William Hayes summed up his recollection of Max's personal aspects in a biographical memoir for the National Academy of Sciences.

> How can I begin to describe what Max was like to those who did not know him, for he was all things to those who did? His profound intelligence and scholarship in so many fields of science, philosophy and the arts, all of which he regarded as a cultural unity; his blend of critical and quiet spoken aloofness with outgoing gregariousness, affection, and sense of fun; his basic seriousness and childish love of practical joking; all could be seen as the essence of paradox, or as the embodiment of "natural man." This, perhaps, has been best expressed by a close colleague of Max who always thought of Max as a human archetype, perhaps the best way to convey an impression of Max's individuality is through a kaleidoscope of reminiscences and impressions by various friends who knew him well.

GANDHI OF BIOLOGY

Delbruck had been called a "Gandhi of biology," who, without possessing any temporal power at all, was an ever-present and sometimes irksome spiritual force. "What will Max think of it?" had become the central question of the molecular biology psyche.

The Caltech period was noted for the extraordinary and informal hospitality of Max and Manny in their home in Pasadena, which was "open house" to all and sundry, and the famous weekend camping trips to the desert, organized by Manny, that might include undergraduates, graduates, post-docs, staff, visitors, children, and dogs, with long treks up and down the hills and canyons, on which Max might unexpectedly block the path by stopping abruptly to ponder a sudden thought. After returning to camp and a welcome siesta, Manny would prepare dinner over the campfire. "Evening brought a big fire and wild stories until each wandered into the dark to find his own bag and pile of clothes under the sky freckled with stars. One would occasionally wake up to see Max balancing his binoculars against the car."

Delbruck's wit and humor were very much a part of his image because they accentuated the depth and seriousness of his personality in such a striking way. His wit was light, amusing, and self-deprecating, as when he told Jean Weigle that he supposed the *Festschrift* in honor of his (Max's) 60th birthday would be an opportunity for everyone to publish papers that had been rejected repeatedly by many journals.

Finally, to show how intimidating Max might at first appear to those who didn't know him well, it was not uncommon for him, with a rather serious expression, to say to a lecturer after his performance, "Well, that was the worst seminar I have ever heard."

CORRESPONDENCE WITH J.B.S. HALDANE

Delbruck was interested in scientific creativity. There was an exchange of letters between Max Delbruck and J.B.S. Haldane, who was one of the founders of population genetics, shortly before Haldane's death in 1964. Haldane died in India. It was mainly concerned about the nature of scientific research. In his letter dated April 29, 1964, Delbruck wrote: "I agree entirely with you that in India an enormous amount of first-rate biological research that costs little money could be done and should be done. The same might even be true in the Western Worlds but it just does not appeal to the imagination of the ambitious to do anything that lies off the main highway of science. I have, for a number of years, tried to do such one-man research on a problem which lies off the main highway, but I found it very difficult to find anybody interested in it."

In his reply dated May 8, 1964, Haldane wrote: "Judging from your letter, ambition means something different in U.S.A. and Europe. A really ambitious European looks out for something which lies off the main highway of science, like Mendelian genetics, radioastronomy, or radioactivity. Their bye-road may become a main highway. Whereas the main highway can lead to little but teaching courses after a generation or so. Sometimes it does not even do that.

Here people follow the highway without even being ambitious for anything beyond a professorship. I have generated something more lively in a few of them, but it tends to flutter and go out. I shall try to make one of them find out how much earth is brought up per hectare per year by termites in my garden.

We don't have many worms, but termites bring up a lot to make shelters over dried herbage. But that is off the main stream, the sort of thing Darwin did in his spare time, Similarly the next number of the Journal of Genetics will contain a paper by Reber, one of the pioneers of radioastronomy, on the effect of forcing beans to coil in the sense which is 'unnatural' for them. It seems to improve the yield. I think such people should be encouraged."

NOBEL PRIZE

Delbruck received the 1969 Nobel Prize in Physiology or Medicine, shared with Salvatore Luria and Alfred Hershey, "for their discoveries concerning the replication mechanism and the genetic structure of viruses." The Nobel committee also noted that "The honour in the first place goes to Delbrück who transformed bacteriophage research from vague empiricism to an exact science. He analyzed and defined the conditions for precise measurement of the biological effects. Together with Luria he elaborated the quantitative methods and established the statistical criteria for evaluation which made the subsequent penetrating studies possible. Delbrück's and Luria's forte is perhaps mainly theoretical analysis, whereas Hershey above all is an eminently skillful experimenter. The three of them supplement each other well also in these respects."

EVOLUTIONARY EPISTEMOLOGY

In the winter of 1972, Max gave an extensive course of 20 lectures at Caltech on "Evolutionary Epistemology." He later condensed these into a long but elegant essay entitled "Mind from Matter??" presented as a single lecture to the XIIIth Nobel Conference in 1977. The essay ranges from cosmology and the beginning of life, through the evolution of prokaryotes and perception, higher organisms and behavior, the nervous system, consciousness, language and culture, to cognitive ability. He then goes on to ask if mind evolved and was selected merely for its survival value, "to let us get along in the cave, how can it that (it) permit(s) us to obtain deep insights into cosmology, elementary particles, molecular genetics, number theory? To this question I have no answer.... The feeling of absurdity that attaches to the notion 'Mind from Matter' is perhaps of a similar nature to the feeling of absurdity we have learned to cope with when we permit relativity to reorganize time and space and quantum theory to reconcile waves and corpuscles. If so, then there may yet be hope for developing a formal approach permitting a Grand Synthesis."

The essay begins with a brief recapitulation of Schrödinger's book, *What Is Life?*, and outlines Bohr's subtle complementarity argument. Thus, Max's thinking continued to be swayed by Bohr's ideas, but in a new dimension, after 45 years.

Delbruck taught regularly at Caltech, and his method of learning was to teach, and every year, he would assign himself the task of teaching a course in some new subject that he wanted to learn. This ranged all the way from statistical mechanics to epistemology. Long after he had been officially retired, he volunteered to teach freshman physics here at Caltech as a sort of refresher course for himself. He never lost his interest or skill in theoretical physics and mathematics and, as late as 1980, published a paper on Bose–Einstein statistics.

LAST DAYS

When Delbruck reached the normal age of retirement in 1977, the Caltech trustees appointed him to the special position of Board of Trustees Professor of Biology, Emeritus, so that he could continue the research of his *Phycomyces* Group at the Institute. Early in 1978, he learned that he was suffering from multiple myeloma. This responded well to chemotherapy, and he was able to travel to Paris with his daughter Nicola in the spring of 1979 to be inducted as a Foreign Member of the French Academie des Sciences.

He retained the interest of a scientist toward his disease from its beginning, never complained, and, from first to last, retained the upper hand. A few months before his death, he suffered a mild stroke, which impaired his vision on one side; he found this interesting and smilingly said, "The students need me as a guinea pig; they are setting up some tests they cannot do with the monkeys."

FURTHER READING

Benzer, S. (1955) Fine structure of a genetic region of bacteriophage. *Proc. Natl. Acad. Sci. USA, 41*: 344–354.

Benzer, S. (1961) On the topography of the genetic fine structure. *Proc. Natl. Acad. Sci. USA, 47*: 403–415.

Bohr, N. (1933) Light and life. *Nature, Lond., 131*: 421–423, 457–459.

Bohr, N. (1965) Light and life revisited. In: *Essays in Atomic Physics and Human Knowledge.* New York: John Wiley.

Cairns, J., G.S. Stent, and J.D. Watson (1968) *Phage and the Origins of Molecular Biology.* Cold Spring Harbor, NY: Cold Spring Harbor Press.

Delbruck, M. (1977) The life of a Nobel wife. *Eng. Sci. (California Institute of Technology), March–April*: 14–24.

Delbruck, M. (1978) The arrow of time—beginning and end. *Eng. Sci. (California Institute of Technology), 42.*

Delbruck, M. (1980) Oral history: how it was. *Eng. Sci. (California Institute of Technology), March–April*: 21–26; *May–June*: 21–27.

Doermann, A.H. (1952) The intracellular growth of bacteriophages. I. The liberation of intracellular bacteriophage T4 by premature lysis with another phage or with cyanide. *J. Gen. Physiol., 35*: 646–656.
Hershey, A.D. (1946) Spontaneous mutations in bacterial viruses. *Cold Spring Harb. Symp. Quant. Biol., 11*: 67–77.
Hershey, A.D., and M. Chase (1952) Independent functions of viral protein and nucleic acid in growth of bacteriophages. *J. Gen. Physiol., 36*: 39–56.
Lipson, E.D. (1980) Sensory transduction in Phycomyces photoresponses, In: H. Senger (ed.), *The Blue Light Syndrome.* New York: Springer.
Mullins, N.C. (1972) The development of a scientific speciality: the Phage Group and the origins of molecular biology. *Minerva X, 1*: 51–82.
Schrödinger, E. (1944) *What Is Life?* Cambridge: Cambridge University Press.
Stent, G.S. (1968) That was the molecular biology that was. *Science, 160*: 390–395.
Stent, G.S. (1981) Obituary: Max Delbruck. *Trends Biochem. Sci., 6(5)*: III–IV.
Stent, G.S., and R. Calender (1978) *Molecular Genetics: An Introductory Narrative.* 2nd ed. San Francisco: W.H. Freeman & Co.
Watson, J.D. (1968) The double helix. p. 217; facsimile of letter following p. 226. London: Weidenfield & Nicolson.
Watson, J.D., and F.H.C. Crick (1953) Genetic implications of the structure of deoxyribonucleic acid. *Nature, Lond., 171*: 964–967.

Selected Publications of Max Delbruck

(With N.W. Timofeef-Ressovsky and K.G. Zimmer.) Ueber die Natur der Genmutation und der Genstruktur. *Nachr.* 1935.
Ges. Wiss. Gottingen 6 N.F. Nr. 13: 190–245. 1935.
(With N.W. Timofeef-Ressovsky.) Strahlengenetische Versuche tiber sichtbare Mutationen und die Mutabilitat, einzelner Gene bei *Drosophila melanogaster. Z. indukt. Abstamm.—u. VerebLehre, 71*: 322–334. 1936.
(With N.W. Timofeef-Ressovsky.) Cosmic rays and the origin of species. *Nature, Lond., 137*: 358–359. 1936.
(With E.L. Ellis.) The growth of bacteriophage. *J. Gen. Physiol., 22*: 365–384. 1939.
The growth of bacteriophage and lysis of the host. *J. Gen. Physiol., 23*: 643–660. 1940.
Adsorption of bacteriophage under various physiological conditions of the host. *J. Gen. Physiol., 23*: 631–642. 1940.
(With L. Pauling.) The nature of the intermolecular forces operative in biological processes. *Science, 92*: 77–79. 1940.
A theory of autocatalytic synthesis of polypeptides and its application to the problem of chromosome reproduction. *Cold Spring Harb. Symp. Quant. Biol., 9*: 122–124. 1941.
(With S.E. Luria.) Interference between bacterial viruses. I. Interference between two bacterial viruses acting upon the same host and the mechanism of virus growth. *Archs Biochem., 1*: 111–141. 1942.
(With S.E. Luria.) Interference between bacterial viruses. II. Interference between inactivated bacterial virus and active virus of the same strain and of a different strain. *Archs Biochem., 1*: 207–218. 1942.
Bacterial viruses (bacteriophages). *Adv. Enzymol., 2*: 1–32. 1942.
(With S.E. Luria.) Mutations of bacteria from virus sensitivity to virus resistance. *Genetics, 28*: 491–511. 1943.
(With S.E. Luria and T.F. Anderson.) Electron microscope studies of bacterial viruses. *J. Bacteriol., 46*: 57–76. 1943.

(With S.E. Luria.) A comparison of the action of sulphadrugs on the growth of a bacterial virus and its host. *Proc. Indiana Acad. Sci., 53*: 28–29 (Abstract). 1943.
Spontaneous mutations of bacteria. *Ann. Mo. Bot. Gdn., 32*: 223–233. 1945.
The burst size distribution in the growth of bacterial viruses. *J. Bacteriol., 50*: 131–135. 1945.
Effects of specific antisera on the growth of bacterial viruses. *J. Bacteriol., 50*: 137–150. 1945.
Interference between bacterial viruses. III. The mutual exclusion and the depressor effect. *J. Bacteriol., 50*: 151–170. 1945.
Bacterial viruses or bacteriophages. *Biol. Rev., 21*: 30–40. 1946.
Experiments with bacterial viruses (bacteriophages). *Harvey Lectures, 41*: 161–187.
(With W.T. Bailey.) Induced mutation in bacterial viruses. *Cold Spring Harb. Symp. Quant. Biol., 11*: 33–37. 1946.
Biochemical mutants of bacterial viruses. *J. Bacteriol., 56*: 1–10. 1948
(With M.B. Delbruck.) Bacterial viruses and sex. *Sci. Am., 179*: 46–51. 1948.
Genetique du bacteriophage. *Colloques Int. C.N.R.S., 8*: 91–103. 1949.
A physicist looks at biology. *Trans. Conn. Acad. Arts Sci., 38*: 173–190. 1949.
(With J.J. Weigle.) Mutual exclusion between an infecting phage and a carried phage. *J. Bacteriol., 62*: 301–318. 1952.
(With N. Visconti.) The mechanism of genetic recombination in phage. *Genetics, 38*: 5–33. 1953.
On the replication of DNA. *Proc. Natl. Acad. Sci. USA, 40*: 783–788.
Current views on the reproduction of bacteriophage. *Scientia, 91*: 118–126. 1956.
(With W. Reichardt.) System analysis of the light growth reactions of *Phycomyces*. In: D. Rudnick (ed.), *Cellular Mechanisms in Differentiation and Growth*. Princeton: Princeton University Press, pp. 3–44. 1956.
(With G.S. Stent.) On the mechanism of DNA replication. In: W.D. McElroy and B. Glass (eds.), *The Chemical Basis of Heredity*. Baltimore: Johns Hopkins Press. 1957.
Bacteriophage genetics. *Proc. IVth. Int. Poliomyelitis Congr.* New York: Lippincott. 1958.
Knotting problems in biology. In: Mathematical problems in the biological sciences; *Proc. Symp. App. Math., 14*: 55–68. 1962.
Primary transduction mechanisms in sensory physiology and the search for suitable experimental systems. *Isr. J. Med. Sci., 1*: 1363–1365. 1965.
(With others.) General discussion. *Radiat. Res. Suppl., 6*: 227–234. 1966.
Molecular aspects of genetics. In: R.A. Brink (ed.), *Heritage from Mendel*. Madison, WI: University of Wisconsin Press, pp. 49–50. 1967.
Molecular biology—the next phase. *Eng. Sci., 32*: 36–40. 1968.
A physicist's renewed look at biology: twenty years later. *Science, 168*: 1312–1315. 1970; also in Les Prix Nobel en 1969. Stockholm: The Nobel Foundation.
Lipid bilayers as models of biological membranes. In: F.O. Schmitt (ed.), *The Neurosciences: Second Study Program*. New York: Rockefeller University Press, pp. 677–684. 1970.
Aristotle-totle-totle. In: J. Monod and E. Borek (eds.), *Of Microbes and Life (Andre Lwoff Festschrift)*. New York: Columbia University Press. 1971.
Homo scientificus according to Beckett. In: W. Beranek (ed.), *Science, Scientists and Society*. New York: Bogden & Quigley Inc. 1972.
Signal transducers: *Terra incognita* of molecular biology. *Angew. Chem. Int. Ed. Engl., 11*: 1–7. 1972.
Out of this world. In: F. Reines (ed.), *Cosmology, Fusion and Other Matters, George Gamow Memorial Volume*. Boulder, CO: Colorado Associated University Press. 1972.
Light and life III. *Carlsberg Res. Commun., 41*: 299–309. 1976.
How Aristotle discovered DNA. In: K. Huang (ed.), *Physics and Our World; A Symposium in Honor of Victor F. Weisskopf*. New York: American Institute of Physics. 1976.

Virology revisited. In: M. Chakravorty (ed.), *Proceedings of International Symposium on Molecular Basis of Host Virus Interaction*. Princeton, NJ: Science Press. 1978.

Mind from matter?? In: N.H. Heidcamp (ed.), *The Nature of Life. XIII Nobel Conference*. Baltimore, MD: University Park Press. 1985. (Shorter version in *The American Scholar*, *47*: 339–353.)

Was Bose-Einstein statistics arrived at serendipity? *Chem. Educ.*, *57*: 467–474. 1980.

15 Francis Crick (1916–2008) and James Watson (1928–)

CRICK–WATSON COLLABORATION

They were a strange unlikely pair. Francis Crick was an English physicist and Jim Watson was a young American biologist. Crick was older than Watson by 12 years. He was still working for his doctorate, whereas Watson had already obtained his PhD from Indiana University at the age of 22. Their first meeting was in the fall of 1951, when Jim Watson arrived at the Cavendish Laboratory by a circuitous route which involved Indiana, Copenhagen and Naples.

In his best-selling autobiographical account, *The Double Helix*, Watson wrote: "From my first day in the lab I knew I would not leave Cambridge for a long time. Departing would be idiocy, for I had immediately discovered the fun of talking to Francis Crick. Finding someone in Max [Perutz]'s lab who knew that DNA was more important than proteins was real luck. Moreover, it was a great relief for me not to spend full time learning X-ray analysis of proteins. Our lunch conversations quickly centered on how genes were put together. Within a few days after my arrival, we knew what to do: imitate Linus Pauling and beat him at his own game" (p. 37).

FRANCIS CRICK

Francis grew up in a household of business and religion. His birthplace was Holmgarth, the family home on the outskirts of the village of Weston Favell, close to the town of Northampton in England. The Cricks were not as wealthy as some of their neighbors, but they occupied a comfortable middle-class home in a pleasant neighborhood. They had no car, so the garage was conveniently available for amateur theatricals and various chemical experiments, which Crick later recalled with much pleasure. The experiments often involved electrical circuits and glass bottles containing explosive mixtures, strangely echoing what he would do later, designing electrical circuits to activate mines during the Second World War. Francis never knew his grandfather Walter Sr. but admired his interest in science. A prominent member of the

Northampton Natural History Society, Walter Sr. was the author of several papers on geology and paleontology. He was proud of corresponding with Charles Darwin, who mentioned him in a note sent for publication at the end of his life (Darwin, C. 1882. On the dispersal of freshwater bivalves. *Nature 25*: 529–530). However, as Francis's interest in science grew, there was no one in the family with whom he could talk about science. Francis found some answers in an old Children's Encyclopedia from which he "absorbed great chunks of explanation." As he learned more and more about science, it was also the beginning of his disaffection with the Church and religion. As Francis got older, growing skepticism about religion gradually led to a full-blown acceptance of atheism and agnosticism in later life.

Crick chose University College London (UCL) for his university education, mainly because, in contrast to Oxbridge, it was totally devoid of any religious associations. Any form of religious teaching or requirement for religious conformity was forbidden in UCL's statutes. One of his friends later wrote: "They were very interesting times from 1933 to 1939, with plenty of opportunity for political comment…. He…was moving from the Protestant Christianity of his youth…towards his later firm atheism." There was scandal; the abdication of Edward VIII on December 10, 1936, because of his relationship with Mrs. Wallis Simpson, with the accompanying ribald jokes and cartoons. Odile Crick later recalled: "Hark the herald angels sing. Mrs. Simpson has stole our King." Crick was not impressed by the undergraduate physics lectures, which did not stimulate him. It was outdated physics that did not include the exciting developments in quantum theory and relativity. Crick continued his graduate studies at UCL, investigating a "dull" problem that he did not choose but was given to him, to build an apparatus to measure the viscosity of water under pressure at temperatures above 100°C. However, wartime events, especially the London blitz, which destroyed his apparatus, put an end to his doctoral research. He spent the years 1940–1947 in wartime research.

Genetics

A major stimulus for Crick's introduction to biology and genetics was a small book titled *What Is Life?* by the Austrian scientist and Nobel laureate Erwin Schrödinger. It conveyed an exciting idea that in biology, molecular explanations would not only be extremely important but also that they were just around the corner. It also focused attention on the molecular structure of the genetic material. Crick had already followed the earlier work of the physicist John Desmond Bernal, who produced, with Isidore Fankuchen, beautiful X-ray pictures of the crystals of tobacco mosaic virus. Crick found this subject appealing to his interests. For the first time in his life, he found a subject that turned his thoughts to biology and his own future in an entirely new direction.

When Crick began studying biology in 1947, he was one of several physical scientists who ventured into biological research. Crick had to adjust from the "elegance and deep simplicity" of physics to the "elaborate chemical mechanisms that natural selection had evolved over billions of years." He described this transition as, "almost as if one had to be born again." According to Crick, the experience of learning physics had taught him something important—hubris—and the conviction that since physics was already a success, great advances should also be possible in other sciences such as biology. Crick felt that this attitude encouraged him to be more daring than typical biologists who tended to concern themselves with the daunting problems of biology and not the past successes of physics.

Fortuitously, when he was visiting Edward Mellanby, the Secretary of the Medical Research Council (MRC), to report progress on his research with the physical properties of cytoplasm at the Strangeways Laboratory, he learned of a new proposal to establish an MRC unit at the Cavendish laboratory to study the structure of proteins using the method of X-ray diffraction. It was going to be headed by Max Perutz. To his great surprise, Crick's opinion about this proposed unit was asked. In his autobiographical book, *What Mad Pursuit: A Personal View of Scientific Discovery*, Crick wrote: "To my surprise (because I was still very junior), he asked me what I thought about it. I said I thought it was an excellent idea. I also told Mellanby that now that I had a background in biology, I would like to work on protein structure, since I felt my abilities lay more in that direction. This time he raised no objection, and the way was cleared for me to join Max Perutz and John Kendrew at the Cavendish" (p. 23).

The Cavendish Laboratory at Cambridge was under the general direction of Sir Lawrence Bragg, who shared a Nobel Prize with his father, in 1915 at the age of 25. Bragg was influential in the effort to beat a leading American chemist, Linus Pauling, to the discovery of DNA's structure (after having been piped at the post by Pauling's success in determining the alpha helix structure of proteins). At the same time, Bragg's Cavendish Laboratory was also effectively competing with King's College London, whose Biophysics Department was under the direction of Prof. John Randall. (Randall had earlier refused Crick's application to work at King's College.)

Meeting Max Perutz

About their first meeting, Perutz recalled meeting the tall Englishman with bushy eyebrows who would soon "put us all in high spirits by his laughter." As one biographer put it, "two years younger than Perutz and certainly not shy, the hopeful Crick joked and interacted in a lively manner with the restrained and quiet Perutz. One can imagine this meeting of the voluble and extroverted Crick at 6 foot 2 inches tall with Perutz, a mere 5 foot 6 inches, slight of frame, quietly spoken, dressed in subdued colors, his brown eyes glinting beneath a prominent forehead."

Crick, in his autobiography, *What Mad Pursuit*, described his position, "I was still, at this time, a beginning graduate student." But that did not stop Crick from dispensing advice to a meeting of experienced X-ray crystallographers. He wrote: "Bragg was furious. Here was this newcomer telling experienced X-ray crystallographers, including Bragg himself, who had founded the subject and been in the forefront of it for almost forty years, that what they were doing was most unlikely to lead to any useful result. The fact that I clearly understood the theory of the subject and indeed was apt to be unduly loquacious about it did not help. A little later I was sitting behind Bragg, just before the start of a lecture, and voicing to my neighbor my usual criticism of the subject in a rather derisive manner. Bragg turned around to speak to me over his shoulder, 'Crick', he said, 'you're rocking the boat.'" (Crick 1988, pp. 50–51).

Early Steps

Crick was interested in two fundamental problems: how molecules make the transition from the nonliving to the living and how the brain makes a conscious mind. His background made him more qualified for research on the first topic and the field of biophysics. It was at this time of Crick's transition from physics to biology that he was influenced by both Linus Pauling and Erwin Schrödinger. As Crick saw it, Charles Darwin's theory of evolution and Gregor Mendel's genetics and knowledge of the molecular basis of genetics, when combined, revealed the secret of life. It was clear that some macromolecule such as a protein was likely to be the genetic molecule. However, it was well known that proteins are structural and functional macromolecules, some of which carry out enzymatic reactions of cells. In the 1940s, some evidence had been found pointing to another macromolecule, DNA, the other major component of chromosomes, as a candidate genetic molecule. In 1944, Avery et al. showed that a heritable phenotypic difference could be caused in bacteria by providing them with a particular DNA molecule.

However, other evidence was interpreted as suggesting that DNA was structurally uninteresting and possibly just a molecular scaffold for the apparently more interesting protein molecules. Crick was in the right place, in the right frame of mind, at the right time (1949), to join Max Perutz's project at the University of Cambridge, and he began to work on the X-ray crystallography of proteins. X-ray crystallography theoretically offered the opportunity to reveal the molecular structure of large molecules like proteins and DNA, but there were serious technical problems then preventing X-ray crystallography from being applicable to such large molecules.

In the year preceding his meeting with Watson (1949–1950), Crick taught himself the mathematical theory of X-ray crystallography. At that time, researchers in the Cambridge lab were attempting to determine the most stable helical conformation of amino acid chains in proteins (the alpha helix). Linus Pauling was the first to identify the 3.6 amino acids per helix turn ratio of the alpha helix. Crick observed the kinds of errors that his coworkers made in their failed attempts to make a correct molecular model of the alpha helix. These turned out to be valuable lessons that proved helpful later when he was working with Watson on the helical structure of DNA. For instance, he learned the importance of the structural rigidity that double bonds confer on molecular structures, which is relevant both to peptide bonds in proteins and the structure of nucleotides in DNA.

1951–1953: DNA Structure

In 1951 and 1952, together with William Cochran and Vladimir Vand, Crick assisted in the development of a mathematical theory of X-ray diffraction by a helical molecule. This theoretical result matched well with X-ray data for proteins that contain sequences of amino acids in the alpha helix conformation. Helical diffraction theory turned out to also be useful for understanding the structure of DNA.

First Meeting with Watson

In his autobiographical account, *What Mad Pursuit*, Crick wrote that he first heard of Jim Watson from his wife Odile. One day, when he came home, Odile told him, "Max (Perutz) was here with a young American he wanted you to meet and—you know what—he had no hair!" Jim had a crew cut, which was still a novelty in Cambridge.

Crick wrote that he and Watson hit it off immediately, partly because their research interests were similar. But he added that they shared some other qualities as well—"a certain youthful arrogance, a ruthlessness, and an impatience with sloppy thinking came naturally to both of us." However, their background knowledge was quite different. Crick knew more about proteins and X-ray diffraction, while Watson knew a lot more about the experimental work on phages, as well as bacterial genetics. Their knowledge of classical genetics was about the same.

They were given a separate room at the Cavendish, where they could converse and argue with each other without disturbing the others. Watson wrote of his first impressions of Crick in his memoir, *The Double Helix*, "I have never seen Francis Crick in a modest mood...It has nothing to do with his present fame. Already he is much talked about, usually with reverence and someday he may be considered in the category of Rutherford or Bohr... Often, he came up with something novel, would become enormously excited, and immediately tell it to anyone who would listen. A day or so later he would often realize that his theory did not work and return to experiments... There was much drama connected with these ideas. They did a great deal to liven up the atmosphere of the lab.... This came partly from the volume of Crick's voice: he talked louder and faster than anyone else and, when he laughed, his location within the Cavendish was obvious. Almost everyone enjoyed these manic moments... But, there was one notable exception. Conversations with Crick frequently upset the Director, Sir Lawrence Bragg... (Watson 1968, pp. 8–9).

Watson and Crick never did any experimental work on DNA (at least until then), but they talked endlessly about the problem. Following Linus Pauling's example, they believed that the way to solve the structure was by model-building. However, their first attempt was a fiasco because they thought, mistakenly, that the structure contained very little water. But, eventually, they were able to overcome this and other mistakes, luckily with some help from Rosalind Franklin and also a Caltech chemist, who was spending a sabbatical year in Cambridge, Jerry Donohue, who advised them that most textbook formulas were erroneous and that each base occurred almost exclusively in one form.

The key discovery was Watson's determination of the exact nature of the two base pairs; A with T, G with C. Crick wrote that Watson reached this conclusion by serendipity ("chance favors the prepared mind"). A curious detail: Linus Pauling could have scooped the Watson–Crick discovery, especially since on the same boat he was travelling on from a conference in Europe to the United States, there was a fellow passenger, chemist Erwin Chargaff, who could have told Pauling about the equivalence of A with T and G with C, but, for some reason, Linus took an immediate dislike toward his shipmate and avoided his company during that entire trip!

Conflict between Francis Crick and Lawrence Bragg

According to Watson's account in *The Double Helix*, there was much rancor and conflict between Francis and the Director Lawrence Bragg. On page 44, Watson wrote that Francis' interest in DNA temporarily fell to almost zero when his attention was drawn to a new manuscript by Bragg and Perutz on the shape of the hemoglobin molecule. As he read that paper, Francis became furious, "for he noticed that part of the argument depended upon a theoretical idea he had propounded some nine months earlier… Yet his contribution had not been acknowledged." When he confronted Bragg, Francis was told that Bragg had no prior knowledge of the matter. Furthermore, Bragg was thoroughly insulted by the implication that he had underhandedly used another scientist's ideas. For Bragg, this latest incident seemed to be the final straw in his relations with Francis. He told Francis that he was considering seriously whether he could continue to give Francis a position in the laboratory after his PhD work was completed! Francis was concerned about his future in science. His career was already delayed by the outbreak of the Second World War. Moreover, he had lost all interest in physics and was considering biology as his new frontier. He was excited to join Perutz and Kendrew and decided to enroll as a PhD student (of Caius College) at the age of 35, with Max as his supervisor. Although he was bored at the thought of the tedium involved in a thesis research, he was happy that he could hardly be dismissed before he completed the formalities for his degree.

Max Perutz and John Kendrew quickly came to Francis' rescue and interceded with Bragg. Everyone agreed to resolve the conflict by acknowledging that Francis and Bragg had the same idea independently. The idea of getting rid of Francis was quietly shelved for the moment, but Bragg was not happy at the thought of Francis continuing at the Cavendish forever!

JIM WATSON

Watson opens his book, *Avoid Boring Other People*, with the following line: "I was born in 1928 in Chicago into a family that believed in books, birds, and the Democratic Party." From an early age, Watson was brought up in an intellectual household with an appreciation of nature. His father took him on bird watching trips, which he continued throughout his college years. He recalled: "I remained all through my college years a fervent ornithologist, especially during the spring and fall migrations…. The birds that fascinated me the most were the shorebirds, ranging from the tiny sandpipers to the much larger curlews. I was always on the lookout for the very rare red Wilson and northern phalaropes that Dad had seen when he was a boy. So I was tremendously thrilled when one day… I spotted three northern phalaropes spinning in the shallow water" (Watson 2007, p. 25).

Watson grew up on the south side of Chicago, attending public schools, including Horace Mann Grammar School and Southshore High School.

When he was only 14, Watson proudly appeared on *Quiz Kids*, a popular radio show that challenged bright youngsters to answer questions. He enrolled at the University of Chicago, where he was awarded a tuition scholarship, at the age of 15. Watson received a fine education, coming into contact with such luminaries as Sewall Wright, who was one of the founders of population genetics, and becoming aware of other prominent scientists such as Herman J. Muller at Indiana, who was just awarded the Nobel Prize in 1946.

In his autobiography, *Avoid Boring Other People*, Watson described the University of Chicago as an idyllic academic institution where he was instilled with the capacity for critical thought. Mainly attracted by the presence of Nobel laureate Muller, in 1947, Watson decided to become a graduate student at Indiana University in Bloomington. He was interested in Muller's research on the nature of the gene, based on his experiments with the fruit fly *Drosophila*. However, Watson wrote of his disappointment with Muller's lectures: "Emanating from a short, heavyset man almost the shape of a *Drosophila* himself, Muller's lectures were streams of consciousness rather than prepared orations. His agitated speech mingled clever genetic reasoning with details of his frustrations over, say, not initially being accepted into T.H. Morgan's lab, and later when finally a member having his ideas given short shrift. Much less absorbing were the lab sessions, in which we were chaotically run through an increasingly complex set of genetic crosses. The insights of such experiments seemed rather arcane, pointing to a truth that could not be avoided: *Drosophila's* days as a model organism were over. Indeed, a new one would soon supplant it as the premier tool for studying the gene." Watson correctly realized even before the fall term was half over that he did not want to do his degree with Muller. Watson further wrote that through Luria's virus course lectures, he saw the genetic wave of the future unfolding; "The key would be microorganisms..." The solution was obvious to Watson, he decided to study with Luria and received his PhD degree from Indiana University in 1950; Salvatore Luria was his doctoral advisor.

Watson's First Visit to Cold Spring Harbor Laboratory: The Phage Group

Watson was initially drawn into molecular biology by the work of Salvatore Luria, who shared a Nobel Prize for his work on the Luria–Delbruck experiment, which concerned the nature of genetic mutations. Luria and Max Delbruck were the leaders of this new "Phage Group." Early in 1948, Watson began his PhD research in Luria's laboratory at Indiana University. That spring, he met Delbrück first in Luria's apartment and again that summer during Watson's first trip to the Cold Spring Harbor Laboratory (CSHL), on Long Island.

Watson's first experience of being an active scientist was with the Phage Group. Importantly, the members of the Phage Group sensed that they were on the path to discovering the physical nature of the *gene*. In 1949, Watson took a course with Felix Haurowitz that included the conventional view of that time: that genes were proteins and able to replicate themselves. The other major molecular component of chromosomes, DNA, was widely considered to be a "stupid tetranucleotide," serving only a structural role to support the proteins. However, even at this early time, Watson, under the influence of the Phage Group, was aware of the work of Avery et al. (1944), which suggested that DNA was the genetic molecule. Watson's research project involved using X-rays to inactivate bacterial viruses. He earned his PhD in Zoology at Indiana University in 1950 (at age 22).

Watson then went to Copenhagen University in September 1950 for a year of postdoctoral research, first heading to the laboratory of biochemist Hermann Kalckar, who was interested in the enzymatic synthesis of nucleic acids, and he wanted to use phages as an experimental system. Watson, however, wanted to explore the structure of DNA, and his interests did not coincide with Kalckar's. After working part of the year with Kalckar, Watson spent the remainder of his time in Copenhagen conducting experiments with microbial physiologist Ole Maaloe, then a member of the Phage Group.

During the previous summer's Cold Spring Harbor phage conference, Watson learned the use of radioactive phosphate as a tracer to determine which molecular components of phage particles actually infect the target bacteria during viral infection. The intention was to determine whether protein or DNA was the genetic material, but their results were inconclusive and could not specifically identify the newly labeled molecules as DNA. Watson never developed a constructive interaction with Kalckar, but he did accompany Kalckar to a meeting in Italy, where Watson attended Maurice Wilkins' talk about his X-ray diffraction data for DNA, which later played a crucial role in the Watson–Crick discovery of DNA molecular structure.

Cambridge University

Upon meeting John Kendrew from the U.K., Luria arranged for a new postdoctoral position for Watson at Cambridge University, where Watson wanted to perform X-ray diffraction experiments to determine the structure of DNA. Linus Pauling published his model of the amino acid alpha helix in 1951, a result that grew out of Pauling's efforts in X-ray crystallography and molecular model building. After his phage work and other experimental research, Watson decided to perform X-ray diffraction experiments to determine the structure of DNA.

1951–1953: WATSON–CRICK COLLABORATION

When Watson came to Cambridge, Crick was a 35-year-old graduate student and Watson was a 23-year-old PhD. They were interested in learning how genetic information might be stored in molecular form. Watson recorded his early impressions of Francis Crick in *Avoid Boring People* (pp. 96–97): "The unit's resident theoretician was by then the physicist Francis Crick, who at thirty-five was two years younger than Max Perutz and one year older than John Kendrew. Francis was of middle-class, nonconformist, Midlands background, though his father's long prosperous shoe factories in Northampton failed during the Great Depression of the 1930s. It was only with the help of a scholarship from Northampton Grammar School that

Francis moved to the Mill Hill School in North London, where his father and uncle had gone. There he liked science but never pulled out the grades required for Oxford or Cambridge. Instead he studied physics at University College London, afterward staying on for a Ph.D. financially sponsored by his uncle Arthur."

Watson and Crick talked and talked about DNA and the idea that it might be possible to guess a good molecular model of its structure. In November 1951, Wilkins came to Cambridge and shared his data with Watson and Crick. Alexander Stokes (another expert in helical diffraction theory) and Wilkins (both at King's College) had reached the conclusion that X-ray diffraction data for DNA indicated that the molecule had a helical structure—but Rosalind Franklin vehemently disputed this conclusion. Stimulated by their discussions with Wilkins and what Watson learned by attending a talk given by Franklin about her work on DNA, Crick and Watson produced and showed off an erroneous first model of DNA. Their hurry to produce a model of DNA structure was driven in part by the knowledge that they were competing against Linus Pauling.

Watson and Crick were not officially working on DNA. Crick was writing his PhD thesis; Watson also had other work such as trying to obtain crystals of myoglobin for X-ray diffraction experiments. In 1952, Watson performed X-ray diffraction on tobacco mosaic virus and found results indicating that it had helical structure. Having failed once, Watson and Crick were now somewhat reluctant to try again, and for a while, they were forbidden to make further efforts to find a molecular model of DNA.

Of great importance to the model-building effort of Watson and Crick was Rosalind Franklin's understanding of basic chemistry, which indicated that the hydrophilic phosphate-containing backbones of the nucleotide chains of DNA should be positioned so as to interact with water molecules on the outside of the molecule while the hydrophobic bases should be packed into the core. Franklin shared this chemical knowledge with Watson and Crick when she pointed out to them that their first model (from 1951, with the phosphates inside) was obviously wrong.

Crick described what he saw as the failure of Wilkins and Franklin to cooperate and work toward finding a molecular model of DNA as a major reason why he and Watson eventually made a second attempt to do so. They asked for, and received, permission to do so from both William Lawrence Bragg and Maurice Wilkins. In order to construct their model of DNA, Watson and Crick made use of information from unpublished X-ray diffraction images of Franklin's (shown at meetings and freely shared by Wilkins), including preliminary accounts of Franklin's results/photographs of the X-ray images that were included in a written progress report for the King's College laboratory of Sir John Randall from late 1952.

It is a matter of debate whether Watson and Crick should have had access to Franklin's results without her knowledge or permission, and before she had a chance to publish the detailed analysis of her X-ray diffraction data, which were included in the progress report. However, Watson and Crick found fault in her steadfast assertion that, according to her data, a helical structure was not the only possible shape for DNA—so they had a dilemma. In an effort to clarify this issue, Max Perutz later published what had been in the progress report and suggested that nothing was in the report that Franklin herself had not said in her talk (attended by Watson) in late 1951. Perutz explained further that the report was to an MRC committee that had been created in order to "establish contact between the different groups of people working for the Council." Randall's and Perutz's laboratories were both funded by the MRC.

X-RAY DIFFRACTION STUDY

Using experimental data collected by Rosalind Franklin (without her permission), who worked in the laboratory of Maurice Wilkins, Watson and Crick deduced, in March 1953, the double helix structure of DNA. Sir Lawrence Bragg, the director of the Cavendish Laboratory, made the original announcement of the discovery at a Solvay conference on proteins in Belgium on April 8, 1953. Watson and Crick submitted a paper to the scientific journal *Nature*, which was published on April 25, 1953, but it was mostly ignored by scientists at large. Bragg gave a talk at the Guy's Hospital Medical School in London on Thursday, May 14, 1953, which resulted in a May 15, 1953, article by Ritchie Calder in the London newspaper *News Chronicle*, entitled "Why You Are You. Nearer Secret of Life."

The Cambridge University student newspaper *Varsity* also ran its own short article on the discovery on Saturday, May 30, 1953. Watson subsequently presented a paper on the double-helical structure of DNA at the 18th Cold Spring Harbor Symposium on Viruses in early June 1953, six weeks after the publication of the Watson and Crick paper in *Nature*. It was the first opportunity for many to see the model of the DNA double helix.

WATSON–CRICK PAPER OF 1953

There were three Watson–Crick papers on DNA in 1953 and a fourth one that was published in the following year. The order of authorship is of interest. Judging by surviving drafts of the papers, an earlier draft of the short note, which announced their achievement (published 25 April), was written by Watson, but all the subsequent corrections were in Crick's handwriting. The first draft of their second collaborative paper on the genetic implications of their model was written by Crick. The manuscript of the third paper, which was written for the 1953 CSHL symposium, was in Watson's handwriting, although it indicated a joint effort by both Crick and Watson. The fourth paper was clearly written by Watson, with minor additions by Crick.

The first paper was entitled "Molecular Structure of Nucleic Acids. A Structure for Deoxyribonucleic Acid." It occupies about one page, beginning with the sentence: "We wish to suggest a structure for the salt of deoxyribonecleic acid (D.N.A.)." The editors suggested the inclusion of periods in D.N.A. The authors continued: "This structure has novel features which are of considerable biological interest" and added at the end: "It has not escaped our notice that the specific pairing we have postulated immediately suggests a possible copying mechanism for the genetic material."

In a seven-page, handwritten letter to his son at a British boarding school on March 19, 1953, Crick explained his discovery, beginning the letter "My Dear Michael, Jim Watson and I have probably made a most important discovery…" The letter was put up for auction at Christie's New York on April 10, 2013, with an estimate of $1 to $2 million, eventually selling for $6,059,750, the largest amount ever paid for a letter at auction.

CENTRAL DOGMA

The adaptor molecules were eventually shown to be tRNAs, and the catalytic "ribonucleic–protein complexes" became known as ribosomes. An important step was later realization (in 1960) that the messenger RNA was not the same as the ribosomal RNA. None of this, however, answered the fundamental theoretical question of the exact nature of the genetic code. In his 1958 article, Crick speculated, as had others, that a triplet of nucleotides could code for an amino acid. Such a code might be "degenerate," with $4 \times 4 \times 4 = 64$ possible triplets of the four nucleotide subunits while there were only 20 amino acids. Some amino acids might have multiple triplet codes. Crick also explored other codes in which, for various reasons, only some of the triplets were used, "magically" producing just the 20 needed combinations. Experimental results were needed; theory alone could not decide the nature of the code. Crick also used the term "central dogma" to summarize an idea that implies that genetic information flow between macromolecules would be essentially one way:

$$\text{DNA} \rightarrow \text{RNA} \rightarrow \text{Protein}$$

Some critics thought that by using the word "dogma," Crick was implying that this was a rule that could not be questioned, but all he really meant was that it was a compelling idea without much solid evidence to support it. In his thinking about the biological processes linking DNA genes to proteins, Crick made explicit the distinction between the materials involved, the energy required, and the information flow. Crick was focused on this third component (information), and it became the organizing principle of what became known as molecular biology. Crick had by this time become a highly influential theoretical molecular biologist.

WATSON'S LIST OF LESSONS DRAWN FROM DNA WORK

The following points were listed by Watson under "Remembered Lessons" (from *Avoid Boring People*).

1. Choose an objective apparently ahead of its time.
   Watson wrote: "Better to leapfrog ahead of your peers by pursuing an important objective that most others feel is not for the current moment. The 3-D structure of DNA in 1951 was such an objective, regarded by virtually all chemists as well as biologists as unripe."
2. Work on problems only when you feel tangible success may come in several years (three to five years).
   Many big goals are truly ahead of their time.

3. Never be the brightest person in a room.

   Getting out of intellectual ruts more often than not requires unexpected intellectual jousts. Nothing can replace the company of others who have the background to catch errors in your reasoning or provide facts that may either prove or disprove your argument of the moment...being the top dog in the pack can work against greater accomplishments.... By the early 1950s, Linus Pauling's scientific interactions with fellow scientists were effectively monologues instead of dialogues. He then wanted adoration, not criticism.
4. Stay in close contact with your intellectual competitors.
5. Work with a teammate who is your intellectual equal.

   Two scientists acting together usually accomplish more than two loners each going their own way.
6. Always have someone to save you.

   In leaving one field for another, it never makes sense to burn your past intellectual bridges at least until your new career has taken off.

Advice for graduate students:

1. Choose a young thesis adviser.

   Watson decided not to stay with H.J. Muller for his PhD work.

   (Because the heyday of *Drosophila* research that Muller then carried on had long since passed.)

   Watson wrote: "I most certainly from being Salva Luria's first Ph.D. student and not having to share his attention with other students".
2. Extend yourself intellectually through courses that initially frighten you.
3. Humility pays off during oral exams.
4. Avoid advanced courses that waste your time.
5. Keep your intellectual curiosity much broader than your thesis objective.

Comments on teaching and research:

1. Teaching can make your mind move on to big problems.
2. Lectures should not be unidimensionally serious.
3. Give your students the straight dope.
4. Encourage undergraduate research experience.
5. Focus departmental seminars on new science.
6. Immediately write up big discoveries.
7. Travel makes your science stronger.

Comments on how to write a best-seller, *The Double Helix*:

1. Be the first to tell a good story.
2. A wise editor matters more than a big advance.

   (An innovative book usually takes more time to write and may cost more money to produce than either you or your producer would guess at the time...of contract.)

3. Find an agent whose advice you will follow.
4. Use snappy sentences to open your chapters
5. Don't use autobiography to justify past actions or motivations.
   (A major reason for writing autobiography is to prevent later biographers getting the basic facts of your life wrong.)
6. Avoid imprecise modifiers
   (Modifiers such as *very*, *much*, *largely*, and *possibly* don't convey useful information and only reduce the impact of otherwise crisp language.)
7. Always remember your intended reader.
8. Read out loud your written words.
   (To make *The Double Helix* read smoothly, I read aloud every sentence to see if it made sense when spoken.)

HARVARD UNIVERSITY

In 1956, Watson joined the Biology department at Harvard University. His work at Harvard focused on RNA and its role in the transfer of genetic information. At Harvard University, Watson achieved a series of academic promotions in quick succession, from assistant professor to associate professor to full professor of biology. Watson stated, however, that he did not receive the customary raise of $1000 in salary after winning the Nobel Prize. In his book *Avoid Boring Other People* (Watson 2007, p. 195), Watson wrote: "From the moment of my Nobel Prize, I took comfort in expecting a larger than ordinary annual salary raise. Over the past two years, I had twice received an annual increase of $1,000, so when I opened the small envelope coming on July 2 from University Hall I expected to see a $2,000 increase. Instead, the historian Franklin Ford, Bundy's successor as dean of the Faculty of Arts and Sciences, informed me that for my first time at Harvard I was to receive no raise at all. Instantly, I went ballistic and let all my friends know my outrage. Was an administrative blunder to blame for Harvard's failure to acknowledge the windfall of prestige that I had provided or did President Pusey want to send a message that celebrities had no place on his faculty and should consider going elsewhere?… Later, Franklin Ford called me to his office to say that no insult had been intended—rather, priority had been given to rewarding other professors whose salaries were particularly low." Nevertheless, Harvard appeared to have shown poor judgment in not rewarding Watson at the time of his Nobel award. His two previous raises were mistimed—given in years when he received no significant award.

Watson argued for a greater emphasis on molecular biology, a shift in focus for the school from classical biology, stating that disciplines such as ecology, developmental biology, taxonomy, physiology, etc., had stagnated and could progress only once the underlying disciplines of molecular biology and biochemistry had elucidated their underpinnings, going so far as to discourage their study by students.

Distinguished Harvard biologist and naturalist Edward O. Wilson had recorded his impressions of young Watson of those early years (Wilson 1994). In *Naturalist* (p. 219–225), Wilson wrote: "Without a trace of irony I can say I have been blessed with brilliant enemies…James Dewey Watson, the codiscoverer of the structure of DNA, served as one such adverse hero for me.… He came to Harvard as an assistant professor in 1956, also my first year at the same rank. At twenty-eight, he was only a year older. He arrived with a conviction that biology must be transformed into a science directed at molecules and cells and rewritten in the language of physics and chemistry..." However, Wilson later wrote: "When Watson became director of the Cold Spring Harbor Laboratory in 1968... He proved me wrong. In ten years he raised that noted institution to even greater heights by inspiration, fund-raising skills, and the ability to choose and attract the most gifted researchers. A new Watson gradually emerged in my mind."

Watson continued to be a member of the Harvard faculty until 1976, even though he took over the directorship of CSHL in 1968.

Views on Watson's scientific contributions while at Harvard are somewhat mixed. His most notable achievements in his two decades at Harvard may be what he wrote about science, rather than anything he discovered during that time. Watson's first textbook, *The Molecular Biology of the Gene*, set a new standard for textbooks, particularly through the use of concept heads—brief declarative subheadings. His next textbook was *Molecular Biology of the Cell*, in which he coordinated the work of a group of scientist–writers. His third textbook was *Recombinant DNA*, which described the ways in which genetic engineering has brought much new information about how organisms function. The textbooks are still in print. Watson has written several other books, which are listed in the references at the end of this chapter. These are mostly autobiographical, containing aspects of his life as well as science.

THE DOUBLE HELIX

In 1968, Watson wrote the best-seller *The Double Helix*, listed by the Board of the Modern Library as number seven in their list of 100 best nonfiction books. The book details the sometimes painful story of not only the discovery of the structure of DNA but also the personalities, conflicts, and controversy surrounding their work. Watson's original title was to have been *Honest Jim*, in that the book recounts the discovery of the double helix from his point of view and included many of his private emotional impressions at the time. The book changed the way the public viewed scientists and the way they work.

Some controversy surrounded the publication of the book. Watson's book was originally to be published by the Harvard University Press, but Francis Crick and Maurice Wilkins, among others, objected. Watson's home university dropped the project and the book was commercially published.

CRICK'S REACTION

Watson's bestseller, *The Double Helix*, brought fame and much irritation as well. To say that Crick was upset by Jim's version of the discovery does not even remotely express his feelings. According to one description, this "fresh, arrogant, catty, bratty and funny account" of how Watson and Crick came to discover the structure of DNA sent Crick into a fury.

In his letter of April 13, 1967, to Watson, Crick expressed his frustration, "The whole exercise of preparing this chatty, personal, gossipy, and egotistical account grossly invades my privacy…a violation of friendship." As a man who always valued his personal privacy, Crick was concerned that Watson wrote critically of his friends' failings and about their personal lives, and did so while they were still alive. He chided Watson in his letter, "It is not customary to write intimate books about your friends without their permission, at least until they are dead. I would remind you that Bertrand Russell delayed the publication of his autobiography till he was over 90, and that Lord Moran's much criticized account of Churchill's health was not published till after the latter's death. The fact that a man is well known does not by itself excuse his friends from respecting his privacy while he is alive. Only if a person himself either gives permission or discusses his own personal affairs in public should his friends feel free to write about them. The only exception is when private matters are of prime and direct concern, as in the case of Mrs. Simpson and King Edward, and even then the British press wrote nothing for months."

Should matters related to a scientific discovery be an exception to this rule? Crick addressed this aspect in his letter:

"But the point of science is what is discovered, not how it was discovered or by whom. It is the results which need to be brought home to the public. It is quite inexcusable to invade someone's privacy to describe how the structure of DNA was discovered to people who don't even know what it is, nor why it is important. I have no objection to a genuine historical description. It is vulgar popularization which is indefensible."

Crick was not alone in expressing concerns about violation of privacy in *The Double Helix*. French scientist Andre Lwoff commented that Watson's "cold objectivity is applied to persons he likes, as it is to crystals or base-pairing. Very few are spared. May God protect us from such friends!"

Watson's manuscript was first submitted to Harvard University Press. However, because of the objections raised by Crick and Maurice Wilkins, Harvard President overruled the press, and it was rejected. It was eventually published by the Atheneum press. During the process, Crick, Wilkins, and Watson's publishers all consulted lawyers to assess their risk. Strangely, the director of the MRC Lab at Cambridge University, Sir Lawrence Bragg, decided to write the preface at Watson's request even though he knew that Crick objected to its publication. It was Bragg who, in May 1965, had urged Watson to write up the discovery for publication. His reasoning that Crick's "greater fame" would otherwise eclipse Watson's role in the discovery is not convincing. It is more likely that the conflicts that had occurred between Bragg and Crick in the preceding years may have left some rancor, leading to Bragg's support of Watson's book. One also wonders about the real motive of Bragg for suggesting to Watson to write up his discovery for publication! Why not ask both Crick and Watson to write up their joint discovery together?

It has been said that, over the years, Crick revised his opinion of *The Double Helix*, recognizing that, for a popular book, it conveys a surprising amount of science. Crick told the *Cambridge Evening News* that they accomplished the work in two months, and it involved mainly theorizing. No experiments were conducted, no animal models were used, and no long hours were spent in the laboratory. It involved many discussions, in the lab, in the Eagle Pub, and walking by the colleges, but the data came from many sources. The X-ray diffraction picture was taken by Rosalind Franklin (who died of ovarian cancer at the early age of 38 years).

It was Watson and Crick who sought out the data from diverse sources, evaluated their significance, and persisted in model building until they discovered the right one. Crick's early years of education and long training prepared him for the task. His experience of earlier protein studies prepared him well for X-ray crystallography.

COLD SPRING HARBOR AGAIN

In 1968, Watson became the director of the CSHL. Watson served as the laboratory's director and president for about 35 years, and later, he assumed the role of chancellor and then Chancellor Emeritus.

In his roles as director, president, and chancellor, Watson performed outstanding service, leading CSHL to articulate its present-day mission, "dedication to exploring molecular biology and genetics in order to advance the understanding and ability to diagnose and treat cancers, neurological diseases, and other causes of human suffering." CSHL substantially expanded both its research and its science educational programs under Watson's direction. He is credited with "transforming a small facility into one of the world's great education and research institutions. Initiating a program to study the cause of human cancer, scientists under his direction have made major contributions to understanding the genetic basis of cancer." In a retrospective summary of Watson's accomplishments there, Bruce Stillman, the laboratory's president, said, "Jim Watson created a research environment that is unparalleled in the world of science."

In October 2007, Watson was suspended following wide criticism of his views on race and intelligence, which he expressed in an interview with *Times* of London, and a week later, he retired at the age of 79 from CSHL from what the lab called "nearly 40 years of distinguished service." In a statement, Watson attributed his retirement to his age and circumstances that he could never have anticipated or desired.

HUMAN GENOME PROJECT

Watson was appointed, in 1990, as the associate director—in charge of the Human Genome Project—at the National Institutes of Health (NIH), a position he held until April 10, 1992. Watson left the Genome Project after conflicts with the new NIH Director, Dr. Bernadine Healy, as he was opposed to Healy's attempts to acquire patents on gene sequences and ownership of the "laws of nature." Two years before stepping down from the Genome Project, he had stated his own opinion on this long and ongoing controversy, which he saw as an illogical barrier to research; he said, "The nations of the world must see that the human genome belongs to the world's people, as opposed to its nations." He left within weeks of the 1992 announcement that the NIH would be applying for patents on brain-specific cDNAs.

It must also be said that neither Watson nor Crick ever sought to benefit commercially from being the codiscoverers of their great discovery.

In 1994, Watson became president of CSHL. Francis Collins succeeded Watson as director of the Human Genome Project at the NIH.

In 2007, James Watson became the second person to publish his fully sequenced genome online. Watson was quoted as saying, "I am putting my genome sequence on line to encourage the development of an era of personalized medicine, in which information contained in our genomes can be used to identify and prevent disease and to create individualized medical therapies."

SOCIOPOLITICAL ACTIVITIES

Watson participated in several political protests, including the following:

- Vietnam: Watson joined 12 other faculty members of the Department of Biochemistry and Molecular Biology, including one other Nobel Prize winner, and spearheaded a resolution for "the immediate withdrawal of U.S. forces from Vietnam."
- Nuclear weapons: In 1975, on the "thirtieth anniversary of the bombing of Hiroshima," Watson, along with "over 2000 scientists and engineers," spoke out against nuclear proliferation to President Ford in part because of the "lack of a proven method for the ultimate disposal of radioactive waste" and because "The writers of the declaration see the proliferation of nuclear plants as a major threat to American liberties and international safety because they say safeguard procedures are inadequate to prevent terrorist theft of commercial reactor-produced plutonium."
- Genetics: In 2007, Watson said, "I turned against the left wing because they don't like genetics, because genetics implies that sometimes in life we fail because we have bad genes. They want all failure in life to be due to the evil system."

For some inexplicable reason, the left classified genetics with other evils such as global warming, although genetics has contributed (and continues to contribute) many benefits to humanity, such as increased food production as well as numerous advances in medicine and public health.

FURTHER READING

Crick, F.H.C. (1988) *What Mad Pursuit: A Personal View of Scientific Discovery.* New York: Basic Books.

De Kruif, P. (1926) *Microbe Hunters.* New York: Harcourt Brace Inc.

Dronamraju, K.R. (1999) Erwin Schrödinger and the Origins of Molecular Biology. *Genetics, 153*: 1071–1076.

Friedberg, E.C. (2005) *The Writing Life of James D. Watson.* Cold Spring Harbor, NY: Cold Spring Harbor Laboratory Press.

McElheny, V.K. (2003) *Watson and DNA: Making a Scientific Revolution.* New York: Perseus Publishing.

Sayre, A. (1975) *Rosalind Franklin and DNA.* New York: W.W. Norton & Company Inc.

Schrödinger, E. (1946) *What Is Life?* Cambridge: Cambridge University Press.

Watson, J.D. (1968) *The Double Helix: A Personal Account of the Discovery of the Structure of DNA.* New York: Atheneum.

Watson, J.D. (2000) *A Passion for DNA: Genes, Genomes and Society*. Cold Spring Harbor, NY: Cold Spring Harbor Laboratory Press.

Watson, J.D. (2001) *Genes, Girls and Gamow: After the Double Helix*. New York: Alfred A. Knopf.

Watson, J.D. (2003) *DNA: The Secret of Life*. New York: Alfred A. Knopf.

Watson, J.D. (2007) *Avoid Boring People*. New York: Alfred A. Knopf.

Watson, J.D. (2012) *The Annotated and Illustrated Double Helix*. New York: Simon & Schuster.

Watson, J.D. (2014) *Molecular Biology of the Gene*. 7th ed. Boston: Pearson.

Watson, J.D., and F.H.C. Crick (1953) Molecular structure of nucleic acids: a structure for deoxyribose nucleic acid. *Nature, 171*: 737–738.

Watson, J.D. et al. (1992) *Recombinant DNA*. 2nd ed. New York: W.H. Freeman and Company.

REFERENCES

Avery, O.T., C.M. MacLeod, and M. McCarty (1944) Studies on the chemical nature of the substance inducing transformation of pneumococcal types. Induction of transformation by a desoxyribosenucleic acid fraction isolated from pneumococcus type III. *J. Exp. Med.*, 79: 137–158.

Wilson, E.B. (1994) *Naturalist*. Washington, DC: Island Press.

Section V

Radiation Genetics

In 1926, while at the University of Texas, Herman J. Muller carried out two experiments with varied doses of X-rays, the second of which used the crossing over suppressor stock ("ClB") of *Drosophila*. He found a clear, quantitative connection between radiation and lethal mutations. Muller's discovery created a media sensation after he delivered a paper entitled "The Problem of Genetic Modification" at the Fifth International Congress of Genetics in Berlin. In 1946, Muller was awarded the Nobel Prize for the discovery that mutations can be induced by X-rays. In his Nobel Prize lecture, Muller argued that there was no threshold dose of radiation that did not produce mutagenesis, which led to the adoption of the linear-no-threshold model of radiation on cancer risks.

He was a signatory (with many other scientists) of the 1958 petition to the United Nations calling for an end to nuclear weapons testing, which was initiated by the Nobel Prize-winning chemist Linus Pauling.

The growth of the nuclear industry has increased the importance of radiation genetics as the theoretical basis for predicting the long-term genetic effects of increased background radiation in human environment. Another trend is space radiation genetics, which studies the patterns of the genetic effects of cosmic rays in relation to other spaceflight factors, including weightlessness and excessive gravitational force.

16 Hermann J. Muller (1890–1967)

Hermann Joseph Muller (1890–1967) was the founder of radiation genetics and one of the few geneticists who exercised enormous influence upon the successful development of genetics in the first half of the twentieth century. Muller won the Nobel Prize in Physiology or Medicine in 1946 for his earlier discoveries on the induction of genetic mutations by radiation. Muller was born in New York City, to a family of German Catholic immigrants on his father's side. On the mother's side, his ancestry was of mixed Episcopalian and Jewish origin. Young Hermann lost his father at the age of 10 and had a difficult life during his early years. Much of his early and middle life was in fact spent in hardship, which was in part due to his inability to get along with others, but also because of his support for the Communist Party as well as his sympathy for the Soviet Union. However, his later years, especially from 1946 onward, when he was awarded the Nobel Prize, were spent in relative peace and stability.

Fortunately, Muller obtained a teaching assistantship in zoology at Columbia University in 1912, leading to his PhD with T.H. Morgan, which was awarded in 1916. The Muller–Morgan relationship was complex and not without some conflict as well as resentment on Muller's part. Muller first met Morgan as a teacher in 1910 after qualifying for his BA degree. Morgan taught courses in experimental embryology and only occasionally mentioned his genetic work during these classes. Muller had recorded that he was allowed to take a course with Morgan only as a graduate in 1910–1911, after his ideas on genetics had already become fairly crystallized. He chafed at not having opportunity to continue work in 1911–1912 but followed its development and helped in the genetic work occasionally.

According to Muller's pupil and biographer, Elof Carlson (1967a), who paraphrased abstracts of Muller's notes: "Muller revealed the personality difference which eventually led to his estrangement with Morgan." Muller wrote: "'It seemed to us students of Lock and Wilson that it was possible to go…further than Morgan did,… and unduly loathe the facts to their logical conclusions. In fact, he (Morgan) really sought in Mendelism and mutation, substitutes for natural selection, and he long refused to adopt generalized, clear-cut ideas of chromosomal Mendelian inheritance'" (Muller 1937, quoted from Carlson 1967a). Morgan hired Calvin Bridges as a laboratory helper to clean his fruit fly bottles and to pursue independent research. And he offered a Fellowship to A.H. Sturtevant, to pursue full-time research and study genetics. Muller resented Morgan's lack of appreciation or recognition of his own talent and interest and temporarily lost any hope of continuing fly research. Upon graduation, Muller taught at Cornell Medical School in midtown Manhattan during the day and hurried to teach English to foreigners in a public school.

Muller resented the extra work he had to seek outside to support himself, which left little time for his research, while Morgan provided full-time support for Sturtevant and Bridges for their research in the fly lab (Carlson 1981, Allen 1978). Nevertheless, Muller occasionally visited the fly lab to join in discussions with the fly group. However, he resented that Morgan did not give him full credit for his ideas and his name did not appear in its early publications.

Muller's PhD thesis (1916) dealt with the phenomenon of crossing over, which was used by A.H. Sturtevant for mapping genes. Muller was particularly skilled in designing complex stocks of *Drosophila* to solve genetic problems. He demonstrated that genes kept heterozygous for many generations still maintained their original characteristics. Through the analysis of small variable or continuous traits and fluctuations, Muller could identify the mechanism for variation, while ruling out the prevailing belief that the main gene itself varied through its association with its alleles in the heterozygous condition.

Muller's classic work, using truncated and beaded wings, demonstrated the complex relationship between the main gene and its genetic as well as its environmental modifiers (Muller and Altenburg 1920). From the time of his PhD dissertation, which was submitted in 1915, until 1920, Muller's entire research was concerned with *Drosophila* work. Muller began a study of the gene itself. He wanted to acquire a thorough knowledge of every genetic and environmental modifier. In each cross, he could predict the percent and intensity of expression of extreme, moderate, or mild wing defect and the percent of flies that showed no abnormality of the wing. Muller himself considered this work as a major contribution to evolutionary theory.

Muller traced the path of mutation after its origin by using developmental models of somatic and germinal distribution of mutant cells. He arrived at the conclusion that mutation could occur at any stage of development, from conception through embryonic stages, leading to gamete formation. He concluded that mutation is a very rare event, affecting a solitary gene rather than both of its alleles in a diploid cell. Muller further resolved two problems (Beaded and Truncate wings) that seem to have baffled Morgan. Muller designed "marker stocks" to search for the suspected genes that modified these character traits. For Beaded wings, Muller demonstrated the first instance of balanced lethals to account for them, and for the Truncate wing analysis, Muller (and Altenburg) located the main gene (later called dumpy), showing that Truncate was a lethal allele whose expression depended on modifiers in other chromosomes. These experiments led to a series of remarkable papers and helped Muller obtain a professorship in 1920 at the University of Texas.

Muller continued his research at the University of Texas, developing tests for measuring spontaneous mutation frequency. While exploring a chemical basis for mutation, Muller attempted to detect mutation frequencies at higher temperatures. Eventually, he turned to testing radiation as a mutagenic agent.

In 1923, in a classic paper on mutation, Muller identified 14 points about gene mutation, restricting the term "mutation" to a change in the individual gene. The gene and its mutations constituted the basis of evolution, providing the basic mechanism for variation that mystified Charles Darwin so much. Muller realized that most mutations are deleterious to the organism, that they were usually recessive, and that lethal mutations were much more frequent than those which have resulted in visible abnormalities in the adult.

In another paper, "The Gene as the Basis of Life," which was presented at the Ithaca Congress of Plant Genetics, preferring *the basis* of life, not *a basis* of life as suggested by the President of the Congress, Muller argued that the evolution of life required no miracles and evolution formed a continuum from the origin of the solar system to the future fate of mankind (Carlson 1967a).

RADIATION RESEARCH

In 1926, using a dentist's X-ray machine, Muller began an intensive series of experiments on the induction of mutations. In 1927, overcoming the setbacks and delays, Muller achieved his objective, he had artificially transmuted the gene. He proved his own earlier proposition that the artificial induction of mutations will be the source for model systems of evolution and the biophysical analysis of the gene (Muller 1927).

In November 1926, Muller carried out two experiments with varied doses of X-rays, the second of which used the crossing over suppressor stock ("ClB"), which he had found earlier in 1919. A clear, quantitative connection between radiation and lethal mutations quickly emerged. Muller's discovery created a media sensation after he delivered a paper entitled "The Problem of Genetic Modification" at the Fifth International Congress of Genetics in Berlin. Muller had effectively founded the new field of radiation genetics. He mapped the radiation-induced mutations and found that their distribution is similar to that of spontaneous mutations. He also applied the new method to study the problem of gene structure and found that the gene or chromatin was no more than a double thread. Other findings, such as visible mutations, and chromosome breaks such as inversions and translocations were found to be similar to the spontaneous anomalies observed in natural populations of *Drosophila*. Muller also found a unique class of mutation, associated with breakage, "eversporting displacements," which were later called variegated position effects. Muller also determined that the size of the gene could not be determined in any way with this procedure.

Earlier studies of that period by Muller include his attempts to define the gene as a unit, concluding that a gene differed from all other classes of molecules in one important respect; when altered by mutation, it did not lose its capacity to reproduce. It not only reproduced, but it also reproduced its defects. Another approach was to determine the biochemical nature of the gene; perhaps they are not unlike the bacteriophage. Although this line of investigation was not followed by Muller or others at that time, some 20 years earlier, English physician Archibald Garrod had already published his work on the genetics of human biochemical variants such as cystinuria. However, Garrod's work was also ignored for a long time!

Collaborating with the physicist Mott-Smith, Muller calculated that virtually none of the spontaneously arising mutations could be explained on the basis of the combined background radiation of cosmic rays and radioactive decay within and outside the environment of flies. Muller concluded that most spontaneous mutations arise from chemical alterations of the gene within the cell. Continuing his collaboration with Mott-Smith, Muller applied target theory to the mutation frequency at the white locus and concluded that the size of the gene could not be determined in any meaningful way by this approach.

PERSONAL DIFFICULTIES

It was a difficult period for Muller both scientifically and personally: his marriage was falling apart, and he was increasingly dissatisfied with his life and political climate in Texas. Public interest in the eugenic movement has been declining, partly because of new work (including his own work on twin studies) on the relation between genetics and evolution.

POLITICAL ACTIVITIES

With the stock market crash and the onset of depression and long unemployment lines during the 1930s, like so many other intellectuals of that period, including J.B.S. Haldane, Muller came to believe that a Capitalist society was incompatible with social justice. He became an active supporter of radical student groups and helped to edit and distribute an underground paper *The Spark*, supporting social causes.

Muller's career reached a crisis point in the 1930s due to the convergence of personal and professional problems. In addition to a failed marriage, Muller's disagreements with his departmental chairman J.T. Patterson compounded his problems. It was hard to continue his work at the University of Texas. Personal conflicts of Muller elsewhere in the United States made it harder to find a position at other major universities. With the aid of a Guggenheim Fellowship, Muller moved to Berlin, where he wanted to collaborate on the biophysical aspects of the gene, with Timofeef-Ressovsky, Zimmer, and Delbruck, but he had to leave Berlin when Hitler rose to power. A timely invitation from N.I. Vavilov provided Muller an opportunity to move to the Soviet Union. At first, matters proceeded smoothly. Muller supervised a large and productive lab and organized work on medical genetics. Most of his work involved further explorations of genetics and radiation. There, he completed his eugenics book, *Out of the Night*, the main ideas of which dated to his student days in 1910. However, that was also his undoing as it was rumored that Stalin read Muller's book and was displeased. Furthermore, difficulties arose with the rise of the anti-Mendelian pseudo-scientist Trofim Lysenko. Vavilov warned Muller that his life was in danger. Once again, Muller was on the move, joining the Canadian Blood Unit in the Spanish civil war, and he sought refuge in

Spain. However, as that war was drawing to a close, Muller was once again in dire need of a university position. Fortunately, his old friend Julian Huxley came to his rescue and recommended Muller to Francis Crew, director of the Institute of Animal Genetics at Edinburgh. Muller developed a thriving graduate program in Edinburgh, where some of his pupils later became famous scientists, including Chrlotte Auerbach, Guido Pontecorvo, and S.P. Ray-Chaudhuri. However, wartime economic constraints and personal reasons forced Muller once again to seek employment in the United States.

EDINBURGH PERIOD

With Julian Huxley's help, Muller obtained a temporary position at the Institute for Animal Genetics at Edinburgh University. He was then 50 years old. Muller attracted several students during that period, including some who became well-known geneticists in their own right, such as Charlotte Auerbach, Guido Pontecorvo, and S.P. Ray Chaudhury. With Pontecorvo, Muller worked out the breakage–fusion–bridge cycle (dicentric chromosome formation), independently of Barbara McClintock, and used it to explain the curious dominant lethals he had observed in large numbers since his X-ray work in 1927. The dicentric chromosomes led to cell death and aborted embryos (Pontecorvo and Muller 1942). With his student S.P. Ray Chaudhury, Muller extended his radiation studies to chronic and acute doses. Another finding was that, for gene mutations, it made no difference whether a dose of 400 R was administered in a few minutes or drawn out over a month-long period. In either case, the mutation rate was the same, confirming Muller's belief that gene mutations were punctiform events (Muller and Chaudhury 1939). The same observation got him embroiled in a dispute with British radiologists who considered it inappropriate it to extrapolate from flies to humans and that Muller's view might unnecessarily alarm the public about the uses of radiation.

MULLER'S CAREER AFTER RETURNING TO THE UNITED STATES

Muller found it difficult to find a faculty position in an American university after his return from Europe, which was mainly due to his past political activities. He took an untenured research position at Amherst College when he returned to the United States in 1940. His *Drosophila* work in this period focused on measuring the rate of spontaneous (as opposed to radiation induced) mutations. His publication rate decreased greatly in this period, from a combination of lack of lab workers and research facilities. Much of his time was spent in teaching. He also worked as an adviser to the Manhattan Project, although he did not know what it was then, as well as a study of the mutational effects of radar. Muller's appointment at Amherst ended after the 1944–1945 academic year. Fortunately, despite his socialist political activities, he was invited by Fernandus Payne, head of the Zoology department at Indiana University, to join their faculty.

In 1946, there was a dramatic shift in Muller's life with the announcement from Stockholm that he was awarded the Nobel Prize for Physiology or Medicine, "for the discovery that mutations can be induced by X-rays." Larger political and social events focused attention on Muller's earlier work on radiation genetics. The atomic bombing of Hiroshima and Nagasaki in 1945 magnified the radiation risks that Muller had observed earlier. Genetics, and especially the physical and physiological nature of the gene, became a central topic in biology. X-ray mutagenesis was playing a key role to understanding many recent advances.

MULLER'S POST-NOBEL YEARS

According to Muller's biographer, Elof Carlson, Muller's post-Nobel years were spent in an immense number of public activities, covering radiation hazards, evolution, peace, humanism, and the guidance of human evolution as a public necessity. In a remarkable address to the Indiana Biology teachers, which was titled "One Hundred Years without Darwinism Is Enough," Muller berated the high school texts that left out the word "evolution," replacing it with a euphemism such as "racial development," to appease conservative members of school boards who were hostile to the concept of evolution.

The Nobel Prize, following the atomic bombings of Hiroshima and Nagasaki, focused public attention on the dangers of radiation, a subject that Muller had been publicizing for two decades. With the increasing number of open air tests of nuclear weapons in the 1950s, nuclear fallout became a public issue. More and more evidence had been leaking out about radiation effects, particularly radiation sickness and deaths caused by nuclear testing. Because of his research on mutagenesis and his newly acquired status as a Nobel laureate, Muller was thrust more and more into public life, a spokesman to warn the public on the evils of open air testing and the nuclear weapons in general. Muller was very much in demand as a public speaker, a spokesman for those who were concerned about the evils of nuclear weapons and a possible nuclear winter caused by a war, where these weapons might be used, at the height of the tensions resulting from a weapons race with the Soviet Union.

The situation became even more urgent after the "Castle Bravo" controversy. Castle Bravo was the code name given to the first U.S. test of a dry fuel hydrogen bomb, detonated on March 1, 1954, at Bikini atoll-Marshall Islands. It was the most powerful nuclear device ever detonated, which led to the most significant accidental radioactive contamination ever caused by the United States. That explosion created an international outcry against atmospheric thermonuclear testing.

RUSSELL–EINSTEIN MANIFESTO

Under the leadership of Bertrand Russell (in which Einstein joined later), a statement highlighting the dangers posed by nuclear weapons that called for a gathering of world leaders was issued on July 9, 1955. Over the years that followed, Russell and Joseph Rotblat (the only scientist who left the Manhattan Project on moral grounds) worked on efforts to curb nuclear proliferation, collaborating with Albert Einstein and other scientists to compose what became known as the Russell–Einstein Manifesto. Several years later, Rotblat himself was honored by the award of the Nobel Peace Prize in 1995.

Muller was one of the signatories, which included 11 prominent intellectuals and scientists. Muller's participation lent credibility as he was the only Nobel laureate–scientist who discovered the mutagenic effects of atomic radiation.

They include Max Born, Percy W. Bridgman, Albert Einstein[1], Leopold Infeld, Frédéric Joliot-Curie, Hermann J. Muller, Linus Pauling, Cecil F. Powell, Joseph Rotblat, Bertrand Russell, and Hideki Yukawa.

A few days after the release, Canadian businessman and philanthropist Cyrus R. Eaton offered to sponsor a conference—called for in the manifesto—in his birthplace Pugwash, Nova Scotia. This conference was to be the first of the Pugwash Conferences on Science and World Affairs, held in July 1957. At first, it was to be held in India at the invitation of Prime Minister Nehru but was cancelled because of the pressure of other international events such as the Suez crisis.

Muller was also a signatory (with many other scientists) of the 1958 petition to the United Nations calling for an end to nuclear weapons testing, which was initiated by the Nobel laureate Linus Pauling.

MULLER AND JULIAN HUXLEY

Herman Muller and Julian Huxley were lifelong friends. Muller was three years younger than Huxley. Their friendship started when they were both in their 20s. In 1914, Muller met Julian Huxley who was visiting the famous "fly" room at Columbia University. In his *Memories*, Huxley wrote of his first visit to New York. "My first visit was to the famous 'fly-room' at Columbia University, where T.H. Morgan and his famous team, Sturtevant, Bridges, Muller and Altenburg, had just completed their first great piece of work on *Drosophila*, the now famous little fruitfly (Morgan et al. 1915). It—led to my inviting H.J. Muller to come to Rice[2] as my assistant—one of the most sensible things I ever did. Muller was a tower of strength, and managed to carry on with his genetical researches as well as helping me with teaching and demonstrating. He was later awarded the Nobel Prize for his splendid work on mutation and its artificial induction by X-rays and other agents" (Huxley 1970, p. 91).

Huxley's friendship with Muller lasted rest of their lives. Several years later, when Muller fled Soviet Union in 1939 and was in dire need for employment, Huxley once again came to his rescue and recommended him to Francis Crew, director of the Institute of Animal Genetics at Edinburgh University.

GERMINAL CHOICE AND EUGENICS

Early Interest

Muller's interest in the possibility of genetic applications to improve the quality of human populations was evident from his childhood and youth. At Columbia University, he joined a progressive undergraduate society called Peithologians in 1910 and presented a talk entitled "Revelations of Biology and Their Significance," projecting the new science of genetics into the future. He saw human evolution as its ultimate goal. These views were evident before he came into contact with Thomas H. Morgan and his "fly room."

These ideas remained dormant as Muller carried on with his research on *Drosophila* genetics. Many years later, in 1950, in his classic paper entitled "Our Load of Mutations," Muller emphasized that accumulated spontaneous mutations in human populations are no longer being selected out through infectious diseases, malnutrition, and a rigorous environment. Muller argued that if this problem is neglected, it would lead to a reversal of evolutionary trends in man, or it would be dysgenic. Muller went further. He believed that the status quo of the genetic load was not the best that genetic research could offer. Beyond preventive genetic medicine (negative eugenics), Muller hoped to launch a eugenics program which avoided the stigma of Aldous Huxley's *Brave New World*. He emphasized "germinal choice," i.e. freedom of choice, by motivated, intelligent, and healthy donors and recipients of germinal material would lead to children, through germinal choice of higher potential for those qualities which Muller identified with humanism: a healthy body, an intelligent mind, and a generous heart" (Carlson 1967b).

Who Will Choose the Values?

Muller responded to the criticism, "Who will choose the values?" He was optimistic. If the married couples, with an infertile husband, are informed of "germinal choice," Muller believed that the more intelligent and concerned couples would want to examine carefully the documented records of the donors and they themselves will determine their choice. It was believed that approximately 5% of married couples belonged to this category. It is hoped that the children, on the average, would have a lower genetic load of mutations and a higher genetic potential, leading one to expect, in due course, a reversal of the dysgenic effects of the accumulation of spontaneous mutation in man.

ENDNOTES

1. There is some irony in Einstein being one of the signatories of the manifesto because the U.S. Atomic Bomb program, which was later termed the "Manhattan Project," was originally initiated at the suggestion of Einstein, who urged President Franklin Roosevelt, in a letter, dated August 2, 1939, to set up such a program. Einstein warned Roosevelt that Germany might have started a similar program as it stopped the export of all uranium from Czechoslovakia (which was then under German occupation) and uranium was a main component of such bombs. Finally, when the atom bomb was exploded, it was on Japan, not Germany, where some suspect that race might have played a part in making that decision.
2. Rice Institute (later Rice University) in Houston, Texas.

REFERENCES

Allen, G.E. (1978) *Thomas Hunt Morgan: The Man and His Science*. Princeton: Princeton University Press. pp. 202–208.

Carlson, E.A. (1967a) The legacy of Hermann Joseph Muller: 1890–1967. *Can. J. Genet. Cytol., 9*: 436–448.

Carlson, E.A. (1967b) H.J. Muller. *Genetics, 56*: s24–s26.

Carlson, E.A. (1981) *Genes, Radiation and Society: Life and Work of H.J. Muller*. Ithaca: Cornell University Press.

Huxley, J.S. (1970) *Memories*. New York: Harper & Row. pp. 90–91.

Morgan, T.H., A.H. Sturtevant, H.J. Muller, and C.B. Bridges (1915) *The Mechanism of Mendelian Heredity*. New York: Henry Holt & Co.

Muller, H.J. (1927) Artificial transmutation of the gene. *Science, 66*: 84–87.
Muller, H.J. and E. Altenburg (1920) The genetic basis of truncate wing, an inconstant and modifiable character in Drosophila. *Genetics*, *5*: 1–59.
Muller, H.J. and S.P. Ray-Chaudhuri (1939) The validity of the Bunsen-Roscoe law in the production of mutations by radiation of extremely low intensity. Proceedings of the Seventh International Congress of Genetics. *J. Genet. Suppl.*, *246*: 194.
Pontecorvo, G. and H.J. Muller (1942) The surprisingly high frequency of spontaneous and induced breakage and its expression through germinal lethals. *Genetics*, *27*: 157–158.

Section VI

Transposons

Barbara McClintock was a cytogeneticist who started her career as the leader in the development of maize cytogenetics. During the 1940s and 1950s, McClintock discovered transposition and used it to demonstrate that genes are responsible for turning physical characteristics on and off. She developed theories to explain the suppression and expression of genetic information from one generation of maize plants to the next. She hypothesized that gene regulation could explain how complex multicellular organisms made of cells with identical genomes have cells of different function. McClintock's discovery challenged the concept of the genome as a static set of instructions passed between generations. Based on the reactions of other scientists to her work, McClintock felt she risked alienating the scientific mainstream and from 1953 stopped publishing accounts of her research on controlling elements. McClintock's work did not receive widespread recognition among scientists until the 1980s. In 1983, McClintock received the Nobel Prize for her experiments detailing the evidence for transposons. According to Evelyn Fox Keller, a historian of biology, McClintock was awarded the Nobel Prize more than 30 years after the publication of her results due to gender inequalities in science. Contrary to Fox Keller's interpretation, historian of science Nathaniel Comfort argued that McClintock had to wait so long before receiving scientific acclaim because some aspects of genetics were still unclear until the early 1980s, and therefore, McClintock's research required a more comprehensive framework to validate her results.

17 Barbara McClintock (1902–1992)

Barbara McClintock was born in Brooklyn, New York, on June 16, 1902. Her father, Thomas Henry McClintock, was a physician. As a young girl, Barbara's parents determined that her given name, Eleanor, was too "feminine" and "delicate" and was not appropriate for her. They chose Barbara instead. From a very early age, Barbara was an independent child, a trait she later identified as her "capacity to be alone." From the age of three onward, Barbara lived with an aunt and uncle in Brooklyn, New York, to reduce the financial burden on her parents while her father was building his medical practice. She was described as a solitary tomboy, who was close to her father but had a difficult relationship with her mother. The entire family moved to Brooklyn in 1908. Barbara completed her education there at Erasmus Hall High School, graduating in 1919. She discovered her love of science and reaffirmed her solitary personality during high school. She wanted to continue her studies at Cornell University's College of Agriculture. Her mother resisted sending McClintock to college, for fear that she would be unmarriageable. McClintock was almost prevented from starting college, but her father intervened just before registration began, and she matriculated at Cornell in 1919.

CORNELL UNIVERSITY

McClintock began her studies at Cornell University's College of Agriculture in 1919, receiving a BSc in Biology in 1923. Her interest in genetics began when she took her first course, taught by plant breeder C.B. Hutchison, in that field in 1921. Hutchison was impressed by McClintock's interest and selected her to participate in the graduate genetics course at Cornell, where she earned her PhD in 1927. During her graduate and postgraduate years as a botany instructor, McClintock gathered a group that studied the new field of cytogenetics of maize, which included Marcus Rhoades and the future Nobel laureate George W. Beadle. She was encouraged by the head of the plant breeding department Rollins A. Emerson.

EARLY WORK

Starting in the late 1920s, McClintock studied how chromosomes change during reproduction in maize. She developed the technique for visualizing maize chromosomes and, by using microscopic analysis, demonstrated genetic recombination and crossing over during meiosis, preparing the first genetic map for maize, linking regions of the chromosome to physical traits. Furthermore, she also demonstrated the role of the telomere and centromere in conserving genetic information.

While working with the geneticist Lewis Stadler at Missouri University during the summers of 1931 and 1932, McClintock was introduced to the use of X-rays as a mutagen. Through her work with X-ray-mutagenesis in maize, she identified "ring chromosomes," which form when the ends of a single chromosome fuse together after radiation damage. She concluded that there must be a structure on the chromosome tip that would normally ensure stability. In 1933, she established that cells can be damaged when nonhomologous recombination occurs. During this same period, McClintock hypothesized that the tips of chromosomes are protected by telomeres.

She showed that the loss of ring-chromosomes at meiosis caused variegation in maize foliage in generations subsequent to irradiation resulting from chromosomal deletion. During this period, she demonstrated the presence of the nucleolus organizer region on a region on maize chromosome 6, which is required for the assembly of the nucleolus.

CYTOGENETICS OF MAIZE

In 1931, McClintock published the first genetic map for maize, showing the order of three genes on maize chromosome 9. She was the first person to describe the interaction of homologous chromosomes during meiosis in 1930. She showed the connection between crossing over during meiosis and the recombination of genetic traits, demonstrating the correlation between recombination of chromosomes seen under a microscope and new traits. In 1938, she produced a cytogenetic analysis of the centromere, describing the organization and function of the centromere, as well as the fact that it can divide.

During the 1940s and 1950s, McClintock discovered transposition and used it to demonstrate that genes are responsible for turning physical characteristics on and off. She developed theories to explain the suppression and expression of genetic information from one generation of maize plants to the next; however, due to skepticism of her research and its implications, she stopped publishing her data in 1953.

McClintock identified specific chromosome groups of traits that were inherited together. McClintock's classic paper on the characterization of triploid maize chromosomes triggered wide interest in maize cytogenetics among biologists.

MUTATIONS

In her classic paper of 1950, McClintock wrote on mutation:

> The mutable loci fall into two major classes: (1) those that require a separate activator factor for instability to be expressed, and (2) those that are autonomous with respect to the factor that controls the onset of mutability. They also may be subdivided on a quite different basis. This is related to the types of expression of the

mutations that occur. The following types are present: (*a*) Changes from the mutant to, or close to, the wild-type expression. After such a mutation, the locus may be permanently stabilized. It may no longer show evidence of the instability phenomenon. (*b*) A second group, similar to (*a*) except that the mutation to wild-type does not produce stability of the locus. The wild-type-producing locus, in turn, may mutate to give the recessive expression. (*c*) A third type where the mutations give rise to a series of alleles of the affected loci. These alleles are distinguished by different degrees of quantitative expression of the normal phenotype. Most of these are relatively stable; only rarely does instability again appear. (*d*) A fourth type, similar to (*c*). Most of the alleles, however, are not stable for they, in turn, can mutate in the direction of a higher or lower grade of quantitative expression of the phenotype. Mutable loci showing these different types of expression of mutation are found in both the major classes, that is, in the activator-requiring class and in the autonomous class.

JUMPING GENES

While continuing her research on the effect of X-rays on maize cytogenetics, McClintock observed the breakage and fusion of chromosomes in irradiated maize cells. In some plants, spontaneous chromosome breakage occurred in the cells of the endosperm. She observed that the ends of broken chromatids were rejoined after the chromosome replication. In the anaphase of mitosis, the broken chromosomes formed a chromatid bridge, which was broken when the chromatids moved toward the cell poles. The broken ends were rejoined in the interphase of the next mitosis, and the cycle was repeated, causing massive mutation, which she could detect as variegation in the endosperm. This breakage–rejoining–bridge cycle was a key cytogenetic discovery for several reasons. First, it showed that the rejoining of chromosomes was not a random event, and second, it demonstrated a source of large-scale mutation. For this reason, it remains an area of interest in cancer research today.

Although her research was progressing at Missouri, McClintock was not satisfied with her position at the University. She recalled being excluded from faculty meetings and was not made aware of positions available at other institutions. McClintock believed she would not gain tenure at Missouri, even though according to some accounts, she knew she would be offered a promotion from Missouri in the spring of 1942. Recent evidence reveals that McClintock more likely decided to leave Missouri because she had lost trust in her employer and in the University administration, after discovering that her job would be in jeopardy if Stadler were to leave for Caltech, as he had considered doing.

In early 1941, she took a leave of absence from Missouri in hopes of finding a position elsewhere. She accepted a visiting professorship at Columbia University, where her former Cornell colleague Marcus Rhodes was a professor. Rhoades also offered to share his research field at Cold Spring Harbor on Long Island. In December 1941, she was offered a research position by Milislav Demerec, the newly appointed acting director of the Carnegie Institution of Washington's Department of Genetics at Cold Spring Harbor Laboratory; McClintock accepted his invitation despite her qualms and became a permanent member of the faculty.

COLD SPRING HARBOR

At Cold Spring Harbor, she was highly productive and continued her work with the breakage–fusion–bridge cycle, using it to substitute for X-rays as a tool for mapping new genes. In 1944, in recognition of her prominence in the field of genetics during this period, McClintock was elected to the National Academy of Sciences—only the third woman to be elected. That same year, she became the first female president of the Genetics Society of America.

In 1944, she undertook a cytogenetic analysis of *Neurospora crassa* at the suggestion of George Beadle, who used the fungus to demonstrate the one gene–one enzyme relationship. He invited her to Stanford to undertake the study. She successfully described the number of chromosomes of *N. crassa* and described the entire life cycle of the species. Beadle said, "Barbara, in two months at Stanford, did more to clean up the cytology of Neurospora than all other cytological geneticists had done in all previous time on all forms of mold." *N. crassa* has since become an ideal species for classical genetic analysis.

DISCOVERY OF CONTROLLING ELEMENTS

At Cold Spring Harbor, McClintock began systematic studies on the mechanisms of the mosaic color patterns of maize seed and the unstable inheritance of this mosaicism in 1944. She identified two new dominant and interacting genetic loci that she named *Dissociator* (*Ds*) and *Activator* (*Ac*). She found that the *Dissociator* did not just dissociate or cause the chromosome to break, it also had a variety of effects on neighboring genes when the *Activator* was also present, which included making certain stable mutations unstable. In early 1948, she made the surprising discovery that both *Dissociator* and *Activator* could transpose, or change position, on the chromosome.

Between 1948 and 1950, she developed a theory by which these mobile elements regulated the genes by inhibiting or modulating their action. She referred to Dissociator and Activator as "controlling units"—later, as "controlling elements"—to distinguish them from genes. She hypothesized that gene regulation could explain how complex multicellular organisms made of cells with identical genomes have cells of different function. McClintock's discovery challenged the concept of the genome as a static set of instructions passed between generations. In 1950, she reported her work on Ac/Ds and her ideas about gene regulation in a paper entitled "The Origin and Behavior of Mutable Loci in Maize," published in the journal *Proceedings of the National Academy of Sciences*. In summer 1951, she reported her work on the origin and behavior of mutable loci in maize at the annual symposium at Cold Spring Harbor Laboratory, presenting a paper of the same name. The paper delved into the instability caused by Dc and As or just As in four genes, along with the tendency of those genes to unpredictably revert to the wild phenotype. She also identified "families" of transposons, which did not interact with one another.

MCCLINTOCK'S VISITS TO MADISON

McClintock's work on "controlling elements" met with disbelief and nonacceptance by some of the leading geneticists of the day, including R.A. Brink and

Sewall Wright, among others, at the University of Wisconsin in Madison. According to the distinguished population geneticist, James F. Crow, "her use of the words 'controlling elements' didn't sit well. And most people here (in Madison), especially—Sewall Wright and R.A. Brink, and by extension me, thought that she was probably wrong on that account" (Crow's interview with Nathaniel C. Comfort, October 13, 1996; quoted in Comfort 2001, pp. 170–171). Comfort (2001) wrote: "Brink, and soon his lunch companions, used more purely descriptive terms such as 'modulating element' and 'transposable element' to strip away the functional connotations of control" (pp. 170–171). However, it is clear that Brink understood McClintock's argument. He wrote that she was arguing that "the various loci are progressively activated during development of the individual by controllers functioning in genetically fixed ways and in a predetermined sequence." But he added that one could "turn McClintock's controller argument around, without compromising the facts, and assume that what she considers primary causes are, themselves, effects." In other words, transposition did not control development, development controlled transposition.

McClintock's visits to Madison came to an abrupt end in 1958, when McClintock gave a seminar to an audience that included Brink, Crow, Kermicle, and Wright. After hearing McClintock's explanation of her developmental hypothesis and watching her diagram on the blackboard, Wright stood up, erased her drawing, and replaced it with an alternative explanation of his own. Wright's argument was similar to Brink's; that she had reversed cause and effect. McClintock then erased Wright's figure and replaced it with her original drawing. After a few rounds of these exchanges, McClintock stormed out of the room, vowing never to return to Madison while Wright was still alive!

However, McClintock found support elsewhere. Edgar Altenburg, a *Drosophila* geneticist and a former colleague of Herman J. Muller at Rice University in Houston, wrote to McClintock in January 1953, asking her to review his manuscript which discussed her mutable-gene work. McClintock read the manuscript and praised him for his insight and objectivity. Altenburg replied that her work heralds a new era in the study of mutation. He believed that the classical gene concept will have to undergo alteration as a result of McClintock's results.

She delivered a series of lectures at Caltech, in 1954, which were arranged by George W. Beadle, who negotiated for her a roving appointment. She started with an overview of maize genetics, followed by a historical account of her work on mutable genes and developmental regulation. The students received the full weight of her evidence and her theoretical speculations. By 1955, it became clear that transposable or controlling elements were not rare, at least in mice. She preferred at first to call them controlling units, similar to the Mendelian unit characters that William Bateson had used long time ago. In June 1955, in a paper presented at the Brookhaven Symposium on Mutation, she wrote that she had called them units because "they may show regular inheritance patterns and thus be followed as 'units' in progeny tests." In other words, they were Mendelian units, but not genes.

McClintock suggested that evidence from bacteriophage, bacteria, protozoa, and several plants and animals indicated that controlling elements were widely distributed in nature. She firmly held the belief that controlling elements are normal components of the chromosome complement and they are responsible for controlling, differentially, the time and type of activity of individual genes. Her talk at the Brookhaven Symposium was followed by extensive formal discussion. Among those present were Ruth Sager, Rollin Hotchkiss, Hermann J. Muller, and others. There was broad skepticism as to the nature of the so-called controlling elements and if they were in fact qualitatively different from genes.

1956 COLD SPRING HARBOR SYMPOSIUM

Shortly afterward, McClintock wrote to her friend Ernst Caspari about the reaction to her work among her colleagues. She thought that they were unable to appreciate her point of view regarding gene mutations and the genetic control of development. She told Caspari that the negative attitude on the part of most geneticists over the past several years made her hesitate to push her account personally. She was tired and weary and hoped that they can take over. However, the 1956 Cold Spring Harbor Symposium, which was titled "Genetic Mechanisms: Structure and Function," presented her yet another opportunity to explain her position once again. Those in attendance included many of her friends: Caspari, Harriet Creighton, Rollin Hotchkiss, Edgar Altenburg, Bentley Glass, Boris Ephrussi, Ruth Sager, Edward Lewis, Jack Schultz, and Charles Metz, as well as George Beadle, Sterling Emerson (son of Rollins), and several other younger geneticists. Once again, she emphasized controlling elements, especially the new element "Suppressor-mutator." She emphasized that controlling elements were different from genes although she did not know their biochemistry. Transposition in maize became a well established fact. However, interpretation of results remained obscure.

McClintock attempted to link her controlling-element work to the most exciting concept of the operon model of genetic regulation proposed by Francois Jacob and Jacques Monod, which drew her back into a debate central to the concerns of bacterial geneticists. However, Jacob and Monod did not accept McClintock's interpretation. Controlling elements are dispensable and are able to move from one chromosomal location to another, which is not the case with the bacterial operators and regulators.

She made an extensive study of the cytogenetics and ethnobotany of maize races from South America. McClintock's research became well understood in the 1960s and 1970s, as other scientists confirmed the mechanisms of genetic regulation that she had demonstrated in her maize research in the 1940s and 1950s. Awards and recognition for her contributions to the field followed, including the Nobel Prize in Physiology or Medicine, awarded to her in 1983 for the discovery of genetic transposition; she is the only woman to receive an unshared Nobel Prize in that category.

EVOLUTIONARY ORIGINS OF MAIZE

In her later years, McClintock was interested in tracing the origins of cultivated maize and the divergence and distribution of the many races grown throughout the world. This project extended her intellectual horizons back to her earlier interest in the cytological techniques and analysis, which occupied her interest in the early years of her career. Her interest was sparked, in part, by a meeting with one of the leading figures in maize genetics, Paul Mangelsdorf, who was visiting her lab at Cold Spring Harbor. She visited Peru to train an individual in maize cytological studies. Her goal was to characterize various strains of maize cytologically and to draw conclusions about their interrelationships if possible. In 9 of 10 highland races from Ecuador, in 11 of 12 highland races from Bolivia, and in all 10 highland races from Chile, she found a consistent pattern of chromosomal knobs. Lowland races differed widely in number, location, and size of knobs. She was able to trace the origin of the races of maize through their chromosomal characteristics. Often, each race had a geographic area of concentration but, it also overlapped into other races. She inferred that present-day maize may have derived from several different races, with migration followed by hybridization.

In her last years, McClintock returned to the idea that environment is the ultimate source of control. The genome responds to the environment. She called for greater attention to the genome, which she described as a highly sensitive organ of the cell, which monitors unexpected events and responds to them, restructuring and evolving as necessary. By the end of her career, McClintock developed a sweeping, integrated view that combined development and evolution.

FURTHER READING

Comfort, N.C. (2001) *The Tangled Field. Barbara McClintock's Search for the Patterns of Genetic Control.* Cambridge: Harvard University Press.

Creighton, H.B., and B. McClintock (1931) A correlation of cytological and genetical crossing-over in *Zea mays. Proc. Natl. Acad. Sci. USA, 17*: 492–497.

Fedoroff, N. (1994) Barbara McClintock. 16 June 1902–2 September 1992. *Biogr. Memoirs Fellows R. Soc., 40*: 266–280.

Fedoroff, N. (1995) *Barbara McClintock, Biographical Memoirs V.68.* Washington, DC: National Research Council/The National Academies Press, pp. 211–236.

Kass, L.B., and C. Bonneuil (2004) Mapping and seeing: Barbara McClintock and the linking of genetics and cytology in maize genetics, 1928–1935. In H.-J. Rheinberger, J.P. Gaudilliere (eds.), *Classical Genetic Research and Its Legacy: The Mapping Cultures of 20th Century Genetics.* London: Routledge, pp. 91–118.

Keller, E.F. (1983) *A Feeling for the Organism. The Life and Work of Barbara McClintock.* New York: Henry Holt.

McClintock, B. (1929) A cytological and genetical study of triploid maize. *Genetics*: 180–222.

McClintock, B. (1931) The order of the genes C, Sh and Wx in *Zea mays* with reference to a cytologically known point in the chromosome. *Proc. Natl. Acad. Sci. USA, 17(8)*: 485–491.

McClintock, B. (1941) The stability of broken ends of chromosomes in *Zea mays. Genetics, 26(2)*: 234–282.

McClintock, B. (1950) The origin and behavior of mutable loci in maize. *Proc. Natl. Acad. Sci. USA, 36(6)*: 344–355.

McClintock, B. (1953) Induction of instability at selected loci in maize. *Genetics, 38*: 579–599.

McClintock, B. (1961) Some parallels between gene control systems in maize and in bacteria. *Am. Nat., 95(884)*: 265–277.

McClintock, B., T.A. Kato Yamakake, and A. Blumenschein (1981) *Chromosome Constitution of Races of Maize. Its Significance in the Interpretation of Relationships between Races and Varieties in the Americas.* Chapingo, Mexico: Escuela de Nacional de Agricultura, Colegio de Postgraduados.

Archives and Research Collections

The Barbara McClintock Papers—Profiles in Science, National Library of Medicine
Barbara McClintock Papers, 1927–1991, at the American Philosophical Society
Cold Spring Harbor Laboratory Archives

Section VII

Applications of Genetics

GREEN REVOLUTION

A remarkable application of genetics is in agricultural food production, especially wheat and rice. Agronomist Norman Borlaug developed a strain of wheat that could resist diseases, was short, reduced damage by wind, and could produce large seed heads and high yields. He introduced this variety of wheat in Mexico, and within 20 years, the production of wheat had tripled. Geneticist M.S. Swaminathan helped to introduce this high-yield variety of wheat to India. Borlaug eventually won a Nobel Peace Prize for his work with developing high-yield crops and for helping prevent starvation in many developing countries.

In addition to producing larger quantities of food, the Green Revolution was also beneficial because it made growing more crops on roughly the same amount of land with a similar amount of effort possible. This reduced production costs and also resulted in cheaper prices for food in the market. The ability to grow more food on the same amount of land was also beneficial to the environment because it meant that less forest or natural land needed to be converted to farmland to produce more food.

The main development of Green Revolution in India was higher-yielding varieties of wheat, for developing rust-resistant strains of wheat. The introduction of high-yielding varieties of seeds and the increased quality of fertilizers and irrigation technique led to the increase in production to make the country self-sufficient in food grains, thus improving agriculture in India. Dr. Mankombu S. Swaminathan played a leading role in this transformation.

MEDICAL GENETICS

Another major application of genetics is in the field of Medicine and human health. Victor Almon McKusick has been the founder of this field. He is widely known as the "father of medical genetics." He was a proponent of the mapping of the human genome due to its use for studying congenital diseases. He is well known for his studies of the Amish and what he called "little people." He was the original author and, until his death, remained chief editor of *Mendelian Inheritance in Man* (*MIM*) and its counterpart *Online Mendelian Inheritance in Man* (*OMIM*).

MIM was the first published catalog of all known genes and genetic disorders, in 1966. The complete text of *MIM* was made available online free of charge beginning in 1987. The 12th and final print edition was published in 1998. At the time of McKusick's death in 2008, *OMIM* contained 18,847 entries. He also led the Annual Course in Medical Genetics at the University of Bologna Residential Center in Bertinoro di Romagna, Italy, in 1987. McKusick was founding president of the Human Genome Organization in 1989. In 1960, McKusick founded and codirected the Annual Short Course in Medical and Experimental Mammalian Genetics at the Jackson Laboratory in Bar Harbor, Maine.

18 Norman Ernest Borlaug (1914–2009), M.S. Swaminathan (1925–), and Green Revolution

Norman Borlaug was awarded the Nobel Peace Prize in 1970 for his advances in plant breeding, which led to spectacular success in increasing food production in Latin America and Asia. In the 1960s and 1970s, Mankombu S. Swaminathan, working closely with Borlaug, implemented and expanded Borlaug's methods in India, which established "green revolution" as a highly successful program, preventing famine.

On the occasion of his Nobel Prize award, Norman Borlaug wrote to Swaminathan: "The Green Revolution has been a team effort and much of the credit for its spectacular development must go to the Indian officials, organization scientists and farmers. However, to you, Dr. Swaminathan, a great deal of the credit must go for first recognizing the potential value of the Mexican dwarfs. Had this not occurred, it is quite possible that there would not have been a Green Revolution in Asia."

POPULATION MONSTER

In his Nobel Prize speech, titled "Green Revolution, Peace and Humanity" and delivered on December 11, 1970, Norman Borlaug cautioned that increasing the food supply is only a temporary measure. The continuing birth rate and population explosion will pose a greater threat in the long term. He wrote:

> The green revolution has won a temporary success in man's war against hunger and deprivation; it has given man a breathing space. If fully implemented, the revolution can provide sufficient food for sustenance during the next three decades. But the frightening power of human reproduction must also be curbed; otherwise the success of the green revolution will be ephemeral only. Most people still fail to comprehend the magnitude and menace of the "Population Monster".... The ticktock of the clock will continually grow louder and more menacing each decade. Where will it all end?

> Malthus signaled the danger a century and a half ago. But he emphasized principally the danger that population would increase faster than food supplies…. We must recognize the fact that adequate food is only the first requisite for life. For a decent and humane life we must also provide an opportunity for good education, remunerative employment, comfortable housing, good clothing, and effective and compassionate medical care. Unless we can do this, man may degenerate sooner from environmental diseases than from hunger. And yet, I am optimistic for the future of mankind, for in all biological populations there are innate devices to adjust population growth to the carrying capacity of the environment. Undoubtedly, some such device exists in man, presumably Homo sapiens, but so far it has not asserted itself to bring into balance population growth and the carrying capacity of the environment on a worldwide scale. It would be disastrous for the species to continue to increase our human numbers madly until such innate devices take over. It is a test of the validity of sapiens as a species epithet.
>
> Since man is potentially a rational being, however, I am confident that within the next two decades he will recognize the self-destructive course he steers along the road of irresponsible population growth and will adjust the growth rate to levels which will permit a decent standard of living for all mankind.

Regrettably, we are still waiting for that optimism which Borlaug had predicted. Human population is increasing in size at an even greater pace than in Borlaug's time!

EARLY LIFE AND EDUCATION

Norman Ernest Borlaug was born on March 25, 1914, in his grandfather's farmhouse near the tiny Norwegian settlement of Saude, in northeastern Iowa. As a boy, he trudged across snow-covered fields to a one-room country school. He was described as a high-spirited boy of boundless curiosity. His sister, Charlotte Culbert, recounted in an interview in 2008 in Cresco, Iowa, that he would whistle aloud as he milked the cows and pester his parents and grandparents with questions about the surrounding plants and animals. His grandfather Nels Borlaug, regretting his own lack of education, urged his grandson to continue his education; "better fill your head now, Norman, so you can fill your belly later!"

Norman attended the University of Minnesota during the years of Great Depression, often encountering starving families, which sharpened his interest in the problems of food production.

Borlaug received his BSc in Forestry in 1937 and PhD in plant pathology and genetics from the University of Minnesota in 1942. He took up an agricultural research position in Mexico, where he developed semidwarf, high-yield, disease-resistant wheat varieties. During the mid-twentieth century, Borlaug led the introduction of these high-yielding varieties combined with modern agricultural production techniques to Mexico and India. As a result, Mexico became a net exporter of wheat by 1963. Between 1965 and 1970, wheat yields nearly doubled in India, greatly improving its food security. Later in his life, Borlaug helped apply these methods of increasing food production in other countries in Asia and Africa.

RESEARCH CAREER

In the 1940s, Borlaug was involved in research for the U.S. Armed Forces. One of his first projects was to develop glue that could withstand the warm salt water of the South Pacific. The Imperial Japanese Navy had gained control of the island of Guadalcanal and patrolled the sky and sea by day. The only way for U.S. forces to supply the troops stranded on the island was to approach at night by speedboat and jettison boxes of canned food and other supplies into the surf to wash ashore. The problem was that the glue holding these containers together disintegrated in saltwater. Within weeks, Borlaug and his colleagues had developed an adhesive that resisted corrosion, allowing food and supplies to reach the stranded Marines.

In 1940, U.S. Vice President-Elect Henry Wallace persuaded the Rockefeller Foundation to work with the Mexican government in agricultural development. He saw Mexico's ambitions as beneficial to U.S. economic and military interests. The Rockefeller Foundation contacted Dr. E.C. Stakman and two other leading agronomists and they developed a proposal for a new organization, the Office of Special Studies, as part of the Mexican Government, but directed by the Rockefeller Foundation. It was staffed with both Mexican and U.S. scientists and focused on soil development, maize and wheat production, and plant pathology.

The project leader was Dr. Jacob George "Dutch" Harrar, who wanted to hire Borlaug as head of the newly established Cooperative Wheat Research and Production Program in Mexico. However, Borlaug initially declined, choosing to finish his war service, but later in 1944, after rejecting DuPont's offer to double his salary, he flew to Mexico City to head the new program as a geneticist and plant pathologist.

CIMMYT

In 1964, Borlaug was made the director of the International Wheat Improvement Program at El Batán, Texcoco, on the eastern fringes of Mexico City, as part of the newly established Consultative Group on International Agricultural Research's (CGIAR's) International Maize and Wheat Improvement Center (Centro Internacional de Mejoramiento de Maíz y Trigo, or CIMMYT). Borlaug retired officially from the position in 1979 but remained a CIMMYT senior consultant. In addition to taking up charitable and educational roles, he continued to be involved in plant research at CIMMYT with wheat, triticale, barley, maize, and high-altitude sorghum.

Borlaug joined the faculty of Texas A&M University in 1984 and began a teaching and research career. Shortly afterward, he was given the title of Distinguished Professor of International Agriculture at the university and the holder of the Eugene Butler Endowed Chair in Agricultural Biotechnology. Borlaug remained at A&M until his death in September 2009.

WHEAT RESEARCH

The Cooperative Wheat Research Production Program, a joint venture by the Rockefeller Foundation and the Mexican Ministry of Agriculture, involved research in genetics, plant breeding, plant pathology, entomology, agronomy, soil science, and cereal technology. The goal of the project was to boost wheat production in Mexico, which at the time was importing a large portion of its grain. Plant pathologist George Harrar recruited and assembled the wheat research team in late 1944. The four other members were soil scientist William Colwell, maize breeder Edward Wellhausen, potato breeder John Niederhauser, and Norman Borlaug, all from the United States. During the 16 years Borlaug remained with the project, he bred a series of remarkably successful high-yield, disease-resistant, semidwarf wheat.

Borlaug's first few years in Mexico were difficult. There were no trained scientists and equipment. Local farmers were hostile toward the wheat program because of serious crop losses from 1939 to 1941 due to stem rust. "It often appeared to me that I had made a dreadful mistake in accepting the position in Mexico," he wrote in the epilogue to his book, *Norman Borlaug on World Hunger.* He spent the first 10 years breeding wheat cultivars resistant to disease, including rust. In that time, his group made 6000 individual crossings of wheat. It was back-breaking and slow, involving long hours of meticulous hybridization work under the Mexican sun!

TWO SEASONS

Initially, Borlaug's work had been concentrated in the central highlands, in the village of Chapingo near Texcoco, where the problems with rust and poor soil were most prevalent. He realized that he could speed up breeding by taking advantage of the country's two growing seasons. In the summer, he would breed wheat in the central highlands as usual, then immediately take the seeds north to the Yaqui Valley research station near Ciudad Obregon, Sonora. The difference in altitudes and temperatures would allow more crops to be grown each year.

However, Borlaug's boss, George Harrar, opposed this expansion. Besides the extra costs of doubling the work, Borlaug's plan went against a then-held principle of agronomy that has since been disproved. It was believed that to store energy for germination before being planted, seeds needed a rest period after harvesting. Despite Harrar's opposition, Borlaug persisted with the support of his mentor Elvin Stakman. From 1944 onward, wheat was bred in Mexico, at locations 700 miles (1000 km) apart, 10 degrees apart in latitude, and 8500 ft (2600 m) apart in altitude. This was called "shuttle breeding."

An unexpected benefit of the double wheat season was that the new breeds did not have problems with photoperiodism. Normally, wheat varieties cannot adapt to new environments, due to the changing periods of sunlight. Borlaug later recalled, "As it worked out, in the north, we were planting when the days were getting shorter, at low elevation and high temperature. Then we'd take the seed from the best plants south and plant it at high elevation, when days were getting longer and there was lots of rain. Soon we had varieties that fit the whole range of conditions." It was contrary to the expected outcome. This meant that the project would not need to start separate breeding programs for each geographic region of the planet.

MULTILINE VARIETIES

Genetically identical plant varieties often only have one or a few major genes for disease resistance, and plant diseases such as rust are continuously producing new races that can overcome a pure line's resistance. Multiline varieties were developed to resolve this problem. Multiline varieties are mixtures of several phenotypically similar pure lines, which each have different genes for disease resistance. Because they have similar heights, flowering and maturity dates, seed colors, and agronomic characteristics, they remain compatible with each other and do not reduce yields when grown together on the field.

BACKCROSSING

In 1953, Borlaug extended this technique by suggesting that several pure lines with different resistance genes should be developed through backcross methods using one recurrent parent. Backcrossing involves crossing a hybrid and subsequent generations with a recurrent parent. As a result, the genotype of the backcrossed progeny becomes increasingly similar to that of the recurrent parent. Borlaug's method would allow the various different disease-resistant genes from several donor parents to be transferred into a single recurrent parent. To make sure each line has different resistant genes, each donor parent is used in a separate backcross program. Between 5 and 10 of these lines may then be mixed depending upon the races of pathogen present in the region. As this process is repeated, some lines will become susceptible to the pathogen. However, these lines can easily be replaced with new resistant lines.

DWARFING

Borlaug worked with varieties that had tall, thin stalks. Taller wheat grasses better compete for sunlight but tend to collapse under the weight of the extra grain—a trait called lodging—from the rapid growth spurts induced by nitrogen fertilizer Borlaug used in the poor soil. To prevent lodging, he bred wheat to grow shorter, stronger stalks that could better support larger seed heads. Dwarfing is an important agronomic quality for wheat; dwarf plants produce thick stems.

In 1953, Borlaug acquired a Japanese dwarf variety of wheat called *Norin 10* developed by Orville Vogel, that had been crossed with a high yielding American cultivar called *Brevor 14*.

Norin 10/Brevor 14 is semidwarf (one-half to two-thirds the height of standard varieties) and produces more stalks and, thus, more heads of grain per plant. Borlaug crossbred the semidwarf Norin 10/Brevor 14 cultivar with his disease-resistant cultivars to produce wheat varieties that were adapted to tropical and subtropical climates.

Borlaug's new semidwarf, disease-resistant varieties, called Pitic 62 and Penjamo 62, changed the potential yield of spring wheat dramatically. By 1963, 95% of Mexico's wheat crops used the semidwarf varieties developed by Borlaug. That year, the harvest was six times larger than in 1944, the year Borlaug arrived in Mexico.

Mexico had become fully self-sufficient in wheat production and a net exporter of wheat. Four other high-yield varieties were also released, in 1964: Lerma Rojo 64, Siete Cerros, Sonora 64, and Super X.

Borlaug's dwarf spring wheat strains were sent for multilocation testing in the International Wheat Rust Nursery, organized by the U.S. Department of Agriculture in 1961–1962. In March 1962, a few of these strains were grown in the fields of the Indian Agricultural Research Institute (IARI) in New Delhi, where M.S. Swaminathan was the geneticist.

GREEN REVOLUTION IN INDIA

M.S. Swaminathan

Monkombu Sambasivan Swaminathan (born August 7, 1925) is an Indian geneticist and international administrator, renowned for his leading role in India's Green Revolution, a program under which high-yield varieties of wheat and rice seedlings were planted in the fields of poor farmers. Swaminathan is known as "Father of Indian Green Revolution" for his leadership and success in introducing and further developing high-yielding varieties of wheat in India. He is the founder and chairman of the M.S. Swaminathan Research Foundation. His stated vision is to rid the world of hunger and poverty. He is an advocate of moving India to sustainable development, especially using environmentally sustainable agriculture, sustainable food security, and the preservation of biodiversity, which he calls an "evergreen revolution."

From 1972 to 1979, he was director general of the Indian Council of Agricultural Research. He was Principal Secretary, Ministry of Agriculture, from 1979 to 1980. He served as Director General of the International Rice Research Institute (IRRI) (1982–1988) and became president of the International Union for the Preservation of Nature in 1988.

Early Life and Education

Dr. M.S. Swaminathan was born in Kumbakonam, Tamil Nadu, India. His father was Dr. M.K. Sambasivan, a surgeon and follower of Mahatma Gandhi. M.S. Swaminathan learnt from his father "that the word 'impossible' exists mainly in our minds and that given the requisite will and effort, great tasks can be accomplished." After his father's death when he was 11, young Swaminathan was looked after by his uncle, M.K. Narayanaswami, a radiologist. He attended the local high school and later the Catholic Little Flower High School in Kumbakonom, from which he matriculated at age 15. Coming from a family of doctors, he naturally took admission in a medical school. But when he witnessed the Great Bengal Famine of 1943, he decided to devote his life getting rid of hunger from India. He simply switched from the medical field to the agricultural field. He then went on to finish his undergraduate degree at Maharaja's College in Trivandrum, Kerala. He studied there from 1940 to 1944 and earned a bachelor of science degree in zoology in 1944.

Early Career

Swaminathan enrolled in Madras Agricultural College (now the Tamil Nadu Agricultural University), where he graduated as valedictorian with another bachelor of science degree, this time in Agricultural Science. He explained this career decision thus: "My personal motivation started with the great Bengal famine of 1943, when I was a student at the University of Kerala. There was an acute rice shortage, and in Bengal about 3 million people died from starvation." In 1947, the year of Indian independence from British colonial rule, he moved to study at the IARI in New Delhi as a postgraduate student in genetics and plant breeding. He obtained a postgraduate degree in cytogenetics with high distinction in 1949.

He was awarded the UNESCO Fellowship to continue his research on potato genetics at the Wageningen Institute of Genetics in the Netherlands. Here, he succeeded brilliantly, standardizing procedures for transferring genes from a wide range of wild species of *Solanum* to the cultivated potato, *Solanum tuberosum*. In 1950, he moved to study at the Plant Breeding Institute of the University of Cambridge School of Agriculture. He earned a PhD in 1952, for his thesis, "Species Differentiation, and the Nature of Polyploidy in certain species of the genus *Solanum*—Section Tuberarium." His work presented a new concept of the species relationships within the tuber-bearing genus *Solanum*. His Cambridge college, Fitzwilliam, made him an Honorary Fellow in 2014.

Swaminathan then accepted a postdoctoral research associateship in genetics at the University of Wisconsin in Madison, to help set up a U.S. Department of Agriculture potato research station. Despite his strong personal and professional satisfaction with the research work in Wisconsin, he declined the offer of a full-time faculty position, returning to India in early 1954.

Professional Achievements

Swaminathan has worked worldwide in collaboration with colleagues and students on a wide range of problems in basic and applied plant breeding, agricultural research and development, and the conservation of natural resources.

His professional career began in 1949:

- 1949–1955—Research on potato (*S. tuberosum*), wheat (*Triticum aestivum*), rice (*Oryza sativa*), and jute genetics.
- 1955–1972—Field research on Mexican dwarf wheat varieties. Taught cytogenetics, radiation genetics, and mutation breeding and build up the wheat and rice germplasm collections at IARI.
- 1972–1979—Director-General, Indian Council of Agricultural Research (ICAR), established the National Bureau of Plant, Animal, and Fish Genetic Resources of India. Established the International Plant Genetic Resources Institute (changed in 2006 to Biodiversity International).
- 1979–1980—Principal Secretary in the Ministry of Agriculture, Government of India, Transformed the Pre-investment Forest Survey Programme into the Forest Survey of India.

- 1981–1985—Independent chairman, Food and Agriculture Organization (FAO) Council, Rome, played a significant role in establishing the Commission on Plant Genetic Resources.
- 1983—Developed the concept of Farmers' Rights and the text of the International Undertaking on Plant Genetic Resources. President of the International Congress of Genetics.
- 1982–1988—Director General, IRRI; organized the International Rice Germplasm Centre, now named International Rice Genebank.
- 1988–1991—Chairman of the International Steering Committee of the Keystone International Dialogue on Plant Genetic Resources, regarding the availability, use, exchange, and protection of plant germplasm.

1988–1996—President, World Wide Fund for Nature–India; organized the Indira Gandhi Conservation Monitoring Centre; organized the Community Biodiversity Conservation Programme.

1988–1999—Chairman/Trustee, Commonwealth Secretariat Expert Group, organized the Iwokrama International Centre for Rainforest Conservation and Development, for the sustainable and equitable management of tropical rainforests in Guyana. The President of Guyana wrote in 1994 that "there would have been no Iwokrama without Swaminathan."

1990–1993—Founder/President, International Society for Mangrove Ecosystems.

1988–1998—Chaired various committees of the Government of India to prepare draft legislations relating to biodiversity (Biodiversity Act) and breeders' and farmers' rights (Protection of Plant Varieties and Farmers' Rights Act).

In 1993, Dr M.S. Swaminathan headed an expert group to prepare a draft of a national population policy that would be discussed by the Cabinet and then by Parliament. In 1994, it submitted its report.

1994—Chairman of the Commission on Genetic Diversity of the World Humanity Action Trust. Established a Technical Resource Centre at M.S. Swaminathan Research Foundation (MSSRF) for the implementation of equity provisions of CBD and FAO's Farmers' Rights.

1994 onward—Chairman of the Genetic Resources Policy Committee of the CGIAR, development of policies for the management of the ex situ collections of International Agricultural Research Centers.

1995–1999—Chairman, Auroville Foundation.

1999—Introduced the concept of trusteeship management of Biosphere reserves. Implemented the Gulf of Mannar Biosphere Reserve Trust, with financial support from the Global Environment Facility.

2001—Chairman of the Regional Steering Committee for the India–Bangladesh joint Project on Biodiversity Management in the Sundarbans World Heritage Site, funded by the UN Foundation and United Nations Development Program (UNDP).

2002—President of the Nobel Peace Prize-winning Pugwash Conferences on Science and World Affairs, which work toward reducing the danger of armed conflict and to seek solutions to global security threats.

2002–2005—Cochairman with Pedro Sanchezof the UN Millennium Task Force on Hunger, a comprehensive global action plan for fighting poverty, disease, and environmental degradation in developing countries.

2004–2014—Chairman, National Commission on Farmers.

Books by M.S. Swaminathan

An Evergreen Revolution. Delhi: Konark Publishers. 2006.

I Predict: A Century of Hope Towards an Era of Harmony with Nature and Freedom from Hunger. Delhi: Konark Publishers. 1999.

Gender Dimensions in Biodiversity Management. Delhi: Konark Publishers. 1998.

Implementing the Benefit Sharing Provisions of the Convention on Biological Diversity: Challenges and Opportunities. Delhi: Konark Publishers. 1997.

Agrobiodiversity and Farmers' Rights. Delhi: Konark Publishers Limited. 1996.

Farmers' Rights and Plant Genetic Resources: A Dialogue. 1995.

Wheat Revolution: a Dialogue. Madras: Macmillan India Limited. 1993.

(With S. Jana.) *Biodiversity: Implications for Global Food Security.* Madras: Macmillan India Limited. 1992.

Recommended for Further Reading

Mathur, G.C. (2015) *A Living Legend: Prof. M.S. Swaminathan, Leader of the Ever Green Farm Revolution Movement.*

Swaminathan, M.S. *50 Years of Green Revolution—An Anthology of Research Papers,* 2017.

Swaminathan, M.S. (2017) *Legend in Science and Beyond.*

NORMAN BORLAUG INTERNATIONAL SYMPOSIUM

Dr. Swaminathan was the featured speaker at the 2006 Norman E. Borlaug International Symposium in Des Moins, Iowa, on October 19, 2006.

He currently holds the UNESCO—Cousteau Chair in Ecotechnology at the M.S. Swaminathan Research Foundation in Chennai, India.

GREEN REVOLUTION IN INDIA

In May 1962, Dr. M.S. Swaminathan, a member of IARI wheat program, requested of Dr. B.P. Pal, director of IARI, to arrange for the visit of Borlaug to India and to obtain a wide range of dwarf wheat seed possessing the Norin 10 dwarfing genes. The letter was forwarded to the Indian Ministry of Agriculture, headed by Shri C. Subramaniam, which arranged with the Rockefeller Foundation for Borlaug's visit. In March 1963, the Rockefeller Foundation and the Mexican government sent Borlaug and Dr. Robert Glenn Anderson to India to continue his work. He supplied 100 kg (220 lb) of seed from each of the four most promising strains and 630 promising selections in advanced generations to the IARI in October 1963, and test plots were subsequently planted at Delhi, Ludhiana, Pant Nagar, Kanpur, Pune, and Indore. Anderson stayed as head of the Rockefeller Foundation Wheat Program in New Delhi until 1975.

During the mid-1960s, India experienced minor famine and starvation, which was limited partially by the U.S. shipping a fifth of its wheat production to India in

1966 and 1967. In 1965, as a response to food shortages, Borlaug imported 550 tons of seeds for the government.

In 1965, after extensive testing, Borlaug's team, under Anderson, began its effort by importing about 200 tons of Lerma Rojo and Sonora 64 semidwarf seed varieties to India. Transportation delays prevented Borlaug's group from conducting the germination tests needed to determine seed quality and proper seeding levels. A week later, Borlaug discovered that his seeds were germinating at less than half the normal rate. It was found later that the seeds had been damaged in a Mexican warehouse by overfumigation with a pesticide. He immediately ordered all locations to double their seeding rates. The initial yields of Borlaug's crops were higher than any ever harvested in India. William Gaud of the U.S. Agency for International Development called Borlaug's work a "Green Revolution." High yields led to a shortage of various utilities—labor to harvest the crops, bullock carts to haul it to the threshing floor, jute bags, trucks, rail cars, and grain storage facilities. Some local governments were forced to close school buildings temporarily to use them for grain storage.

WHEAT YIELDS IN INDIA SINCE 1961

In India, yields increased from 12.3 million tons in 1965 to 20.1 million tons in 1970. By 1974, India was self-sufficient in the production of all cereals. By 2000, India was harvesting a record 76.4 million tons (2.81 billion bushels) of wheat. Since the 1960s, food production has increased faster than the rate of population growth. India's use of high-yield farming has prevented an estimated 100 million acres (400,000 km^2) of virgin land from being converted into farmland—an area about the size of California, or 13.6% of the total area of India. The use of these wheat varieties has also had a substantial effect on production in six Latin American countries, six countries in the Near and Middle East, and several others in Africa.

Borlaug's work with wheat contributed to the development of high-yield semidwarf indica and japonica rice cultivars at the IRRI and China's Hunan Rice Research Institute. Borlaug's colleagues at the CGIAR also developed and introduced a high-yield variety of rice throughout most of Asia. Land devoted to the semidwarf wheat and rice varieties in Asia expanded from 200 acres (0.8 km^2) in 1965 to over 40 million acres (160,000 km^2) in 1970. In 1970, this land accounted for over 10% of the more productive cereal land in Asia.

WORLD FOOD PRIZE

The World Food Prize is an international award recognizing the achievements of individuals who have advanced human development by improving the quality, quantity, or availability of food in the world. The prize was created in 1986 by Norman Borlaug as a way to recognize personal accomplishments and as a means of education by using the Prize to establish role models for others. The first prize was given to M.S. Swaminathan, in 1987, for his work in India. The next year, Swaminathan used the US$250,000 prize to start the MS Swaminathan Research Foundation in Chennai for research on sustainable development.

NOBEL PEACE PRIZE

For his contributions to the world food supply, Borlaug was awarded the Nobel Peace Prize in 1970. Norwegian officials notified his wife in Mexico City at 4:00 am, but Borlaug had already left for the test fields in the Toluca Valley, about 40 miles (65 km) west of Mexico City. A driver took her to the fields to inform her husband. According to his daughter, Jeanie Laube, "My mom said, 'You won the Nobel Peace Prize,' and he said, 'No, I haven't',... It took some convincing... He thought the whole thing was a hoax." He was awarded the prize on December 10. In his Nobel Lecture the following day, he speculated on his award: "When the Nobel Peace Prize Committee designated me the recipient of the 1970 award for my contribution to the 'green revolution', they were in effect, I believe, selecting an individual to symbolize the vital role of agriculture and food production in a world that is hungry, both for bread and for peace." His speech repeatedly presented improvements in food production within a sober understanding of the context of population.

BORLAUG HYPOTHESIS

Borlaug continually advocated increasing crop yields as a means to curb deforestation. The large role he played in both increasing crop yields and promoting this view has led to this methodology being called by agricultural economists the "Borlaug hypothesis," namely, that "increasing the productivity of agriculture on the best farmland can help control deforestation by reducing the demand for new farmland." According to this view, assuming that global food demand is on the rise, restricting crop usage to traditional low-yield methods would also require at least one of the following: the world population to decrease, either voluntarily or as a result of mass starvations; or the conversion of forest land into crop land. It is thus argued that high-yield techniques are ultimately saving ecosystems from destruction.

But other land uses exist, such as urban areas, pasture, or fallow, so further research is necessary to ascertain what land has been converted for what purposes, to determine how true this view remains. Increased profits from high-yield production may also induce cropland expansion in any case, although as world food needs decrease, this expansion may decrease as well.

CRITICISMS

Borlaug's name is nearly synonymous with the Green Revolution. Throughout his years of research, Borlaug's programs often faced opposition by people who consider genetic crossbreeding to be unnatural or to have negative effects. Borlaug's work has been criticized for bringing large-scale monoculture, input-intensive farming techniques to countries that had previously relied on subsistence farming.

These farming techniques often reap large profits for U.S. agribusiness and agrochemical corporations and have been criticized for widening social inequality in the countries owing to uneven food distribution while forcing a capitalist agenda of U.S. corporations onto countries that had undergone land reform.

Other concerns of his critics and critics of biotechnology in general include the following: that the construction of roads in populated third-world areas could lead to the destruction of wilderness; the crossing of genetic barriers; the inability of crops to fulfill all nutritional requirements; the decreased biodiversity from planting a small number of varieties; the environmental and economic effects of inorganic fertilizer and pesticides; the amount of herbicide sprayed on fields of herbicide-resistant crops.

Borlaug dismissed most claims of critics but did take certain concerns seriously. He stated that his work has been a change in the right direction, but it has not transformed the world into a Utopia. He said the critics had never experienced the physical sensation of hunger. They do their lobbying from comfortable air-conditioned office suites in Washington or Brussels. He said if they lived just one month amid the misery of the developing world, as I have for 50 years, they'd be crying out for tractors and fertilizer and irrigation canals and be outraged that fashionable elitists back home were trying to deny them these things.

LATER YEARS

Borlaug continued to participate in teaching and research. He spent much of the year based at CIMMYT in Mexico, conducting research, and four months of the year serving at Texas A&M University, where he had been a distinguished professor of international agriculture since 1984. In 1999, the university's Board of Regents named its US$16 million Center for Southern Crop Improvement in honor of Borlaug.

AFRICA

In the early 1980s, environmental groups that were opposed to Borlaug's methods campaigned against his planned expansion of efforts into Africa. They prompted the Rockefeller and Ford Foundations and the World Bank to stop funding most of his African agriculture projects. Western European governments were persuaded to stop supplying fertilizer to Africa. In 1984, during the Ethiopian famine, Ryoichi Sasakawa, the chairman of the Japan Shipbuilding Industry Foundation, contacted the semiretired Borlaug, wondering why the methods used in Asia were not extended to Africa and hoping Borlaug could help. Borlaug helped with the new effort and subsequently founded the Sasakawa Africa Association (SAA) to coordinate the project.

The SAA is a research and extension organization that aims to increase food production in African countries that are struggling with food shortages. At first, Borlaug wanted to do a few years of research, but after he saw how terrible their circumstances were, Borlaug and the SAA started projects in seven countries. Yields of maize in developed African countries tripled. Yields of wheat, sorghum, cassava, and cowpeas also increased in these countries. Program activities were extended later to several African countries.

GLOBAL FOOD SITUATION

Borlaug was greatly concerned about the limited potential for land expansion for cultivation. In March 2005, he stated that, "we will have to double the world food supply by 2050." With 85% of future growth in food production having to come from lands already in use, he recommends a multidisciplinary research focus to further increase yields, mainly through increased crop immunity to large-scale diseases, such as the rust fungus, which affects all cereals but rice. His dream was to "transfer rice immunity to cereals such as wheat, maize, sorghum and barley, and transfer bread-wheat proteins (gliadin and glutenin) to other cereals, especially rice and maize."

Borlaug believed that genetic modification of organisms (GMOs) was the only way to increase food production as the world runs out of unused arable land. GMOs were not inherently dangerous "because we've been genetically modifying plants and animals for a long time. Long before we called it science, people were selecting the best breeds." In a review of Borlaug's 2000 publication entitled "Ending World Hunger: The Promise of Biotechnology and the Threat of Antiscience Zealotry," the authors argued that Borlaug's warnings were still true in 2010.

GM crops are as natural and safe as today's bread wheat, opined Dr. Borlaug, who also reminded agricultural scientists of their moral obligation to stand up to the anti-science crowd and warn policy makers that global food insecurity will not disappear without this new technology and ignoring this reality global food insecurity would make future solutions all the more difficult to achieve.

According to Borlaug, "Africa, the former Soviet republics, and the cerrado are the last frontiers. After they are in use, the world will have no additional sizable blocks of arable land left to put into production, unless you are willing to level whole forests, which you should not do. So, future food-production increases will have to come from higher yields. And though I have no doubt yields will keep going up, whether they can go up enough to feed the population monster is another matter. Unless progress with agricultural yields remains very strong, the next century will experience sheer human misery that, on a numerical scale, will exceed the worst of everything that has come before."

Borlaug remained on the advisory board of Population Media Center, an organization working to stabilize world population, until his death.

DEATH

Borlaug died of lymphoma at the age of 95, on September 12, 2009, in his Dallas home. Borlaug's children released a statement saying, "We would like his life to be a model for making a difference in the lives of others and to bring about efforts to end human misery for all mankind.

As we celebrate Dr. Borlaug's long and remarkable life, we also celebrate the long and productive lives that his achievements have made possible for so many millions of people around the world…we will continue to be inspired by his enduring devotion to the poor, needy and vulnerable of our world."

Among many other honors, he was given an honorary doctorate by the Agricultural University of Norway in 1970.

FURTHER READING

Andrews, A. (2010) *The Kid Who Changed the World*. Nashville, Tennesse: Thomas Nelson Publisher.

Barber, D. (2014) *The Third Plate: Field Notes on the Future of Food*. New York: Penguin Books.

Bickel, L. (1974) *Facing Starvation; Norman Borlaug and the Fight Against Hunger*. New York: Reader's Digest Press.

Capshaw, N.C. (1970) *Heroes for Peace: Winners of the Nobel Peace Prize*. Norman C. Capshaw, Privately published.

Dil, A. (ed.) (1997) *Norman Borlaug on World Hunger*. Essex (UK): Bookservice International.

Hesser, L. (2006) *The Man Who Fed the World*. Dallas, Texas. Durban House Publishing Co.

Mann, C.C. (2018) *The Wizard and the Prophet*. New York: Knopf.

Paarlberg, R. and N. Borlaug (2009) *Starved for Science: How Biotechnology Is Being Kept out of Africa*. Cambridge, Mass.: Harvard University Press.

Pence, G.E. and R. Bailey (2002) *The Ethics of Food: A Reader for the Twenty-First Century*. Lanham, MD: Rowman and Littlefield Publishers Inc.

Swanson, L. (2011) *Norman Borlaug: Hero in a Hurry*. Booksurge Publishing (Amazon).

Vietmeyer, N. (2009) *Borlaug; Volume 1, Right off the Farm 1914–1944*. Lorton, VA: Bracing Books.

Vietmeyer, N. (2011) *Our Daily Bread, The Essential Norman Borlaug*. Lorton, VA: Bracing Books.

19 Victor McKusick (1921–2008)—Father of Medical Genetics

One of the spectacular successes of Mendelian genetics was its application to investigate human genetic diseases and defects as well as complex syndromes.

Victor McKusick is widely considered to be the founding father of medical genetics. An innovative clinician, medical educator, and researcher, he established the first medical genetics program and clinic at Johns Hopkins in 1957, conceived and compiled *Mendelian Inheritance in Man* (*MIM*), an annually updated catalog of human phenotypes (first published in 1966 and now published online), and conducted landmark studies of hereditary disorders in the Amish. He was an early advocate of mapping the human genome, was closely involved in the early years of the Human Genome Project, and served as founding president of the Human Genome Organization (HUGO). In 1997, in recognition of his lifelong contributions, he received the Lasker Award for Special Achievement in Medical Science.

EARLY LIFE AND EDUCATION

Victor Almon McKusick was born in Parkman, Maine, on October 21, 1921. He and his identical twin brother, Vincent, were the youngest of five children. Raised on the family dairy farm, the twins attended a one-room schoolhouse for eight years and graduated from a small local high school (in a class of 28). Like their older siblings, they were expected to attend college—their father was a college graduate and had been a school principal in Vermont before returning to his home town and the farm. In his teens, Victor considered becoming a minister, but a close encounter with medicine during the summer of 1937 changed his plans: he developed an abscess under one arm, and the infection (caused by an unusual streptococcus strain) spread and would not heal. McKusick spent 10 weeks in hospitals, both in Maine and at Massachusetts General in Boston, where he was treated successfully with the new antibiotic sulfanilamide. Inspired by the experience, he chose to pursue a career in medicine and entered a premedical program at Tufts University in 1940. (His brother Vincent chose to practice law and later became chief justice of Maine's Supreme Court.)

JOHNS HOPKINS UNIVERSITY

McKusick left Tufts in 1943 without finishing his degree because Johns Hopkins University School of Medicine (JHUSM) had begun admitting advanced premed students into an accelerated program, due to a wartime shortage of physicians. McKusick began his medical studies in March 1943, received his MD in 1946, and remained at Johns Hopkins for the rest of his career. McKusick's original plan was to complete his internship at Johns Hopkins and then return to Maine to establish a general practice. But circumstances conspired to keep him in Baltimore. He received the prestigious Osler Medical Service internship, which emphasized training for academic medicine, and then served two years as chief of the cardiology unit at the Baltimore U.S. Marine Hospital to fulfill his military obligation (a U.S. Army program had paid for his medical education in exchange for two years in the U.S. Public Health Service). In 1949, he married a fellow Hopkins physician, Anne Bishop, who was then completing her training in rheumatology. In 1950, he returned to Johns Hopkins and the Osler Medical Service, as senior and then chief resident. He joined the faculty in 1951 as an instructor and was promoted to full professor in 1960. From 1973 to 1985, he served as physician-in-chief of Johns Hopkins Hospital and chairman of the Department of Medicine. From 1985, he was university professor of Medical Genetics and remained active in teaching and research.

McKusick took his first step into medical genetics in 1947 during his internship. A teenager with intestinal polyps and curious pigmented spots on his lips became McKusick's patient; within two years, McKusick had seen four other patients with this combination, three of them in the same family, indicating a hereditary condition. Hearing that a Boston physician, Harold Jeghers, had seen five such cases, McKusick contacted him, and the two wrote up their cases for the *New England Journal of Medicine*. This became McKusick's first medical publication. Because the syndrome had been noted many years earlier by J.L.A. Peutz, a Dutch physician, it was later named the Peutz–Jeghers syndrome (PJS) (Jeghers 1949; Bruwer et al. 1954). McKusick had been intrigued by genetics since taking a premed genetics course at Tufts and wondered if PJS was a case of genetic linkage, i.e., individual genes causing separate symptoms, that were inherited together due to their proximity on a given chromosome. His mentor, Bentley Glass, persuaded him that pleiotropism—a single mutant gene causing diverse symptoms—was a more likely cause. (Later research confirmed Glass's clinical intuition about PJS.)

Medical genetics was not a specialty at the time, however, and in the course of his residency, McKusick was drawn to cardiology. Subsequently, he worked for two years (1948–1950) on the cardiovascular unit at the U.S. Public Health Service Marine Hospital under Luther L. Terry who subsequently became famous as the U.S. Surgeon General who blamed cigarette smoking for lung cancer and other ailments. He was doing cardiac catheterizations and studying the movement in the heart borders with a new imaging method called electromyography. During his early faculty years at JHUSM, he pursued a study of heart sounds and murmurs using sound spectrography, which had been developed at the Bell Telephone Laboratory for studying speech sounds. Adapted to cardiac studies, the technology could pick up and record the frequency spectrum of heart sounds, allowing physicians to visualize what they were hearing with their stethoscopes. McKusick renamed it spectral phonocardiography and used the studies as the basis for a comprehensive treatise on heart sounds titled *Cardiovascular Sound in Health and Disease*, published in 1958.

As a cardiologist, McKusick encountered many patients with congenital heart problems such as defective valves or deformed aortas. These were sometimes accompanied by other symptoms such as the skeletal and eye abnormalities that characterized a condition known as Marfan syndrome. He thought the various features of this inherited disorder, like those of PJS, were pleiotropic rather than linked and probably due to a mutation-determined defect in one element of connective tissue wherever it occurred in the body. In parallel with his cardiographic studies, he also carried out a comprehensive study of Marfan syndrome and four similar disorders, collecting patients and family histories from his own practice and from many other clinical departments at Johns Hopkins. This work produced his first book, *Heritable Disorders of Connective Tissue*, first published in 1956.

DIVISION OF MEDICAL GENETICS

Medical genetics as a distinct clinical and academic discipline at Johns Hopkins began in 1957, when McKusick was invited by the Chairman of the Department of Medicine, A. McGehee Harvey, to serve as director of the multifaceted chronic disease clinic developed at Johns Hopkins by J. Earle Moore in 1952. McKusick accepted the position, on the condition that he could develop a Division of Medical Genetics within the Department of Medicine, based in this clinic. Arguing that genetic disease is the ultimate chronic disease, he envisioned the new division carrying out teaching, research, and patient care related to hereditary disorders, much as other subspecialty divisions did for, e.g., cardiac or endocrine disorders. Increased understanding of rare genetic disorders would vastly improve differential diagnosis and treatment and enable physicians to better counsel affected patients and their families.

At the Moore Clinic (it was renamed Division of Medical Genetics in 1957), the research focused on nosology (defining the multiple distinct forms of genetic diseases) and on gene mapping (tracing the chromosomal locations of inherited disease genes and their linkages to other genes.) The program soon became a premier postdoctoral training ground for specialists from many areas of medicine, including dentistry and veterinary medicine, and its alumni helped propagate the new field of medical genetics in the United States and many other countries.

The Division of Medical Genetics was a beehive of activity. McKusick's energy and enthusiasm were infectious. Many Fellows came from Great Britain, which was mainly due to McKusick's friendship with Cyril Clarke in Liverpool. In his obituary of Cyril Clarke, McKusick wrote: "The passing...of Cyril Clarke brought back memories of pleasant and productive interchanges between Liverpool and the Johns Hopkins Hospital over a 20 year period or more beginning in 1957. Cyril was a central figure in those transatlantic collaborations in training and clinical research.... We at Johns Hopkins are much indebted to Cyril for the able protégés he sent Baltimore in the formative years of medical genetics here."

Among the Liverpudlians who came to the Moore Clinic were David A. Price Evans, Peter Brunt, J. Michael Connor, Brian Hanley, F. Michael Pope, Brian Walker,

David Weatherall, J.C. Woodrow, and Ronald Finn (the latter worked with Julius R. Krevens). Malcolm Ferguson-Smith came from the University of Glasgow to establish the cytogenetics laboratory. Digambar Borgaonkar had earlier joined Moore Clinic and started work on his distinguished compilation of Chromosomal variation in Man, which was published later. Eventually, it was expanded under the title "Repository of Human Chromosomal Variants and Anomalies"—a most valuable registry, and later The International Registry Project, which was funded by the World Health Organization. This work received international recognition involving several countries, including China and Russia.

McKusick was ably assisted by two colleagues in the Division of Medical Genetics; one was Edmond A. (Tony) Murphy in statistical and population genetics and Samuel H. ("Ned") Boyer in biochemical genetics. David Weatherall worked first with Ned Boyer and later with C. Lockart Conley in the study of hemoglobinopathies and with others in the Department of Biophysics directed by Howard Dintzis.

PLEIOTROPISM

For the genetic interpretation of the polyps-and-spots syndrome, Victor sought the advice of Bentley Glass, who was by then a member of the Biology faculty of Johns Hopkins University. It was Glass who impressed upon Victor that pleiotropism of a single mutant gene was a more likely explanation for the association between polyps and spots rather than genetic linkage.

McKusick was greatly interested in historical studies in genetics; his writings include the initial chapter of successive editions of Emery and Rimoin's *Principles and Practice of Medical Genetics*, biographical articles on Jonathan Hutchinson, Frederick Parkes Weber, Walter Stanborough Sutton, and Marcella O'Grady Boveri. McKusick was especially eloquent on the contributions of the London multispecialist Jonathan Hutchinson, dedicating two volumes of *Clinical Delineation of Birth Defects* to the memory of Hutchinson (1896).

MARFAN'S SYNDROME

While still on the junior faculty of Johns Hopkins, McKusick pursued a detailed study of Marfan syndrome as well as four others, grouped under the category, "heritable disorders of connective tissue": Ehlers–Danlos syndrome, osteogenesis imperfect, pseudoxanthoma elasticum, and Hurler syndrome (the prototype of the mucopolysaccharidoses). This work was ultimately published as a book entitled *Heritable Disorders of Connective Tissue*. He was greatly impressed by the multiple organ systems that were affected by the pleiotropic action of the mutation involved. He focused attention on all the Marfan patients he could, fully realizing that Johns Hopkins was a "superb site" for such studies because of its excellent departments of ophthalmology, pediatric and adult cardiology, orthopedics, and other specialties. He preferred to say "the Marfan syndrome" (rather than Marfan's syndrome) because it makes it clear that the surname is merely a tag. He wrote: "After all, Marfan described the skeletal features only".

AMISH POPULATIONS

During the 1960s, McKusick expanded his genetic investigations beyond the clinic to study inherited disorders in Amish populations in Pennsylvania, Ohio, and Indiana. There, he identified a number of rare recessive disorders, some of them new. He and his colleagues also expanded their knowledge of skeletal abnormalities by establishing ties with Little People of America, a support organization for people with dwarfism.

Soon after the medical genetics program was founded, McKusick and his colleagues began doing comprehensive annual reviews of the relevant medical and scientific literature, which were published in the *Journal of Chronic Diseases* from 1958 to 1963. Concurrently, McKusick compiled and published an annotated catalog of traits known to be carried on the X chromosome in humans (McKusick 1962). As McKusick began his genetic studies of Amish populations—which promised to reveal new inherited disorders—he and his colleagues decided to replace the annual reviews with a comprehensive catalog of known Mendelian phenotypes (inherited physical or physiological traits, including disorders and deformities), to which additions could be made as new phenotypes were described and more information became available on older phenotypes. With such a reference list, clinicians could readily find information on rare inherited diseases they encountered in their patients. Genetics researchers would have current information about the state of the art in gene mapping. The catalog, titled *Mendelian Inheritance in Man*, was first published in 1966; in collaboration with the National Library of Medicine, the *Online Mendelian Inheritance in Man* (*OMIM*) became available in 1987 and remains a standard reference work.

INBREEDING IN THE AMISH

Victor discussed nosology in relation to inbreeding, for instance, the delineation of "new" genetic entities with recessive inheritance is enhanced in inbred communities. It was, in fact, his interest in inbreeding in the Amish populations that brought us together because of my research on inbreeding in India with J.B.S. Haldane (Bowman et al. 1965, Dronamraju 1963). In his genetic studies of the Amish, Victor combined multiple research interests. He wrote: "in the 1960s, skeletal dysplasias became an area of both clinical and research nosologic interest because of their relationship to the heritable disorders of connective tissue, because of studies of dwarfism in the Amish…which we initiated in 1963, and because of collaboration with Little People of America, Inc., which began in 1965 when I first attended the annual national convention of this fraternal organization of persons of short stature. Skeletal dysplasias became a major interest of many who were my fellows in that period including David Rimoin, Judith G. Hall, and Charles Scott, who like me became honorary life members of Little People of America."

McKusick's expectation that "new" recessive disorders would be found in the Amish population was "richly fulfilled" (McKusick 2006). The first phenotype that was studied in detail in the Amish was dwarfism, two forms of which were found in the Amish of Lancaster County, Pennsylvania. One was the Ellis-van Creveld syndrome, or six-fingered dwarfism. The second form of recessive dwarfism, a previously unknown entity, was designated cartilage–hair–hypoplasia, later called metaphyseal chondrodysplasia, McKusick type. A large number of other "new" disorders in the Amish were described by several investigators, including the Hopkins group. The resulting publications on the Amish were compiled and edited by Victor in a convenient volume published by the Johns Hopkins University Press in 1978.

From 1968 to 1972, Victor arranged one week-long conferences each year, entitled "Clinical Delineation of Birth Defects," which were held at the Johns Hopkins Hospital and were supported by the National Foundation–March of Dimes (MOD). They were centered on genetic nosology and covered all areas of medicine. The conferences and the resulting publications enabled colleagues from all departments of the Johns Hopkins Hospital as well as elsewhere an opportunity to participate and share their experiences with rare or not so rare genetic disorders.

> Soon after the medical genetics program was founded, McKusick and his colleagues began doing comprehensive annual reviews of the relevant medical and scientific literature, which were published in the *Journal of Chronic Diseases* from 1958 to 1963. Concurrently, McKusick compiled and published an annotated catalog of traits known to be carried on the X chromosome in humans (McKusick 1962). As McKusick began his genetic studies of Amish populations—which promised to reveal new inherited disorders—he and his colleagues decided to replace the annual reviews with a comprehensive catalog of known Mendelian phenotypes (inherited physical or physiological traits, including disorders and deformities), to which additions could be made as new phenotypes were described, and more information became available on older phenotypes. With such a reference list, clinicians could readily find information on rare inherited diseases they encountered in their patients. Genetics researchers would have current information about the state of the art in gene mapping. The catalog, titled *Mendelian Inheritance in Man*, was first published in 1966; in collaboration with the National Library of Medicine, the *OMIM* became available in 1987 and remains a standard reference work.

BAR HARBOR COURSE

An important aspect of McKusick's contribution to medical genetics was the highly successful annual course in medical genetics, which he started in 1960. The genesis of the course was narrated by his wife Anne in a chapter she contributed to the memorial tribute, which Dronamraju had edited shortly after his death; "In July 1959, following a visit to his birthplace and family home in Parkman Maine, Victor, our daughter Carol, and I were having lunch at Testa's restaurant in Bar Harbor, ME. We had been joined by Dr. John Fuller of the Jackson Laboratory. It was then that the idea of a jointly sponsored genetics course occurred to Victor. It was discussed with Dr. Fuller and either that night or the next, with Dr. Earl Green, director of the Jackson Laboratory, and his wife Dr. Margaret Green, a mouse geneticist. There was agreement that the course should be tried."

> One third of the faculty of the course came from Johns Hopkins, another third came from the Jackson Laboratory, and another third came from various other institutions. For two successive years, 1964 and 1965, I was a member of the faculty. Participants of

the first year were mostly department chairmen who were there to evaluate if genetics should be a part of their curricula.

The location of the Jackson Laboratory on Mt. Desert Island, adjacent to Acadia National Park, made it a particularly desirable site for students and their families to spend the two weeks of the course. Bar Harbor is an ideal holiday resort. Mornings of the course were spent in attending lectures, but the afternoons were left free for climbing mountain trails, swimming, or other social activities and relaxation. There were workshops and discussion groups in the evenings.

NOVA SCOTIA HOUSE

Victor's wife Anne Bishop provided some information about their summer house in Nova Scotia:

> Following The Bar Harbor Course our family customarily went on to our old farmhouse in the Annapolis Valley of Nova Scotia, Canada. The Nova Scotia house played a significant role in our lives. It was a simple country house, purchased in 1842 by my great, great grandfather, Elias Bishop. My forebears were from Connecticut, coming as one of the 'planter families' to hold the land for Britain following the eviction of the French Acadians. Victor and I owned this family retreat and took our children there annually. Later, some Geneticists frequently visited us there. Among them were John Edwards, the Professor of Genetics at Oxford and his wife, Felicity; Rod Howell and Sally, Judy Hall, and Clarke Fraser and Marilyn.
>
> Other Nova Scotia activities included picking blueberries or wild blackberries, the latter on our own land, or visiting the zoo our neighbors, Gail and Ron Rogerson ran. In the evenings, we played various games with our children. Particular favorites were the Dictionary game, Scrabble, Mastermind, Password, Charades, Careers and Chronology. We tried Bridge but Victor was a somewhat frustrating partner in this, since he tried to be dummy so that he could read the stack of journals on the floor beside him. We also organized beach parties at the Bay of Fundy, where the tides are the highest in the world, and cooked on a grating perched on the rocks.

MCKUSICK'S PUBLICATIONS

Victor's major research contributions and their publications are described and discussed by several colleagues. These include his work on Marfan syndrome and related disorders; his extensive work with the Amish populations (McKusick 1978); his contributions to genetic nosology; his great interest in gene mapping; his massive compilation of Mendelian disorders and the online version (*OMIM*); his role in founding and guiding the HUGO; his contribution to education and training in medical genetics through a series of summer courses in Bar Harbor (United States), Bologna (Italy), and Beijing (China); and his most valuable studies in the history of science and medicine.

Victor was a prolific writer (see Table 19.1). During his lifetime, he published an enormous output of 772 research papers, books, reviews, and miscellaneous reports, which are listed in the Appendix. These are analyzed in Table 19.1. During the years 1949–1973, he published 538. Compared to that, he published

TABLE 19.1
Publications of Victor McKusick

Years	Number of All Publications
1949–1953	37
1954–1958	112
1959–1963	125
1964–1968	122
1969–1973	142
1974–1978	74
1979–1983	38
1984–1988	32
1989–1993	25
1994–1998	29
1999–2003	23
2004–2008	13
Total	772

much less in the remaining years 1974–2008, only 234! Once he became chairman of the department of medicine in 1973, there was a sharp drop in the number of publications. For example, in the five preceding years (1969–1973), his publications numbered 142, as compared to 74 during the years 1974–1978. This is understandable because of additional administrative work and other responsibilities, which he assumed in 1973. Also, the work on the catalogs of *MIM*, Human Gene Mapping (HGM), and HUGO took up more and more time in later years. He was publishing until the very end; during the last two years, 2007–2008, there were nine publications—altogether an impressive record of lifetime performance.

MISCONCEPTIONS IN HUMAN GENETICS

Writing in 1971, McKusick found it necessary to list 14 genetic misconceptions (McKusick 1971). His article has a timeless quality. This was extremely useful then (and still is today) as we come across these false ideas. Certain fundamental truths need to be repeated often.

1. Congenital is synonymous with genetic: Congenital merely means present at birth. Exceptions are found in both directions. Some genetic disorders are not congenital in the usual sense, and many congenital malformations do not have a predominantly genetic cause. For instance, Huntington's chorea is genetic but not congenital, whereas rubella embryopathy is congenital but not genetic.
2. If a disorder is inherited, a chromosome analysis will show abnormality: Most Mendelian disorders have no chromosome abnormality that can be demonstrated with existing techniques.

3. A buccal smear provides full information on the chromosomes: The buccal smear only tells the maximal number of X chromosomes per cell.
4. If a genetic disorder is dominant, all children of an affected person will be affected; conversely, if all children of normal parents are affected by a genetic disorder, this is evidence of dominant inheritance. This indicates ignorance of Mendel's laws and the meaning of "dominance."
5. When individuals of only one sex are affected in a family, it indicates sex-linkage of the disorder: Once again, this indicates ignorance of basic principles of genetic transmission.
6. A disorder that occurs in multiple siblings with normal parents is not hereditary: Recessive inheritance is characterized by affected sibs with normal parents, and a recessive disorder is as genuinely hereditary as a dominant one.
7. Consanguinity brings out sex-linked disorders; for instance, hemophilia was frequent in the inbred royal families of Europe: Consanguinity increases the occurrence of homozygous affected females but has no effect on the frequency of affected males.
8. The occurrence of a hereditary syndrome composed of two or more manifestations is the result of close linkage on the same chromosome of separate genes, each resulting in one of the individual manifestations: All "mendelizing" syndromes studied in full detail to date have been found to have their basis in the pleiotropic effect of a single mutant gene. Linkage produces no permanent association of traits because even closely linked genes become separated through the process of crossing over.
9. Dominantly inherited disorders tend to increase in severity with transmission from generation to generation, a process called anticipation. For example, age of onset, an expression of severity, tends to be lower in affected children than in their affected parents: Ascertainment bias is responsible for apparent anticipation. On the average, only in milder cases of many dominant disorders, the offspring are expected to be more severely affected than their parents.
10. Dominant disorders are common; recessive disorders are rare: There is no correlation between the frequency of a disorder and its mode of inheritance.
11. Inbreeding causes a build-up of "bad genes" in populations: Inbreeding does not directly change the gene frequencies. It does change genotype frequency; it increases the frequency of homozygotes. If the homozygote is at a disadvantage, inbreeding actually results in a decrease in the deleterious genes.
12. Dominant disorders are more severe than recessive disorders: As a general rule, just the opposite is the case; however, severity may be associated with either type.
13. If a couple has had three children born with a given recessive disorder, the chance that the fourth child will also be affected is vanishingly small: The risk of an affected child from two carrier parents remains 1 in 4, regardless of previous family history.
14. Genetic disease is not treatable: False. Many genetic diseases are treatable.

GENETIC NOSOLOGY: ON LUMPERS AND SPLITTERS

McKusick believed that certain fundamental facts are worth repeating. He emphasized the concepts of pleiotropism and genetic heterogeneity in his now classic paper, "On Lumpers and Splitters, or the Nosology of Genetic Disease," which appeared in the *Birth Defects, Original Article Series*, in 1969 (McKusick 1969). This paper is noted for delineating several fundamental problems, for instance, what constitutes a "genetic entity," what is the impact of inbreeding on nosology, and what are the problems in naming genetic entities. He mentioned examples from his own work, pleiotropism as a leading concern with respect to Marfan syndrome whereas genetic heterogeneity became increasingly the focus in studies of the genetic mucopolysaccharidoses and the separation of homocystinuria from the Marfan syndrome. This became evident in successive editions of *Heritable Disorders of Connective Tissue*.

McKusick compared the classification of disease with the taxonomy of plants and animals. Taxonomists like nosologists can be either lumpers or splitters. But there is one important difference. The nosologist's major concern is whether syndromes A and B are one and the same entity or distinct entities. The taxonomist, on the other hand, constructs a branching classification based on the phylogeny, the components in his classification bearing varying degrees of genetic relationship to each other, based on their descent from a common ancestor.

The fundamental questions tackled by McKusick include the following: What constitutes a genetic entity? How does one identify genetic heterogeneity (including clinical, genetic, and biochemical methods)? Practical difficulties include rarity of the phenotype in the population, and small size of the families, and the naming of genetic entities. The phenotype resulting primarily from a specific and unitary factor is an entity. He wrote that delineation of genetic entities is on safe ground if a fundamental biochemical defect or a specific chromosomal anomaly is identified.

GENE MAPPING

Victor once wrote: "I am not certain why, in the late 1950s, I became enthralled with mapping genes on human chromosomes" (McKusick 2006). One of his first attempts at gene mapping was in collaboration with Ian Porter, which involved the map distance between the loci for G6PD deficiency and colorblindness on the X chromosome. Linkage studies in the Moore Clinic continued in the 1960s in collaboration with James Renwick, who (with Jane Schulze and David Bolling) wrote one of the first computer programs for linkage analysis. It was used in the analysis of possible linkage of XG and CB. Renwick was also involved in several other Moore Clinic linkage studies, including assignment of the Duffy blood group locus to chromosome 1 in 1968, first assignment of a specific locus to a specific autosome. Other linkage studies from the Moore Clinic included mapping of the interval between nail-patella syndrome and the adenylate kinase locus.

The HGM workshops were a forerunner of the HUGO. The first HGM workshop was planned in New Haven in 1973 by Frank Ruddle, but the funding from the MOD Foundation was organized by Victor. These workshops served an important function in the compilation of gene map information, both published and unpublished, over a period of two decades. HGM3 was organized by Victor in Baltimore in 1975.

Victor played a key role in convincing the National Foundation that birth defects fall into the domain of medical genetics. He sent a copy of the first edition of his *Heritable Disorders of Connective Tissue* to Thomas M. Rivers, who was Vice President of Research at the MOD. As Victor recalled later, Rivers was a loyal Hopkins medical alumnus, with training in pediatrics at the Johns Hopkins Hospital. Shortly afterward, Victor was invited to join the MOD Medical Advisory Committee in 1959. The MOD was the sole funder of the Bar Harbor Short Course for its first 25 sessions and continued to provide some support afterward.

GENOMICS

At the Cold Spring Harbor symposium in May 1986, which was entitled "The Human Genome," Victor presented a paper on the status of the human gene map (McKusick 1986). He thought it was an "eye-opener" to molecular biologists. It was about that time that Victor was approached by Brian Crawford of Academic Press, who invited Victor to edit a new journal on mapping and sequencing. Victor was at first reluctant to undertake the new task but was encouraged to do so when Frank Ruddle agreed to join him as coeditor-in-chief. They were about to call the new journal *Genome*, but it was already preempted by the *Canadian Journal of Genetics and Cytology*, which had renamed their journal *Genome*. Victor later recalled that a long evening session of the editorial board was held, over beer, in July 1986, and Thomas Roderick of the Jackson Laboratory in Bar Harbor, Maine, came up with the name *Genomics* as the title.

FOUNDER PRESIDENT OF HUGO

From the beginning, HUGO was supported by funds from the Howard Hughes Medical Institute (HHMI), which also supported the development of *OMIM* beginning in 1985 and for several years afterward. HHMI, in fact, played the same supporting role as had the MOD with reference to the Bar Harbor courses, the conferences on the Clinical Delineation of Birth Defects, and most of the HGM workshops.

HUGO was founded in 1988 as a coordinating agency for the global effort. It was proposed by Sydney Brenner, James Watson, Lee Hood, and others at the Cold Spring Harbor Symposium, and Victor was invited to serve as organizing president (Victor's Hugo)! Victor later reminisced that he was invited because of his active role in the HGM workshops and the ongoing record he had maintained of the status of the human gene map. The first task was to establish a founding council to write a mission statement and bylaws for HUGO. The first meeting of the council was held at Montreaux, Switzerland, at the suggestion of James Watson, and HUGO was incorporated in Geneva. Victor noted that there were five Nobelists at that meeting, including Watson, Dausset, Gilbert, Dulbecco, and Jacob, as well as several other scientists from 19 countries. It was incorporated in Delaware to facilitate handling of U.S. funds.

The objectives of HUGO involved intellectual property rights, ethical issues, research materials, information sharing, and standardization of nomenclature. It was never conceived as a funding agency. After the completion of his presidency of HUGO, Victor continued his work as co-chair of the HUGO ethics committee and was a member of the ethics committee for the National Center for Human Genome Research (NCHGR). From the beginning, Jim Watson wisely decided that funding should be set aside in the budget of the NCHGR for the study of what he called ELSI, for ethical, legal, and societal issues.

The funding of the Human Genome Project by the federal government was mainly based on the National Research Council (NRC)/National Academy of Sciences (NAS) committee report on mapping and sequencing the human genome in 1988. That committee was chaired by Bruce Alberts. The report concluded that the project could be completed in 15 years at an annual budget of $200 million and that "map first, sequence later" was the correct approach for several reasons. These are as follows: (a) mapping of DNA markers and cloned fragments would provide a useful scaffolding for sequencing, (b) sequence technology at that time needed further improvement, and (c) the sequence of genomes of other species should also be continued along with that of the human species, etc. The National Institutes of Health (NIH) project was initiated on October 1, 1990, with James Watson as the first director of the NCHGR. After his resignation, Francis Collins took over on January 1, 1993, and the center was given institute status within the NIH.

CELERA ADVISORY BOARD

Victor was invited by Craig Venter to join the Celera Scientific Advisory Board in 1998, when he founded that company to undertake the complete sequencing of the human genome. Victor later recalled that his participation and association with Hamilton ("Ham") Smith and Craig Venter was very stimulating. He thought that the "completion" of the Human Genome Project was appropriately accelerated by the competition that was introduced by the private effort. The motto of Celera, "Discovery cannot wait," added the right touch and excitement. Both the Venter map and the map created by the international consortium led by Francis Collins were announced at the White House on June 26, 2000.

Victor told me that he attended that ceremony and enjoyed it immensely. Later, I had learned all the details of that White House ceremony from my friend and neighbor in Houston, Neal Lane, who was then assistant to the President for Science and Technology and director of the White House Office of Science and Technology Policy under President Bill Clinton.

MIM AND OMIM

Starting in the early 1960s, Victor compiled a catalog of human genes and genetic disorders. The basic idea of *MIM* (McKusick 1966) originated in the annual reviews of medical genetics, which were undertaken by Victor and several associates during the years 1958–1963. With the onset of the genetic program in the Moore Clinic in 1957, a journal club was initiated and its contributions became eventually a part of the *MIM* catalog, although the annual reviews of the literature were first published each year as one month's issue of the *Journal of Chronic Diseases*.

From the very beginning, the phenotypic nature of the mendelizing catalogs was emphasized. The subtitle of the first ten editions published was "Catalogs of Autosomal Dominant, Autosomal Recessive and X-Linked Phenotypes." Initially, the focus was on X-linked traits, and a monograph entitled "On the X Chromosome of Man" was published in the *Quarterly Review of Biology* in 1962 (McKusick 1962b). The second, the recessive catalog, was meant to be a resource in connection with study of recessive disorders in the Amish. The autosomal dominant catalog was then undertaken "for the sake of completeness" (McKusick 2006).

The first print edition of *MIM* appeared in 1966, and it was maintained on the computer since 1964. For the 11th (1994) and 12th (1995) editions, the subtitle was changed to "Catalogs of Human Genes and Genetic Disorders," thus indicating the progress in the field. In the 1980s, *MIM* was prepared for online presentation, and a search engine enhanced its usefulness. Online access, *OMIM*, was first provided from the Welch Medical Library at Johns Hopkins beginning in 1987 and from the National Center for Biotechnology Information of the National Library of Medicine from 1995 onward.

Victor once wrote that he reveled in the teaching of medical students and residents; "From the distinguished careers in genetics of several of those students and residents, it appears that through teaching in a framework of classic Oslerian medicine I may have helped bring genetics into mainstream medicine."

HUMAN GENOME PROJECT

By 1986, advances in gene mapping and in molecular biology had generated much discussion about projects to map and sequence the entire human genome. Not surprisingly, McKusick played a leading role on the NRC committee charged with assessing the feasibility of what became the Human Genome Project. In 1988, he became founding president of the HUGO, an international coordinating agency for the global mapping and sequencing efforts. During the next few years, he would also serve on the ethics committees for the Human Genome Project and HUGO. In the early 1990s, he chaired another NRC committee charged with establishing guidelines for the use of DNA analysis in law enforcement.

McKusick was a prolific writer throughout his career and published over 500 medical articles and seven books in addition to the ongoing compilation of *MIM*. He also pursued a lifelong interest in the history of medicine; many of his clinical publications included historical components, and he wrote a history of medical genetics for a standard textbook in the field. Other projects included a study of hemophilia in colonial New England, and an annotated edition of selections from William Osler's classic medical text *Principles and Practice of Medicine*.

McKusick's work earned him many honors and awards, including the John Phillips Award of the American College of Physicians for distinguished contributions in internal medicine (1972), election to the National Academy of Sciences (1973), the Gairdner International Award (1977), the William A. Allan Award of the American Society of Human Genetics (1977), the James Murray Luck Award from the National Academy of Sciences (1982), the Sanremo International Prize for Genetic Research (1983), the American Association of Physicians' George M. Kober Award (1990), the Lasker Award for Special Achievement in Medical Science (1997); the National Medal of Science (2002), and the Japan Prize in Medical Genomics and Genetics (2008), along with more than 20 honorary doctorates.

Victor McKusick died on July 22, 2008, at the age of 86.

BRIEF CHRONOLOGY

- 1921—Born in Parkman, Maine, October 21 (with identical twin Vincent)
- 1940–1943—Attended Tufts University
- 1943—Entered JHUSM, without undergraduate degree, as part of special war time policy
- 1946—Received MD from Johns Hopkins
- 1946–1947—Internship at Osler Medical Clinic, Johns Hopkins Hospital
- 1947–1948—Residency at Osler Medical Clinic, Johns Hopkins Hospital
- 1947–1951—Assistant in Medicine, JHUSM
- 1948–1950—Executive chief, Cardiovascular Unit, Baltimore Marine Hospital
- 1949—Married Anne Bishop, a fellow JH physician
- 1951–1954—Instructed in Medicine, JHUSM
- 1954–1956—Assistant Professor of Medicine, JHUSM
- 1956—Published *Heritable Disorders of Connective Tissue*
- 1956–1960—Associate professor of medicine, JHUSM
- 1957–1973—Physician in charge, Joseph Earle Moore Clinic, Johns Hopkins Hospital, and chief, Division of Medical Genetics, Department of Medicine, JHUSM
- 1958—Published *Cardiovascular Sound in Health and Medicine*
- 1960—With John Fuller of Jackson Lab, organized and directed the first Short Course in Medical Genetics, held in Bar Harbor, Maine
- 1960–2008—Professor of medicine, JHUSM
- 1962—Began population studies with Amish
- 1966—Published first edition of *MIM*
- 1970—Elected honorary Life Member of Little People of America
- 1973—Elected to National Academy of Sciences
- 1973–1985—Chairman, Department of Medicine, and physician in chief, Johns Hopkins Hospital
- 1977—Received Gairdner International Award
- 1980—Founding member, American Board of Medical Genetics
- 1985–2008—University professor of medical genetics
- 1986–1988—Member, National Academy of Science Committee on Mapping and Sequencing the Human Genome
- 1987—*Online Mendelian Inheritance in Man* became available, through a collaboration between the National Library of Medicine and the William H. Welch Library at JHUSM
- 1988–1990—Founder president of HUGO

- 1990–1992—Chairman, National Academy of Sciences Committee on DNA Technology in Forensic Science
- 1997—Received Lasker Award for Special Achievement in Medical Science
- 2002—Received National Medal of Science
- 2008—Received Japan Prize in Medical Genomics and Genetics
- 2008—Died in Baltimore, Maryland, July 22

REFERENCES

Bowman, H.S., V.A. McKusick, and K.R. Dronamraju (1965) Pyruvate kinase deficient hemolytic anemia in an Amish isolate. *Am. J. Hum. Genet., 17*: 1–8.

Bruwer, A., J.A. Bargen, and R.R. Kierland (1954) Surface pigmentation and generalized intestinal polyposis (Peutz–Jeghers syndrome). *Mayo Clin. Proc., 29*: 168–171.

Dronamraju, K.R. (1963) Genetic studies of the Andhra Pradesh population. In: E. Goldschmidt (ed.), *The Genetics of Migrant and Isolate Populations.* New York: Williams & Wilkins, pp. 154–159 (Proceedings of an international conference on Human Population Genetics held at Hebrew University, Jerusalem, Israel).

Jeghers, H., V.A. McKusick, and K.H. Katz (1949) Generalized intestinal polyposis and melanin spots of the oral mucosa, lips and digits. *N. Engl. J. Med., 211*: 993–1005, 1031–1036.

Hutchinson, J. (1896) Pigmentation of lips and mouth. *Arch. Surg., 7*: 290.

McKusick, V.A. (1956) *Heritable Disorders of Connective Tissue.* 1st ed. St. Louis: Mosby.

McKusick, V.A. (1958) *Cardiovascular Sound in Health and Disease.* Baltimore: Williams & Wilkins.

McKusick, V.A. (1962) On the X Chromosome of man. *Q. Rev. Biol., 37*: 69–175.

McKusick, V.A. (1966) *Mendelian Inheritance in Man: Catalogs of Autosomal Dominant, Autosomal Recessive and X-linked Phenotypes.* 1st ed. Baltimore: Johns Hopkins University Press.

McKusick, V.A. (1969) On lumpers and splitters, or the nosology of genetic disease. *Birth Defects Original Article Series V*: 23–32.

McKusick, V.A. (1971) Fourteen genetic misconceptions. *Ann. Int. Med., 75*: 642–643.

McKusick, V.A. (1978) *Medical Genetic Studies of the Amish.* Baltimore: Johns Hopkins University Press.

McKusick, V.A. (1986) The gene map of Homo sapiens: status and prospectus. *Cold Spr. Harb. Symp. Quant. Biol., 51*: 15–27.

McKusick, V.A. (2002) History of medical genetics. In: D.L. Rimoin, J.M. Connor, R.E. Pyeritz, B.R. Korf (eds.), *Emery-Rimoin Principles and Practice of Medical Genetics.* 4th ed. Edinburgh: Churchill Livingstone, pp. 3–36.

McKusick, V.A. (2006) A 60-year tale of spots, maps, and genes. *Annu. Rev. Genom. Hum. Genet., 7*: 1–27.

Index

H

I

J

K

L

M

T

U

V

W

X